The Mind of an Engineer

Purnendu Ghosh · Baldev Raj
Editors

The Mind of an Engineer

Springer

Editors
Purnendu Ghosh
Birla Institute of Scientific Research
Jaipur, Rajasthan
India

Baldev Raj
National Institute of Advanced Studies
Bangalore, Karnataka
India

ISBN 978-981-10-0118-5 ISBN 978-981-10-0119-2 (eBook)
DOI 10.1007/978-981-10-0119-2

Library of Congress Control Number: 2015955356

Springer Singapore Heidelberg New York Dordrecht London

Printed on acid-free paper

Springer Science+Business Media Singapore Pte Ltd. is part of Springer Science+Business Media (www.springer.com)

Preface

The Indian National Academy of Engineering (INAE) advocates, encourages and practices excellence in all endeavours in engineering, technology and related sciences to meet the challenges and find solution to the problems of national importance. INAE provides an active and interactive eminent forum to augment and expedite developmental activities in India, especially those related to engineering and technology. Its activities include formulation of technology policies, promotion of quality engineering education and R&D. The INAE represents India at the International Council of Academies of Engineering and Technological Sciences (CAETS), and currently the INAE leads the Council.

The initiative, the first of its kind, taken by the INAE to publish this book—The Mind of an Engineer—is an attempt to reflect the personal experiences of some of its Fellows in the fields of science, technology and engineering. The book deals with the reminiscences, eureka moments, inspirations, challenges and opportunities in the journey of the professionals towards achievements and self-realisation. The book contains 58 chapters on diverse subjects that depict the beauty and diversity of the mind. It truly reflects the way the mind of an engineer works. Engineers have meaningful and constructive minds. The meaning of the mind is understood in the context of the basic needs, orientations, experiences, emotions and consciousness of an individual. Meaning is interactive, selective and value-driven. It is said that meaning is like a large map or Web, gradually filled in by the cooperative work of countless generations. The world is changing, and to keep pace with the new world, new creative minds are evolving. Some new realities are arising. Some old issues shall remain and some old ones will be forgotten. Truly as the Millennium Project of the UN says "Some future issues are further in the future than others". One of the future issues of engineering relates to how technology would influence a growing mind, and what would be the consequential ethical obligations of engineers to the society.

We thank Dr. B.N. Suresh, President of the INAE, the Editorial Board of the INAE, and our Fellow colleagues of the INAE for their valuable contributions in publishing this compendium. Their expertise, views and perspectives, we hope, will be meaningful for the engineers, particularly the young professionals. This book would not have been possible without the generous support of the officials of INAE, ably led by the Executive Director Brig. Rajan Minocha and the former Executive Director, Brig. S.C. Marwaha. We thank all of them most sincerely. We gratefully acknowledge the editorial help and support of Dr. Subimal Sinha Roy, Mr. Aninda Bose, Ms. Kamiya Khatter, and Mr. P.K. Manoharan. Seeing the response of our colleagues, we hope to develop a sequel of the book. We keenly look forward to receiving your impressions and response to enrich the future volumes of *The Mind of an Engineer*.

Purnendu Ghosh
Baldev Raj

Contents

Editors and Contributors

About the Editors

Purnendu Ghosh is the Executive Director, Birla Institute of Scientific Research, Jaipur. Prior to this he was a Professor at Indian Institute of Technology, Delhi. He worked at the Swiss Federal Institute of Technology (ETH), Zurich, Switzerland; University of Melbourne, Australia; and National Research Centre for Biotechnology (GBF), Braunschweig, Germany. He has taken a lead role in augmenting bioprocess engineering and biotechnology activities in the country as a member of several Task Forces of the Department of Biotechnology, Government of India. He is currently the Chief Editor of Publications of Indian National Academy of Engineering. He is a columnist of a leading newspaper group. Besides technical publications, he has written books of different genre.

Baldev Raj is the Director of National Institute of Advanced Studies, Bangalore. His pursuits in basic science and its interdisciplinary application have found expressions of merit and distinction in materials science and technologies, cultural heritage and medical technologies providing solutions to challenges in varied fields of energy, water, health care and manufacturing. He is known for his leadership and experience to overcome the barriers between academic research and industry. His work on non-destructive evaluation and different aspects of materials is well recognised. He is the Fellow of 4 Academies of Sciences and Engineering in India, German Academy of Science, The World Academy of Sciences, International Medical Sciences and Member of Indian Institute of Metals.

Contributors

A.B. Pandit is a Professor at Institute of Chemical Technology, Mumbai. He initiated a major research programme on cavitationally induced transformations and developed the novel technique of hydrodynamic cavitation. He has designed many industrial systems in addition to over 300 research publications. His research group is internationally recognised, and has many industry and academic collaborations. Currently, he is involved in the development of cavitating water disinfecting hand pump and the prototypes are undergoing field trials.

A.K. Anand was the Director, Technical Coordination and International Relations Group, Bhabha Atomic Research Centre; the Director, Reactor Projects Group, leading the programme for nuclear propulsion for the Indian nuclear submarine programme; and also the Executive Director, Kudankulam Project, in Nuclear Power Corporation. At present, he is the Director Technical, Microtrol Sterilisation Services Private Ltd., Mumbai.

Ahindra Ghosh was with the Department of Materials and Metallurgical Engineering, IIT Kanpur. He was also associated with Tata Research Development and Design Centre, Pune, and Tata Steel, Jamshedpur. He has to his credit 5 books and about 90 research publications. For the last three decades, his principal interest has been the theory of metallurgical processes in iron and steelmaking with specific emphasis on sponge iron, steelmaking, solidification, ingot and continuous casting. He had significant interactions with industry, including several R&D collaborations with steel plants.

Aloknath De is the Chief Technology Officer of Samsung R&D Institute India in Bangalore and also the Corporate VP for Digital Media and Communication activities pertaining to mobile and consumer appliances. He has more than twenty-five years of industrial and research experiences including the most recent one in ST-Ericsson, where he was the Director and Country Manager. He is a recipient of "Alexander Graham Bell Prize" in Canada and received IETE Memorial Award in India. He has served as an Adjunct Professor at IIT Delhi and an AICTE-INAE Distinguished Visiting Professor at IIT Roorkee. He is a senior member of IEEE, and a Fellow of IE, IETE and INAE.

Anil Kumar Sinha is the Head, Centre for Design & Development Section, BARC, Trombay. He has published/presented nearly 100 technical papers and has received BARCOA Award for the design of a special cutting m/c to remove the coolant channel from one of the power reactors, special Group Achievement Award for cutting and removing coolant channel from Reactor, HomiBhabha Science & Technology Award for design and development of various types of high-precision instruments.

Arun Kumar De embraces the characteristics and qualities of a professional engineer, a researcher, an educationist and an administrator. After he had worked as a design and development engineer in a sewing machine factory in Kolkata, he switched over to teaching and joined IIT Bombay. He then went to CSIR-Central Mechanical Engineering Institute, Durgapur, as Director and then returned to IIT Bombay as Director. He was the first Chairman of Atomic Energy Regulatory Board. He is a Fellow of INAE and Institution of Engineers. He has been accorded with the Lifetime Achievement Award by the Institution of Engineers (India) and by IIT Bombay. He has published over thirty research and other technical papers. He has penned a book entitled "Meandering Streams".

Ashutosh Sharma is the Secretary, Department of Science and Technology (DST), Government of India. He is the founding PI and Coordinator of Nanosciences Centre and Advanced Imaging Centre at IIT Kanpur. He has served on the Governing Boards/Councils of over 15 prominent scientific institutions in India and has had a broad international experience as a research faculty at SUNY Buffalo School of Medicine, visiting faculty at University of Texas at Austin, University of Western Ontario, University of Erlangen-Nuremberg and the World Class University Program of South Korea and as a Member of the European Research Commission. He has published around 290 papers, filed over 10 patents and mentored a nanotechnology start-up. He is a recipient of numerous honours and awards including the inaugural Infosys Prize in Engineering and Computer Science, TWAS Science Prize of the World Academy of Sciences, Bessel Research Award of the Humboldt Foundation, J.C. Bose Fellowship, Bhatnagar Prize, Homi J. Bhabha Award of UGC, The Syed Husain Zaheer Medal of INSA, Distinguished Alumnus Award of IIT Kanpur and the Lifetime Achievement Award of the Indian Science Congress. His other interests are in ancient history and philosophy, poetry and art.

B.N. Suresh is presently the President of Indian National Academy of Engineering. He is Vikram Sarabhai Distinguished Professor at ISRO HQ, Bangalore. He is a member of Board of Governors, IIT Madras, and the Chairman, Research Board for Aeronautical Development Establishment, DRDO. He was the Director, Vikram Sarabhai Space Centre, Trivandrum, and Founder Director of Indian Institute of Space Science and Technology, Trivandrum. He was the Member of Space Commission and also the Distinguished Professor at IIT Bombay, and MIT Manipal. He has won several awards and honours including Lifetime Contribution Award in Engineering by INAE, Karnataka State Rajyotsava Award. In recognition of his meritorious contributions for Science and Technology, Government of India conferred on him Padma Shri and Padma Bhushan.

C.G. Krishnadas Nair held many positions in HAL and contributed significantly to the indigenous development of aerospace materials and process technologies and led the design and development of Advanced Light Helicopter, new Jet trainer and Light Armed Helicopter. He founded Society of Indian Aerospace Technologies and Industries. He served as Vice Chancellor of Mahaveer Academy of Technology and Science, Raipur. He is currently a Professor and a Chairperson of the Governing Council of IIAEM and also the Chancellor of the Jain deemed University. He has authored 24 books and about 200 research and technical papers. He is the recipient of Lifetime Achievement Award from INAE and has received several national and international awards, including Padma Shri, Scope Award for Best Chief Executive, Birla Gold Medal Award, National Aeronautical Prize, Vasvik Research Award, Omprakash Bhasin Award, National Metallurgist Award, IIM Platinum Award, Swadeshi Science Puraskara and Gold Cross Award from ComiteD'honneurdu Meriteet Devouement Francais, France.

D.N. Singh is an Institute Chair Professor at the Department of Civil Engineering, IIT Bombay. His main research area is Environmental Geotechnology. He is the Editor-in-Chief of Environmental Geotechnics, UK. He is the recipient of Young Teachers' Award, Shamsher Prakash Foundation Award, John R. Booker Excellence Award and Richard Feynman Prize. He is a fellow of American Society of Civil Engineers.

Dhruba Basu has spent the major part of his career in ISRO designing and developing important electronic systems for the country's satellite launch vehicle programs and was responsible for the total R&D activities in Electronics in VSSC. He led a team of engineers in the development of on-board computer systems for India's satellite launchers. He is the recipient of several awards, including Calcutta University Silver medal and best article award of the Space Engineers' Association. He is the Fellow of INAE.

Dipak Mazumdar currently holds the Ministry of Steel Chair Professorship at IIT Kanpur. He has been teaching and conducting research in steelmaking for over three decades.

Ganti Prasada Rao worked at PSG College of Technology, Coimbatore, and IIT Kharagpur. He was also at UMIST, Manchester, England and Lehrstuhlfür Elektrische Steuerung und Regelung, Ruhr-Universität Bochum, Germany, University Henri Poincare, Nancy, France, and at Brunel University, UK. He was the Scientific Advisor to the Directorate of Power and Desalination Plants, Water and Electricity Department, Government of Abu Dhabi and the International Foundation for Water Science and Technology. He is presently a member of the UNESCO-EOLSS Joint Committee. He has authored 4 books and 150 research papers. He is the coeditor of Identification of Continuous Systems—Methodology and Computer Implementation. He received the IIT Kharagpur Silver Jubilee Research Award, The Systems Society of India Award and Honorary Professorship of the East China University of Science and Technology, Shanghai. He is the Fellow of the IEEE. The International Foundation for Water Science and Technology has established, in his honour, Systems and Information Laboratory at IIT Kharagpur.

Indranil Manna currently the Director of IIT Kanpur is an educationist and materials engineer with wide-ranging research interests covering structure–property correlation and modelling in nanometric metals and ceramics, laser/plasma-assisted surface engineering, nanofluid and bainitic steel. Before moving to IIT Kanpur, he headed CSIR-CGCRI in Kolkata. Earlier, he was the Professor at IIT Kharagpur. While serving at IIT Kharagpur, he worked as a guest scientist in different renowned Institutions/Universities abroad such as Max Planck Institute at Stuttgart, Technical University of Clausthal, Liverpool University, Nanyang Technological University and University of Ulm. He has over 250 journal publications and received several awards. He is a Fellow of all the national academies of science (INSA, IASc, NASI) and engineering (INAE) in India. He is a J.C. Bose Fellow of DST and a recipient of TWAS prize for engineering sciences. He is currently the Vice-President of the Indian Institute of Metals and the INAE.

J. Raghava Rao is presently working as the Chief Scientist in the Chemical Division of the Central Leather Research Institute, Chennai. His contributions include biomediated tanning, cleaner and greener methods of leather processing, waterless tanning and effective utilisation of tannery solid waste for value addition. He has received several national awards and international recognitions. He has more than 230 publications in reputed journals. Several of his technologies developed have been successfully commercialised.

K.L. Chopra after working in senior R&D positions in Canada, Germany and USA joined IIT Delhi as a Senior Professor and also served as Head and Dean. He served as the Director of IIT Kharagpur. After retirement, he served IITD as IREDA Chair Prof of Solar Energy and is presently a Distinguished Honorary Professor of several institutes. He has authored/coauthored/edited 10 books, published over 430 research papers, held 4 US patents and transferred 10 know-hows to industry in the areas related to thin films and nanomatter. His book "Thin Film Phenomena" is considered as the Bible of the field. Recipient of numerous awards which include S.S. Bhatnagar Prize, Citation Laureate Award of ISI and Lifetime Achievement Awards of Solar Energy Society, Indian Vacuum Society and Indian National Academy of Engineering. Fellow of all Science and Engineering Academies of India, he has been conferred Padma Shri by the Government of India for his contributions to science and engineering.

K. Muralidhar is presently faculty of Mechanical Engineering at Indian Institute of Technology Kanpur. He has published over 100 international journal publications, seven books and monographs, and several book chapters. He has developed optical measurement techniques and numerical algorithms in the area of fluid and thermal sciences. His work has led to better understanding of transport phenomena in porous media, wake dynamics, flow control, crystal growth and biomedical flows. His research finds applications in enhanced oil recovery, gas hydrates, regenerators, nuclear waste disposal and the growth of optical crystals. Presently, he is coordinating a National initiative on Solar Hydrogen Generation, a multi-institutional project on CFD code development on unstructured grids and an Institute-level initiative on futuristic mechanics. He is the immediate past-President of National Society of Fluid Mechanics and Fluid Power (India).

K.V. Raghavan is a Fellow of the INAE, Indian Institute of Chemical Engineers, A P Academy of Sciences and a Distinguished Fellow of University of Grants Commission. He was the Director of Central Leather Research Institute, Chennai, and Indian Institute of Chemical Technology, Hyderabad, Chairman of Recruitment and Assessment Centre of DRDO, and Distinguished Professor of INAE. He is currently the Vice-President (International Cooperation) of the INAE. His areas of specialisation are chemical process development and design, reaction engineering, simulation and modelling and chemical hazard analysis. He published more than 120 papers, filed 45 patents and edited 3 books. His applied research efforts contributed to the successful development of more than 25 chemical processes with high industrial impact in bulk organics, specialities, oilfield chemicals and fluoroorganics.

K.G. Ranga Raju taught at the IIT Roorkee. During his stay at Roorkee, he held several administrative positions including those of Pro-Vice Chancellor and Deputy Director. His research interests include fluvial hydraulics, hydraulic structures, disturbed boundary layers and hydrometry. He is the recipient of several prestigious awards such as the Ippen Award and the Honorary fellowship of the International Association of Hydraulic Engineering and Research. During the last ten years, he has been involved in the resolution of water disputes both at the national and at the international levels.

Kamala Krithivasan was a professor working in the area of theoretical computer science, especially Formal Language Theory and its applications and Unconventional Models of Computing at the Computer Science and Engineering Department, IIT Madras. She has been associated in the area of array grammars, graph grammars, L-systems, application to Kolam patterns, DNA and membrane computing. She has also been involved in computational geometry having worked on optimisation and path planning problems. She has delivered a series of lectures on Discrete Mathematical Structures, Formal Languages, Automata and Computation under the National Programme on Technology Enhanced Learning. She has more than 80 publications in international journals and more than 100 publications in conferences. She visited the University of Maryland, Concordia University, and MPI, Saarbrucken, Germany. She is a Fellow of the INAE and was awarded the Fulbright Fellowship.

K.V. Subramaniam is the President of Reliance Life Sciences, Dhirubhai Ambani Life Sciences Centre, Mumbai.

M.L. Munjal has been on the Faculty of the Indian Institute of Science, Bangalore. Currently, he is the Honorary Professor and INSA Senior Scientist at IISc. He has published over 190 papers and has authored a monograph Acoustics of Ducts and Mufflers, a textbook on Noise and Vibration Control and contributed chapters in a couple of international handbooks. He has over 100 industrial consultancy projects in different aspects of noise control and quieter designs, with particular application to automobiles. He is the Convener of the Facility for Research in Technical Acoustics, Distinguished Member of the Institute of Noise Control Engineering, USA, and an Honorary Fellow of the International Institute of Acoustics and Vibration. He was the Chairman of the National Committee for

Noise Pollution Control and President of the Acoustical Society of India. He is a Member of the Editorial Board of the International Journal of Acoustics and Vibration. He is a Fellow of the INAE and Indian National Science Academy and Member of the European Academy of Sciences.

M.M. Madan has more than 40 years of experience in the field of hydropower. He is the President and CEO (Hydro Business) at Jindal Power Ltd. He is the Director on Board in many hydropower companies. He was the Chief Executive Officer of Malana Power Company Ltd. and AllainDuhangan Hydro Power Ltd. at LNJ Bhilwara Group. He was the Executive Director at NHPC. He is the recipient of CBI&P-I.N.Sinha Award for his contribution in Water Resources Engineering. He has published over 210 technical articles/papers and has also published a Book on Tunnelling. He is a Member of International Society of Rock Mechanics, Tunnelling Association of India, Indian National Hydropower Association, Indian Society of Rock Mechanics and Tunnelling Technology, Chartered Engineer and Fellow, Institution of Engineers (India), Fellow, Institution of Valuers and Fellow of the INAE.

M.O. Garg is an acknowledged expert in petroleum refining and petrochemicals and has been instrumental in developing and commercialising large number of technologies in Indian refineries. He served Engineers India Limited and then Technip KTI before joining IIP. He has been the Director of the Indian Institute of Petroleum and also officiated as the Director General of CSIR. He has won several prestigious awards including OceanTex Excellence Award and CSIR Technology Awards and has published more than 255 papers in refereed journals and 39 national and international patents. He is a Fellow of the INAE.

M.S. Ananth is currently Distinguished Visiting Professor in IIT Bombay. He was the Director of IIT Madras. During this time, he created the largest open-courseware programme in technical education in the world, the National Programme on Technology Enhanced Learning and established the first university-based Research Park in India to promote innovation and entrepreneurship. He has been consistently rated a good teacher. He is a Fellow of the IIChE and the INAE. I&EC Research published the

Ananth-Festschrift Issue in his honour as a "leading researcher, gifted teacher and visionary leader of higher education in India". He was the Member of both the Scientific Advisory Committee to Cabinet and the National Manufacturing Competitiveness Council. He was a Visiting Professor in IIT Kanpur and IISc Bangalore, Princeton University and University of Colorado and RWTH, Aachen (Germany), and a Visiting Scientist in Aspen Tech and in the National Institute of Standards and Technology (USA).

Mahesh C. Chaturvedi was the Professor of Civil Engineering, IIT Delhi. He was also a Visiting Professor in Harvard University.

N.R. Narayana Murthy is the founder of Infosys. He articulated, designed and implemented the Global Delivery Model, which has become the foundation for the huge success in IT services outsourcing from India. He has led key corporate governance initiatives in India. He is an IT advisor to several Asian countries. He serves on the boards of Ford Foundation, United Nations Foundation, Rhodes Trust and the Institute for Advanced Study in Princeton, New Jersey. He has served as a member of the HSBC board and the Unilever board. He has served on the boards of Cornell University, Wharton School and the Graduate School of Business at Stanford University. He has also served as the Chairman of the Indian Institute of Management, Ahmadabad. He is a foreign member of the US National Academy of Engineering and a Fellow of the INAE. He is the recipient of the Hoover Medal, James C. Morgan Global Humanitarian Award, Ernst Weber Medal from the Institute of Electrical and Electronics Engineers, USA. He is the first Indian winner of Ernst and Young's World Entrepreneur of the Year award. He is also a Trustee of the Infosys Science Foundation. He has been awarded the Legion d'honneur by the Government of France, the CBE by the British government and the Padma Vibhushan by the Government of India.

N. Raghavan currently is an Adjunct Professor, IIT Madras. He was the Head, Hydel & Nuclear Projects Sector, Larsen & Toubro Limited, ECC Construction Division, Founder-Chief Executive/Director—L&T-Ramboll Consulting Engineers Limited, Principal Consultant—STUP Consultants Limited; The Cementation Co. Ltd.; PWD Tamil Nadu. His interests include construction management and methods and precast concrete. He received S.B. Joshi Memorial Award for Excellence in Bridge and Structural Engineering, has presented over 180 technical and other papers and talks and has been associated with various Indian and international forums.

Nikhil R. Pal is a Professor at the Electronics and Communication Sciences Unit, Indian Statistical Institute, Calcutta. He was a Visiting Professor at the University of West Florida, USA, RIKEN Brain Science Institute, Japan, and the National Chiao Tung University, Taiwan. His current research interest includes bioinformatics, brain science, fuzzy logic, data mining, machine learning, neural networks and evolutionary computation. He was the Editor-in-Chief of the IEEE Transactions on Fuzzy Systems and has served the Editorial Board of several journals. He is a Fellow of the Institution of Electronics and Telecommunication Engineers, West Bengal Academy of Science and Technology, National Academy of Sciences, India, INAE, Indian National Science Academy, International Fuzzy Systems Association and a Fellow of IETE, USA.

N.S. Mohan Ram was the President, TVS Motor Company, Vice-President, Mukand Limited, the Head, Design Department of Mazagon Docks Ltd., and the Captain at the Indian Navy. He is currently the consultant of TVS Motor Company. His current interest is recycling automobiles at end of life. Among his honours and recognition are included Vishist Seva Medal, Das Gupta and VKRV Rao medals from FMS, FIE Foundation Medal for engineering management, Distinguished Alumnus of IIT Kharagpur, Vaswik Award for engineering excellence. He is a Fellow of the INAE.

P.S. Goel is Hony. Distinguished Professor at ISRO. Formerly, he was the Secretary, Ministry of Earth Sciences and Chairman Earth Commission, Director ISRO Satellite Centre, Chairman, Recruitment and Assessment Board, DRDO. He is a Fellow of the INAE.

Pallab Dasgupta is a Professor at the Department of Computer Science and Engineering, IIT Kharagpur. He works in the area of formal methods for electronic design automation and cyber-physical systems. He has published over 160 research papers, 3 books, 2 US patents and collaborated with leading companies including Intel, General Motors, Synopsys, IBM, Google, National Semiconductors and Semiconductor Research Corporation. He received the young scientist medals of INSA and INAE, the IBM Faculty Award and the IESA Technomentor Award.

Prahlada has served DRDO for 43 years in various capacities and positions. He has worked in ADE, DRDL and finally at the corporate Hqrs in Delhi as Distinguished Scientist and CCR&D. His main contribution is designing, developing, testing, productioning and delivering The Akash Surface to Air Weapon System which has been inducted by the Indian Army and the Indian Air Force. He also held the position of Vice Chancellor, Defence Institute of Advanced Technology, Pune.

P.S. Deodhar is the Chairman, Aplab Limited, Thane. Formerly, he was the Chairman, Electronics Commission, GOI and Advisor (Electronics) to the PM.

R.D. Kale during his initial years worked for the design of the Fast breeder test reactor set-up at Kalpakkam nuclear centre, IGCAR. He sets up the Engineering Development Laboratory. He performed pioneering work in developing liquid sodium technology in the country. He has played a pivotal role in developing large sodium mechanical pumps for the PFBR. He has over 70 research publications. Currently, he is serving as a member of design safety committees of AERB.

R. Krishnan was the Scientific Officer and Head Metallurgy Division, Bhabha Atomic Research Centre, Mumbai; Director, Naval Chemical and Metallurgical Laboratory, DRDO, Mumbai; Chief Controller R&D, DRDO, New Delhi; and Director, Gas Turbine Research Establishment, DRDO, Bengaluru.

R. Sundaravadivelu has 30 years of teaching, research and consultancy experience at IIT Madras in the field of Offshore, Coastal, Ship and Ports & Harbour Structures. He has published 60 journals and 160 conference papers and written two books. His current research interests are International projects on non-ship-shaped FPSO and mobile offshore base and Innovative projects on classic SPAR for offshore wind energy and JACKET SPAR for integrated

OTEC-Offshore desalination system. He is a Fellow of the INAE, Member of the Bureau of Indian Standards on Port & Harbour structures.

Rabindranath Ghosh is a metallurgical engineer. He worked in National Metallurgical Laboratory, Jamshedpur, in various capacities. He is currently a Chair Professor at Metallurgical and Materials Engineering Department, IIT Kharagpur. He is known for his work on Remaining Life Assessment of critical engineering components for power plants, refineries and petro-chemical units; Development of physically based constitutive models for integrity assessment of engineering materials and components; Development of non-destructive tools and techniques for integrity assessment and failure analysis.

Rajaram Bojji is the Director of Rajaram Bojji Inc. He has served Indian Railways for over 35 years. His publications deal with Rail–wheel Interaction. He received Government of India award for efficient running of Konkan Railway, National Award for outstanding contribution to knowledge-based industries by the Federation of Indian Chambers of Commerce & Industries, National Council for Occupational Health and Safety Award.

Rajaram Nagappa worked in Space Science and Technology Centre, Thiruvananthapuram. He retired as the Associate Director, Vikram Sarabhai Space Centre. His major contributions include initiation of microsatellite—Anusat—development and establishment of a high temperature material characterisation laboratory. He is currently with the National Institute of Advanced Studies, Bengaluru, where he anchors the International Strategic and Security Studies Programme. In addition to the fellowship of the INAE, he is a Fellow of the Aeronautical Society of India, Fellow of the High Energy Material Society of India. He is the recipient of awards and honours from the Madras Institute of Technology, the Astronautical Society of India, DRDO and the International Astronautical Federation.

Ram Kanwar Garg was one of the team members engaged in the setting up of the Uranium Metal Plant and producing the first uranium metal ingot used for fabricating fuel for the CIRUS reactor. He then was responsible for process development, design and setting up of the plant for production of nuclear grade Zirconium Oxide at NFC Hyderabad. He led a team to develop the process for Lithium isotope separation and metal production. He developed centrifuge process and sets up the Rare Materials Plant at Mysore for Uranium enrichment. He represented India in the International Nuclear Fuel Cycle Evaluation Study. He served as the Chairman and Managing Director of Indian Rare Earths. Presently, he is the Chairman of a number of committees of the Ministry of Environment, Government of India, Central Pollution Control Board, Maharashtra Pollution Control Board and AERB.

Raja Manuri Venkata Gopala Krishna Rao for over four decades had been intensely involved in the R&D studies pertaining to indigenization and application of polymer composite products for aerospace, non-aerospace and defense sectors. At NAL, he played vital roles in realisation of innumerable composite products based on home-grown technologies, to commercially viable and technology transfer levels. An internationally acknowledged Scientist of repute in Polymer Composite Materials, his motto is, “Cost Effective Appropriate Technologies for Composite Materials”.

S. Kalyanaraman is with the Department of Electronics and Communication, Dayananda Sagar College of Engineering, Bangalore. Formerly, he was the Programme Director, ISRO Satellite Centre, Bangalore.

S.C. Dutta Roy's teaching and research experience includes teaching at an Indian University, two US Universities, one UK University and a long spell of four decades at IIT Delhi. His research has been recognised by several national awards, including the Shanti Swarup Bhatnagar Prize, and through Fellowship of the IEEE, and of all the national academies of science and engineering and Distinguished Fellowship of the IETE. What he values most, however, is the love, affection and appreciation of a very large number of students whom he interacted with personally, and an equally large number of his virtual students, spread throughout the world, through the five video courses he authored, which are available on YouTube. He enjoys listening to classical music and researching on its history and the history makers, no less than the enjoyment he derives out of his professional work.

S.C. Chetal has contributed to Indian sodium cooled fast reactor programme at Indira Gandhi Centre for Atomic Research, Kalpakkam from where he superannuated as Director. He has been the principal design engineer of indigenous designed 500 MWe Prototype Fast Breeder Reactor which is presently under commissioning. He has also contributed to design of zirconium and titanium sponge plant high-temperature retorts. Presently, he is associated with R&D programme of advanced ultra-supercritical power plant technology.

Samir V. Kamat has been working in Defence Metallurgical Research Laboratory, Hyderabad. He has made significant contributions in the area of microstructure–mechanical property correlations in advanced materials such as particulate-reinforced metal matrix composites, ceramic matrix composites, aluminium–lithium alloys, high-strength aluminium alloys and titanium alloys which has led to their development for various defence applications. He has also been instrumental in setting up the state-of-the-art experimental facilities and expertise for characterisation of mechanical behaviour of materials in small volumes, especially for materials used in MEMS. In recent times, he has made significant contributions to the development of Rare Earth Permanent Magnet Technology in DMRL.

Sanak Mishra is presently the Secretary General of the Indian Steel Association and Vice-President (Finance) of INAE. Prior to that he was the Vice-President of Arcelor Mittal. As founder member of the Corporate R&D Centre of the Steel Authority of India Limited (SAIL), he rose to become a full-time Director on the Board of SAIL. He served as the Managing Director of its Rourkela Steel Plant. He is currently the Chairperson of the International Organisation of Minerals, Metals and Materials Societies. He received numerous awards, and among them are JRD Tata Award for Excellence in Corporate leadership in Metallurgical Industries, the National Metallurgist Award from the Ministry of Steel, the Centenary Year Distinguished Alumni Award from the Indian Institute of Science and the Distinguished Merit Alumni Award from the Department of Materials Science & Engineering at the University of Illinois. He is the Fellow of the National Academy of Sciences, INAE, Indian Institute of Metals, Institution of Engineers, Computer Society of India, Institute of Directors and the All India Management Association.

Sanjay Mittal is a Professor in the Department of Aerospace Engineering, IIT Kanpur. He is also the Head of the National Wind Tunnel Facility. His research interests include unsteady aerodynamics, computational fluid dynamics, experimental fluid mechanics, aerodynamic shape optimisation, fluid structure interactions, bluff body flows, linear stability analysis, flows in air intakes, aerodynamics of sports balls and train aerodynamics. He received the Shanti Swarup Bhatnagar Prize in Engineering Sciences and is a Fellow of the Indian National Science Academy, INAE, Indian Academy of Science and National Academy of Sciences India.

Sankar Kumar Pal is a Distinguished Scientist of the Indian Statistical Institute and its former Director. He is widely recognised for his pioneering contributions in Machine Intelligence, Fuzzy Logic, Soft Computing and Pattern Recognition. He worked at the University of California, Berkeley, University of Maryland, NASA Johnson Space Center, Houston, Texas, and US Naval Research Laboratory, Washington DC. He is the Distinguished Visitor of the Institute of Electrical and Electronics Engineers (IEEE) Computer Society, USA, for the Asia-Pacific

Region, and held several visiting positions in Italy, Poland, Hong Kong and Australian universities. He is a coauthor of seventeen books and more than four hundred research publications. He has founded the Machine Intelligence Unit and the nation's first Centre for Soft Computing Research at ISI. He is a Fellow of the IEEE, the Academy of Sciences for the Developing World (TWAS), International Association for Pattern Recognition, International Association of Fuzzy Systems and all the four National Academies for Science and Engineering in India. He has received the S.S. Bhatnagar Prize, Vikram Sarabhai Research Award, Jawaharlal Nehru Fellowship, NASA Tech Brief Award (USA), IETE-Wadhwa Gold Medal, Om Bhasin Award, G.D. Birla Award, Khwarizmi International Award, FICCI Award, INSA-ZaheerMedal, P.C. Mahalanobis Birth Centenary Gold Medal of Indian Science Congress for Lifetime Achievement, J.C. Bose Fellowship of the Government of India and INAE Chair Professorship.

Soumitro Banerjee was in the faculty of the Indian Institute of Technology, Kharagpur, and then moved to the Indian Institute of Science Education & Research, Kolkata. His areas of interest are in the nonlinear dynamics of power electronic circuits and systems, and bifurcation theory for non-smooth systems. He has published three books, served as an Associate Editor of the IEEE Transactions on Circuits & Systems II, and as an Associate Editor of the IEEE Transaction on Circuits & Systems I. He is a recipient of the S.S. Bhatnagar Prize. He is a Fellow of the Indian Academy of Sciences, the INAE, the Indian National Science Academy, The World Academy of Sciences and the IEEE.

Suresh Chandra Gupta assisted the legendry Dr. Vikram Sarabhai in building a rocket research and development team. He was instrumental in promoting technical autonomy for the ab initio development of various subsystems of a rocket. He has held several important positions, before taking over as Director, VSSC. He was a Member of the Space Commission of India and received Dr. Brahm Prakash Distinguished Professorship of ISRO. He continues to contribute actively to the Space Program. He has authored more than 100 technical papers, and a book entitled "Growing Rocket Systems and the Team". He is a Distinguished Fellow of the Astronautical Society of India, and Fellow of the INAE, National Academy of Sciences. He was the President of Systems Society of India. He is a Member of the International Academy of Astronautics, Paris and Member, International Programme Committee on Automatic Control in Aerospace under the International Federation on Automatic

Control. He has received National Systems Award, Shri Hari Om Ashram Prerit Dr. Vikram Sarabhai Research Award, VASVIK Research Award for Electrical Science & Technology, Aryabhata Award of the Astronautical Society of India, for lifetime contributions. He was presented by the Government of India, Department of Space "Outstanding Achievement Award".

Sushmita Mitra is a Professor at the Machine Intelligence Unit (MIU), Indian Statistical Institute, Kolkata. She was in the RWTH, Aachen, Germany, as a DAAD Fellow, Visiting Professor in the Computer Science Departments of the University of Alberta, Edmonton, Canada, Meiji University, Japan and Aalborg University Esbjerg, Denmark. She received the IEEE TNN Outstanding Paper Award for her pioneering work in neuro-fuzzy computing and the CIMPA-INRIA-UNESCO Fellowship. She is the author of three books, beside a host of other edited books. She is an Associate Editor of IEEE/ACM Trans. on Computational Biology and Bioinformatics, Information Sciences, Neurocomputing, and is a Founding Associate Editor of Wiley Interdisciplinary Reviews: Data Mining and Knowledge Discovery (WIRE DMKD). She has more than 100 research publications. She is a Fellow of the IEEE, the INAE, and The National Academy of Sciences, India. She is an IEEE CIS Distinguished Lecturer.

T. Balakrishna Bhat worked at DMRL for more than three decades almost entirely in the area of armour materials for a wide range of applications. He also worked on the theoretical aspects of hot isostatic pressing of powders and on creep and oxidation of two phase materials. As a visiting scientist at JPL, Pasadena, he worked on high-strength honeycomb and cellular solids. He has authored books on explosive compaction, composite armour materials and modules and dynamics of unification.

Tarun K. Ghose initiated the department of food technology and biochemical engineering at Jadavpur University, H B Technological Institute, Kanpur, and Biochemical Engineering Research Centre at IIT Delhi. He was the Senior Fellow at US Army Natick Laboratory, visiting professor at Monash University, ETH Zurich and Laussane, Melbourne University, Rutgers University, Delaware University, National University of Singapore, and Technical University of Denmark. He established international series advances in biochemical engineering. He has served on a number of panel/committees of UN agencies. He received IIChE Ambrose-Congreve Gold Medal as Chemical Engineer of the year. He is a Fellow of INAE, Foreign member of the Academied'Agriculture de France, World Academy of Art and Science.

V.G. Gaikar is Bharat Petroleum Distinguished Professor of Chemical Engineering at the Institute of Chemical Technology, Mumbai. He has over 140 publications, 2 US patents, 1 PCT and 8 Indian Patents. Among major honours and awards are fellowship of INAE and Maharashtra Academy of Sciences, Young Scientist Medal of INSA, Young Associateship of IASc, Best Teacher Award of University of Mumbai, and Herdillia Award of Indian Institute of Chemical Engineers for Excellence in Basic Research in Chemical Engineering. His research interests include Reactive Separations, Engineering Separations through Molecular Simulations, Thermal Conversion of Biomass to Fuels, Clean Process Technology Development and Process Intensification using Microwave. He is also the Director of Aarti Drugs Ltd. and Bharat Oman Refineries Ltd. Currently, he is coordinating the Technical Education Quality Improvement Program for Institute of Chemical Technology, the Innovation Networking of TEQIP Institutes in Maharashtra and the DAE-ICT Centre for Chemical Engineering Education and Research.

Vasant Manohar after working at English Electric Co. Ltd. in England for 2 years joined the Tata Power Company and later the Tata Consulting Engineers, where he rose to the position of the Director-in-Charge and CEO. His work has been mainly in the area of planning, design and management of major engineering projects in India and Malaysia, Nepal, Laos and Iran. His areas of interest include energy matters, technology development and the management of high technology projects and organizations. He was awarded the prestigious Fellowship of the City & Guilds of London Institute, and named an Outstanding Engineer by IEEE Delhi chapter. He is a Fellow of the Institution of Engineers (India), a Life Senior member of IEEE of USA, Past-President of the Society of Power Engineers (India), Past Vice-President and Trustee of the Indian Nuclear Society. He is a Trustee of the Volkart Foundation. He is a Fellow of the INAE.

The Mind of an Engineer

Purnendu Ghosh

Engineering makes the forces of nature work for the good of mankind. Its purpose is to serve our needs and cater to our wants. Need is limited, while want is unlimited. Need is a necessity and defines the limits of enough. Want is an optional and defines the depth of greed. We often do not know where the excess begins. It is important, in this context, to understand how much knowledge is enough. What is more stressful, over-information or under-information? If knowledge increases the way it is increasing, how much intelligence will be enough to contain that knowledge?

We have crossed various ages of engineering. After Stone Age, Bronze Age, Iron Age and Steam Age, we are now in the Information Age. Engineering in the broadest sense relates to the "development, acquisition and application of technical, scientific and mathematical knowledge about the understanding, design, development, invention, innovation and use of materials, machines, structures, systems and processes for specific purposes". An engineer is a composite. "He is not a scientist, he is not a mathematician, he is not a sociologist or a writer; but he may use the knowledge and techniques of any or all of these disciplines in solving engineering problems". Neil Armstrong said, "I am, and ever will be, a white-socks, pocket-protector, nerdy engineer—born under the second law of thermodynamics, steeped in the steam tables, in love with free-body diagrams, transformed by Laplace, and propelled by compressible flow".

The biggest challenge for engineering profession has been its integration with the human needs. On the one hand, engineers are not limited by technology, and on the other, they are worried about the risks to the environment, health, sustenance and safety. The world wants methods that are appropriate to make competition look natural. We are looking for "cooperative competition".

Engineers work under various constraints: nature, cost, safety, environment, ergonomics, reliability, manufacturability and maintainability, among others. Industry wants real-world engineers equipped to deal with the complex interactions

---

P. Ghosh (✉)
Birla Institute of Scientific Research, Statue Circle, Jaipur 302 001, India
e-mail: ghoshiitbisr@gmail.com

P. Ghosh and B. Raj (eds.), *The Mind of an Engineer*,
DOI 10.1007/978-981-10-0119-2_1

across many disciplines. It wants engineers who can handle the unknown and the unexpected problems.

Future engineers are expected to appreciate, more than before, the human dimensions of technology. They are expected to have a grasp of the global issues. There is a need to understand the nuances of working in a culturally diverse space. According to some experts, good engineering designs should not be deprived of the benefits of a broad spectrum of life experiences, as adequate familiarisation with societal demands is essential for practical technological literacy.

Future engineers need to be better equipped to deal with people of diverse backgrounds, such as from social sciences, management and communication. Some of the grand challenges that are enlisted, from an engineer's point of view, are environment protection, hunger, energy and controlling the spread of the diseases. We need developers of responsible technologies and products.

We need managers to manage things, but we also need adequate number of things to manage. There is a need to build faith in the public that engineers are sensitive to their concerns. One of the biggest responsibilities of engineers is to keep themselves updated about professional developments and practices.

This leads us to ask: Do engineers take pride in designing a thing and manufacturing it as they take pride in packaging it?

Convergence of Engineering and Life Sciences

There is optimism in the biotech world. The optimism is due to its ability to manipulate and build living things. In order to retain this optimism, engineering and biology must come closer to each other.

Bioprocess engineers have understood the need for better interaction at the molecular and whole cell level as well as at the whole cell and community level. They want to better understand the behaviour of biological systems when perturbed by genetic, chemical, mechanical, or material interventions or subjected to pathogens or toxins. They have realised that there is a need to evaluate, and if needed, to manipulate metabolic networks. Biocatalysts are also expanding their horizon; they are now being designed for specific uses. New bioinformatics-based strategies are being developed for protein design.

A revolution of sorts is also taking place in the field of biomedical engineering, where the major concern is to establish an interface between man and machine that are relevant for healthcare and medicine. Cell engineering has extended its boundary beyond the cells. It has included the use of cells and part of cells to build structures, such as tissue and ultimately the whole organs. Cell engineering uses, besides cell biology, engineering principles, chemistry, nanotechnology and material science to control cell behaviour. Some of the major application of biomedical engineering is in the areas of development of drug delivery platforms, technologies for imaging the body and its components, devices for replacing neurological function, biomaterials for use in the body, organ replacement systems, artificial blood, etc.

Both bioprocess and biomedical engineering are application based engineering disciplines such as ceramic, agriculture and petroleum engineering. In the changing biotech scenario, it has become essential to bring bioprocess and biomedical engineering disciplines together on the same platform. Biological engineering, the merged discipline, will have extended territory and knowledge base. This broad-based engineering programme needs to be developed in the pattern of science-based engineering disciplines, such as electrical or mechanical engineering. Science-based engineering disciplines are not specific to any particular industry; electrical engineering is not only restricted to electrical industry but also serves several other industries.

Biological engineering has potential to redesign enzymes, genetic circuits and cells to their specifications, or even build biological systems from the scratch. As gene synthesis becomes cheaper and faster, it is believed that in the near future, it will be comparatively easier to synthesise a microbe than to find it in nature or retrieve it from a gene bank.

Life, physical and engineering sciences are converging. This convergence is being hailed as the third revolution in biological sciences. According to a document prepared by the Massachusetts Institute of Technology (MIT), this convergence will be the emerging paradigm for how medical, energy, food, climate and water research will be conducted in the future.

The idea of convergence is to bring closer the technical tools as well as the "disciplined designed approach" traditional to engineering and physics and apply them to life sciences. In the third revolution, it is hoped that engineering and physical sciences will transform biological sciences, and biological models will simultaneously transform engineering and physical sciences. We are hoping for "intellectual cross-pollination". As a result of convergence, several exciting discoveries are shaping into realities. Convergence research in biomedical sciences is possible through the convergence of material science, engineering, chemistry and biology. Another interesting area is developing computational models to understand how the body's immune response operates. Yet another area is prevention of blindness through high-tech imaging technology. The convergence of neuroscience, electrical and computer engineering, and chemistry has made it possible to design brain grafts for treating brain disorders and injury. This is possible due to the coming together of various emerging scientific and engineering disciplines.

Engineers Have the Mind of a Polymath

Someone who knows something about everything and everything about something is a polymath. Engineers have the mind of a polymath. Leonardo Da Vinci, apart from painting studied anatomy, biology, mathematics and engineering. He was known for his intellectual, artistic and physical pursuits. If he was proud of his artistic creation, he was also proud of his ability to bend iron bars.

We all are polymaths, only our qualities are different. There is something else, besides resolve and effort that makes a polymath. Ernst Schumacher writes, "Beethoven's musical abilities, even in deafness, were incomparably greater than mine, and the difference did not lie in the sense of hearing; it lay in mind".

Howard Gardner says that talent and expertise are necessary, but not sufficient to make someone original and creative; "achievement is not just hard work: the differences between performance at time 1 and successive performances at times 2, 3, and 4 are vast, not simply the result of additional sweat". Gardner believes that the answer to the question—why some minds are more beautiful—will come through a combination of several findings. Genetics will give some insight on why highly talented individuals have a distinctive, recognisable genetic profile. Neuroscience will explain why there are differences in structural or functional neural signatures. Cognitive psychologists will tell us more about the psychology of motivation of talented individuals. They will tell us why the talented individuals develop passion to master their art.

Ideas do not magically appear in a genius' head from nowhere. They always build on what came before. Creativity is a chain reaction of many tiny sparks. Creativity, like many other pursuits, is the product of hard work. R. Keith Sawyer says, an idea may seem sudden, but in reality our minds have actually been working on it all along. Sawyer says that insight and execution are inextricably woven together. If the executioner is different from the one who first "saw" the idea, chances of its success are lesser. One of the advices of Sawyer is to develop a network of colleagues and have freewheeling and unstructured discussions with them.

The minds of creative people are special. They are wired for rapid and fluid thinking, as they enjoy the ability to make quick associations. These minds can see things in new ways. They can make connections between old things. These minds are flooded with ideas like all other minds, but they have the ability to control the flight of ideas that many minds do not have. These minds also get bad ideas, but they possess the ability to purge the bad ones. These minds, like all other minds, suffer from mood swings of high and low and are vulnerable and sensitive. But creative people know how to navigate the tides.

Our spiritual and religious traditions tell us that the quieter you become the more you can hear. We are advised to watch the thought, feel the emotion and observe the reaction. Since we have mind, we can make sense of the world in a meaningful way. Since we have mind, we have memory. Because we have memory we can preserve, retain, and subsequently recall, knowledge, information and experience.

The human brain is not complete at the time of birth. Dynamic changes take place in the human brain throughout life, probably for adaptation to our environment. The conventional view was that we are born with a set number of neurons, and we are hardwired in a certain way. But the new understanding about brain functioning is that the human brain can shape, form, eliminate and strengthen new connections throughout life. Neuroscientists say that neurons can change their connectivity, morphology and strength of the connections in their early as well later stages of life in response to environment and experience. Research has also shown that brain has "use it or lose it" approach to neurological maintenance.

The engineering world is changing. A new mind is evolving to deal with this world and with that is changing the future of engineering ethics. Some old ethical questions shall remain, some old issues will be forgotten and some new questions will be asked. As the UN's Millennium Project says, "Some future issues are further in the future than others". One of the future issues of engineering ethics is, if technology grows a mind of its own, what ethical obligations we have for its behaviour.

Engineers have a meaningful mind. Meaning is understood in relation to an individual's basic needs, experiences and emotions. Meaning is interactive, selective and value-driven. A really meaningful mind understands others' mind as much as one's own mind. "Meaning is like a large map or web, gradually filled in by the cooperative work of countless generations". Meaning is thus, more linked to one's cultural identity.

One of the purposes of the universe is to provide and prepare ground for the emergence of an intelligent life. The purpose of the intelligent life is to ask profound questions and probe the nature of the universe itself. We are a miniscule part of a multiverse. Ours is probably the only universe that supports life. Thanks to engineers, we have been able to do what we are supposed to do.

The nature of our questioning mind varies. Milan Kundera says, "It is questions with no answers that set the limit of human possibilities, describe the boundaries of human existence". Questions take us away from our comfort zone and help us to move forward. "The desire to ask a question shows a higher level of thought, one that accepts that your own knowledge of a situation isn't complete or perfect".

Socrates liked to ask questions. Socrates asked questions for clarification, questions that probe assumptions, questions that probe reasons and evidence, questions about viewpoints and perspectives, questions that probe implications and consequences and questioning the questions. Socratic set of questions do not assume you are right or wrong.

Engineers in Innovative Ecosystem

Innovation scenario is changing in the engineering world. Joichi Ito of MIT says, earlier the chain was ideate—propose—raise money—plan—build. Now it should be completely the opposite, build and then figure out the business model. This is unstructured innovation. "Some of the most interesting innovations happen when the person doing it doesn't even know what's going on. True discovery, I think, happens in a very undirected way, when you figure it out as you go along", says Ito. If you are in a discipline, you are worried about peer review, as a consequence you are knowing more and more about less and less, and that is, Ito says, is incremental thing. When you are anti-disciplinary, you have the "freedom to connect things together that aren't traditionally connected".

If you are routine, you are not fit to work in an innovative organisation. Innovative organisations recognise the value of the individuals in the organisation. They provide opportunities, so that one can stay at the bench without forfeiting any of the managerial incentives. Innovative organisations recognise that bureaucracy is the greatest enemy of innovation. The learning cultures that produce innovators include collaboration, problem-based multidisciplinary approach to learning, taking risks and learning from mistakes, creating real products for the real people, and encouraging intrinsic motivation rather than relying on extrinsic motivation (like "carrots and sticks, As and Fs").

In this decade of innovation, our lives and the whole human ecosystem are going through a major transformation. A system collapses when it runs out of resources, and when consumptions are not enough. Innovation is a viable means to avoid this crisis. "As we approach the collapse, a major innovation takes place and we start all over again", explain Geoffrey West and Luis Bettencourt of Santa Fe Institute.

In an innovation ecosystem, like in nature's ecosystem, no single actor can function in isolation. In an innovation ecosystem, "experienced people can mingle with newbies". Getting the right people is the most important ingredient of an innovation ecosystem. Another important element is mixing people in productive ways. The diffusion of ideas, skills and expertise is important for stimulating innovations, and cooperation is essential for the exchange of relationships.

Some of the qualities we expect in an "ideal" engineer are strong analytical skills, creativity, scientific insight, leadership abilities, high ethical standards, dynamism, flexibility, pursuit of lifelong learning, and dedication for public cause. It is a tall order even to conceive such an ideal, but then to create an ideal is always a tall order. Knowledge will never be complete, because perfection is unattainable, and truth is unfathomable. Moreover, we do not want knowledge frontiers to advance so rapidly that we as a society lag behind.

Can a Good Engineer Become a Good Manager?

In the world of engineering, technical skills are not enough to provide leadership. The engineers are expected to better bridge the gap between innovation and manufacturing. Being the "best in the world" in scientific discovery is important. It is also important to learn, as Paul Jacobs says, how to work in interdisciplinary teams, how to iterate designs rapidly, how to manufacture sustainably, how to combine art and engineering and how to address global markets.

People expect from a low-priced car all the good things that make a car a good car. In fact, this kind of expectation tends to push up the bars of engineering excellence. Customer's priorities and demands are most important and that decides what kind of trade-offs can be made to lower the costs. There is nothing like "frugal" nation, as far as expectations are concerned. We all like low-cost-high-value products. Only frugal innovation that works on the principles of "calculated trade-offs" succeeds. The question is—who needs the frugal innovation more—resource rich or resource constrained?

The job profile of and also the expectations from an engineer are different from that of a manager. The skills required to become a good engineer and a good manager are different. The requirement of one job is "focussing", whereas the need of the other job is "overseeing". As an engineer, one is evaluated on the basis of performance. A manager is evaluated on the basis of group performance. For a manager, what matters most is relationship building and conflict resolving skills. Removing bureaucratic hurdles is one of the major responsibilities of a manager. One way is to learn the game of ethical politics to achieve what you want to achieve. The problem is playing politics is not every engineer's cup of tea. They find it difficult to learn and practice it, ethically or otherwise. The job of a manager is like that of a caretaker. A caretaker takes care of what is in place and tries to make it more efficient.

B. Michael Aucoin, author of From Engineer to Manager—Mastering the Transition, says that transition from engineer to manager is possible. His recipe for successful transition includes the following six fundamental principles: mastering relationships, seeing the big picture, getting things done, communicating effectively, using assets wisely and taking things to the next level. It means good engineers must also possess strong interpersonal skills, if they want to become good managers. Engineers are "individualistic" by nature. They are required to develop the gelling capacity that is required in group activities, if they want to become good managers. Aucoin writes, "Engineers are uniquely qualified to be managers and leaders, in large part because they understand systems-thinking so well. Once you understand that organizations are simply systems of people, you've got it made".

We the people have a deep yearning to visualise the big canvas and work together effectively to see the big picture. Perhaps this is one of the major driving forces of becoming a leader. We are prisoners of our own mind. Our world is the size of our mind. We only can free our mind. We only can change our mind. It is our mind where most interesting interdisciplinary conversations happen. It is where the big ideas are generated. So much goes on inside the mind, even when it is supposedly "empty".

Engineering Needs Art

There is an enormous interplay between and mutual dependence of engineering and art. Renaissance engineers were artists. Leonardo Da Vinci was the ultimate engineer and artist. Art, at tandem with engineering, can take us to the blind and imaginary spots. It is said that art deals with incoherence, imprecision, abstraction and contradiction. These attributes are not expected to represent engineering. Isn't incoherence an essential aspect of the human mind? Don't we live in a world full of contradictions? The issue is how to make the "two cultures" merge and move forward together. What would close the communication gap between engineers and artists, since each side has something useful to offer to the other side?

Artists generally are considered more playful. Engineers are considered serious types. Does playful approach diminish seriousness? Are not both engineering and art serious business? Both artists and engineers need imagination. Both get distracted by complex social problems. Both want to come out of the blind alleys in their own way. Engineering gives us perfection. Art brings into the fold of engineering, the unexplainable world of magic. It does not matter if it also brings into the fold some amount of incoherence, imprecision, abstraction and contradiction. Aren't these also the attributes engineers need to be equipped with?

The world of science and technology is changing very fast. Collaboration among diverse disciplines is becoming the norm. More and more people are realising the importance of relatedness among unrelated things; the more unrelated the elements are, the more radical and innovative is the synthesis.

Knowledge sharing is possible when basic values and understanding among the collaborators match. A conceptualist can collaborate with an experimentalist, but their roles in, and their approach to problem-solving are different. This is not to say that others' views should be accepted without critical examination, particularly when they come from entirely diverse sources. In the engineering practices, collaborators share and complement conceptual or experimental approaches. In the world of art, personal taste, vision and style of expression matter. Sharing and complementing two diverse characteristics need better understanding. There is so much truth in the beautiful observation of Murray Gell-Mann, "What is especially striking and remarkable is that in fundamental physics a beautiful or elegant theory is more likely to be right than a theory that is inelegant".

The Future

No one knows where our future lies, still we make projections. The visionaries among us are both utopian and dystopian. Utopian visionaries see in the future world a state of balance and peace, and where all life is valued and sustained. Since the world has achieved its full potential, they see no reason for one to be aggressive. They maintain that man has no enmity or competition with nature. In the future world, they find no difficulty to expect a world of equal opportunity and equitable distribution of goods and services. They see in the future world abolition of cultural, racial and gender-based prejudices. They believe that humanity has solved all its problems with the help of sensibly developed and rightly used technologies. Dystopian visionaries, on the other hand, imagine a future world where life and nature are recklessly exploited and eventually destroyed. They predict catastrophic destruction of our natural environment. They imagine the loss of complete freedom of the mind, due to technological interventions. They believe future generations will depend more upon artificial intelligence than their native intelligence. They believe technology will make them slaves of technology.

Utopian future is projected by the idealist visionaries. Dystopian future, on the other hand, is projected by those who feel oppressed by their environment and are afraid to fight extreme odds. Overly disastrous future projection is definitely not a great idea, as excessive optimism is. The point is to avoid utopian and dystopian extremes and take a conciliatory middle path. If optimism is mixed with some amount of pessimism, it works better. If in certainty, a certain amount of uncertainty is mixed, it generally leads to better end results.

An ideal engineer is an adaptationist. An adaptationist extends or contracts oneself as per the demands of the situation. An adaptationist knows how to manage situations when extrapolations fail.

What should we do, as a country, to become the real player of the future world? Countries not only want exceptional scientists and engineers, but also people who are temperamentally innovators and think they fit into their country's core values. Many would say, prepare young minds and nurture future innovators.

The young innovators want to make a difference and believe that they can live on less. This generation knows how to find support for what they need. They are not afraid to take risks. Mistakes strengthen their self-confidence. For the young innovators, the advice of the wise is "to stay true to what your passion really is and your sense of a larger purpose in life".

We, the people of India, once believed that there was no country like ours, no king like ours, and no science like ours. Can't we regain that spirit, and that confidence? Can't we, a country of 1.2 billion people, get back, literally, to zero?

Directed and Systematic Approaches Towards Sustainability in the Twenty-first Century

Baldev Raj

Life for me has been working on challenges, rejuvenating myself and again working for enhanced purposes. Good proposals and thoughts get non-acceptance, in the hands of a few decision-making persons, but I have learnt that a good proposal and work is accepted sometimes after a multitude of failures. Success happens when divine desires. I have gratitude for my mother, my wife, children, large family of childhood and now, school and higher education teachers, my colleagues and mentors from all walks of life in India and elsewhere in the world.

The human-dominated planet earth ecosystem, which is inclusive of all biological systems and other non-livings, is increasingly being driven by technology and social behaviours. It is, however, sustained by natural life-supporting systems and vitalized by individuals of high merit by bringing in paradigm changes amidst challenges.

The concept of ecological engineering-based engineering management is emerging as a dominant parameter to guide science, technology and innovative approaches towards green and sustainable technologies in the twenty-first century. Effective and comprehensive engineering management is intertwined with socio-economic-nature ecosystem, which is complex and multidimensional. There are systematic but sometimes hidden ecological contexts and contents. Currently, most of the technologies are oriented to mono-objective, namely open loops of material flow, and rigid products and technological processes.

Only when engineering management is placed in the perspective of an ecosystem and is focused on the concept of mutually beneficial coexistence of nature and society, an effective engineering management can be achieved and a goal of sustainable development realized. Ecological engineering can be considered as the bridge between "society economics" and nature-based cosmos. A few unique characteristics and advantages of theory and methods of ecological engineering are proposed by Xu and Li [1] for engineering management. They propose that interdiscipline of engineering and ecology leads to superiority and sustainability of technology. This is to my knowledge a robust approach for integrating engineering

B. Raj (✉)
National Institute of Advanced Studies (NIAS), Indian Institute of Science Campus, Bangalore 560 012, India
e-mail: baldev.dr@gmail.com

P. Ghosh and B. Raj (eds.), *The Mind of an Engineer*,
DOI 10.1007/978-981-10-0119-2_2

management and ecological engineering. Though the approach proposed by them is qualitative at this stage, it has the potency to evolve into an integrated and quantitative approach (based on the weightage of controlling parameters). Further useful models for desired sustainable technologies can be built. It is clearly a direction which needs to be pursued with the help of domain specialists, environment experts and social scientists.

We, human beings, inhabit the earth by occupying self-designated superior hierarchy in biological community [2]. The progress in science and technology without a balance with ecological considerations has resulted in the loss of culture, biodiversity, degradation of the earth and even the universe. This approach has also resulted in building new and hybrid ecosystems which many a times are adverse to the chain of species, evolved over millions of years. These new ecosystems can be visualized as interconnected subsystems defined by parameters such as economic growth, cultures of societies, creativity and sustainability. In this clearly highly nonlinear interconnected system, it is true that activities of human species have considered a huge disconnect with nature (eco-wealth). The basic concept of the ecosystem, for ecological research in places where people either were not in residence, or had relatively modest effect on the habitat, was that such systems were composed of organisms and the physical environment with which these interacted. Plants, animals and microbes constitute one side of the equation and air, soil, water, temperature, light, etc. the other. When these components interact, they constitute an ecosystem. Within ecosystems, man now has a predominant role to play which disturbs the delicate balance of biological system evolved so far, eventually raising a concern of getting engineered in a chaotic fashion.

It is an inescapable reality for human beings to seek "balance" and "sustainability" in the practice of technology within the ecosystem boundary, as human beings are coming to realize that the more the invasion of the nature, the more would be the environmental problems and that sometimes, these will threaten to engulf our achievements meant to improve the quality of life on this earth. The penalties for economics and more importantly for the loss of biodiversity and quality of life can be either immense or irreparable.

It is important to realize that sustainable development calls for paradigm changes in our current modes of production, consumption, decision-making and, of course, engineering management. Ecological engineering-based engineering management practices could be a beneficial attempt to make such effective changes with a systematic approach.

In the twenty-first century, "sustainable development" has become more than ever important concept in the context of economics and environmental intertwined policies. In December 1997, the Kyoto Protocol [3] aimed at addressing the global warming and the deterioration of the ecosystem. In July 2010, countries signed and ratified the Protocol. It is generally argued and accepted that the reason that "sustainability" is important is because the needs of the present are to be met without compromising the ability of the future generations to meet their needs. The sustainability should also push rather than thwart the drivers of continuously improving the quality of life of all living species on this planet.

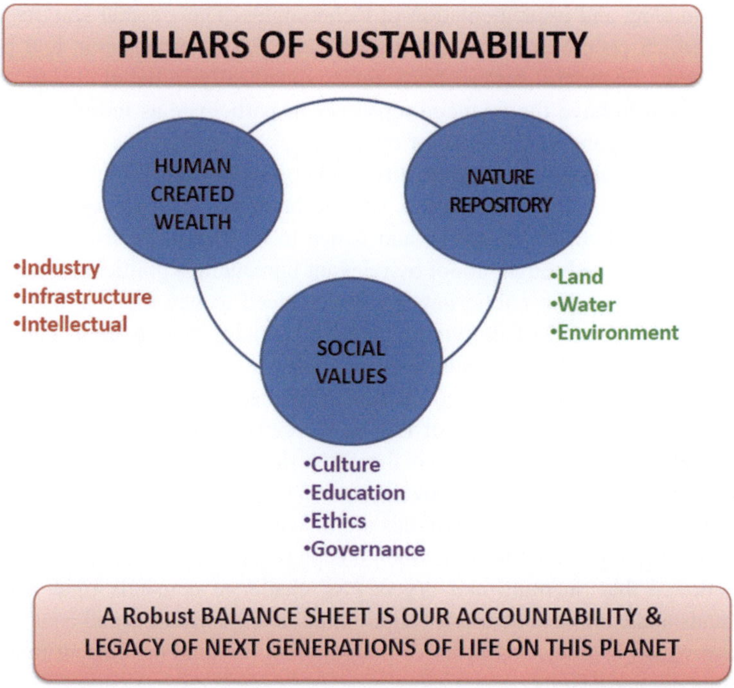

Fig. 1 Three pillars of sustainability

Sustainability in human-dominated ecosystems requires attention to three overlapping realms. An illustration of three pillars of sustainability is given in Fig. 1. The interlinkages and interdependencies are clear, but complex engineering management many a times does not take into account societal and natural wealth considerations. It also ignores cultural strengths and traditions of marginal economical groups, which are vital links in the chain of sustainability. It can be argued that such sustainable technologies are possible with proper merit-based opportunities, national policies, international commitments with clear guidance on issues such as gender equality, child labour, drug trade, low-carbon technologies, sustainable energy, water and agriculture and commitment of professionals in a cohesive and effective way towards developing sustainable technologies in an expeditious manner with clear timelines and promising measurable results.

Sustainability of technologies based on three aspects, i.e. natural wealth, human-created wealth and cultural and intellectual excellence with ethics and equity, is a robust and achievable goal. In simple words, ecological, social and economic sustainability is a challenging proposal to the current generation of professionals. Ecological sustainability is fundamental; it is about ethics and the right directions for social and economic sustainability. Social institutions, communities, culture and quality of life are integral parts of social sustainability. This is often ignored, but it is argued that this vital link needs to be incorporated in decision-making and implementation of technologies. These social features are the ways in which people organize and respond to satisfy their material and psychological needs. Economic sustainability refers specifically to the ability to derive a livelihood and to have the financial resources to participate as individuals, households and institutions in the work of the society.

These three features of sustainability, namely ecological, social and economic components in judicious combination, have to be in place for an ecological–social system to be resilient and adaptive and hence to be able to persist through interlinked changes to be brought about by relevant bureaucrats, politicians, economists and social workers. The public has to be convinced to pay the cost and bear the pains of hardships and of deriving satisfaction arising out of these changes for realizing sustainability. The countries, on their parts, should forge international treaties to realize commerce based on sustainable technologies.

Key to improving the quality of life on the earth through sustainable development is delivering synergy between the three drivers, i.e. economy, society and ecology. The term "sustainable development" recognizes that economic growth, social welfare and environmental issues are linked and must be addressed together, rather than in a fragmented way in which it is pursued and practised currently. It can be mentioned that quantum and urgency of challenge demands right actions of individuals and combined actions of right minds of organizations and countries, and at a pace which is at least an order of magnitude more than the current measure of commitments and results.

I wish to focus on my area of specialization, namely materials science and engineering. Materials are keys for technological advances in engineering management [4, 5]. Added value through materials with higher knowledge content, new functionalities and improved performances is becoming increasingly critical for industrial competitiveness and sustainable development. It can be argued that the materials themselves are the first ingredient in economic and social improvements leading to progress.

Sustainability based on engineering management leads to new demands for materials. Sustainable materials and sustainable materials management (SMM) are needed for sustainable production. The OECD working definition of SMM was developed in 2005, and it states: "Sustainable Materials Management is an approach to promote sustainable use of materials, integrating actions targeted at reducing negative environmental impacts and preserving natural capital throughout the life cycle of materials, taking into account economic efficiency and social equity". Materials technology impacts a manufactured product by its choice of materials (metals, ceramics, glass, polymers, etc.) and processing steps (shaping, joining, finishing), recycling, etc.

I illustrate my perspective with a few examples taken from the literature and private communications [6]. Aluminium is a green metal and a good replacement for wood. Aluminium helps in protecting forest covers. Aluminium industry is continuously facing opposition from environmental ministries and non-governmental organizations. The discovery of East Coast bauxite deposits in the late 1970s has put India on the world aluminium map as the fourth country with more than 3 billion metric tonnes of bauxite reserves. Mineral resources are non-renewable. However, their life can be stretched through adoption of new technologies at various stages of production, processing and use. Moreover, aluminium industry is sensitive from environmental angle, and hence, careful planning is a must to ensure that impact on the environment due to mining, refining and smelting operation is the least.

Researchers in India are pursuing a road map with innovative technologies for alumina and aluminium industry, such as zero waste refinery, water management and red mud management. Prioritized development of science and technologies is enabling India and the world to adopt cleaner and greener technologies. K.K. Khanna [7] recommended a new environmentalism to have acceptable intertwining of technology and management, and the involvement of environmental specialists from the inception of improvement in technologies to have better results of acceptance and competitiveness. This approach is different to that of Xu and Li, mentioned earlier in the article. The two approaches need to be practised in a few case studies to achieve maturity based on validated successes.

India has done well in engineering management towards sustainable green environment-friendly technology in leather processing. There is a need for transition from a "Don't-ecology" to "Do-ecology" approach in leather processing. It has been recognized that end of pipe treatments in isolation is not an adequate strategy to meet the requirements of wastewater norms and standards. J.R. Rao communicated with me to state that an ideal approach is to target the zero or near-zero discharge of waste liquors by appropriately modifying the leather processing. To reach the desired targets, engineering management based on ecological sensitivity has been researched and developed.

Further resources are focused on newer green science and integration with ecological consideration to environmental technology and cleaner technology [8]. Desired approaches, such as "reverse leather processing", "natural colours" and "chrome management", are pathways to realize the goals. Research at Central Leather Research Institute, Chennai, has proved that the innovative approaches enable a significant reduction in chemicals, time, power and cost. Thus, green and ecological engineering is possible and can be made competitive in even small industries with little expertise in advanced science and technology which would be a basis for realizing sustainability and economic growth in a cohesive and inclusive manner.

I would like to dwell in brief on energy security and sustainability, another area on which I have been engaged for almost four decades [9–13]. Adequate availability of energy to all the citizens, high GDP growth along with the concerns and threats of climate change and sustainability doctrines are the corner stones of the energy policy of India. The policy focuses on judicious choice and a mix of fossil, hydro, nuclear and renewables. The policy also balances progress and GDP growth for removing poverty, vis-a-vis climate change concerns, economics of energy and energy security.

Environmental concerns are associated with all forms of energy including fossil fuels, nuclear energy and renewables, throughout the energy chain from exploration, mining, transportation and generation to end-use. Life cycle considerations

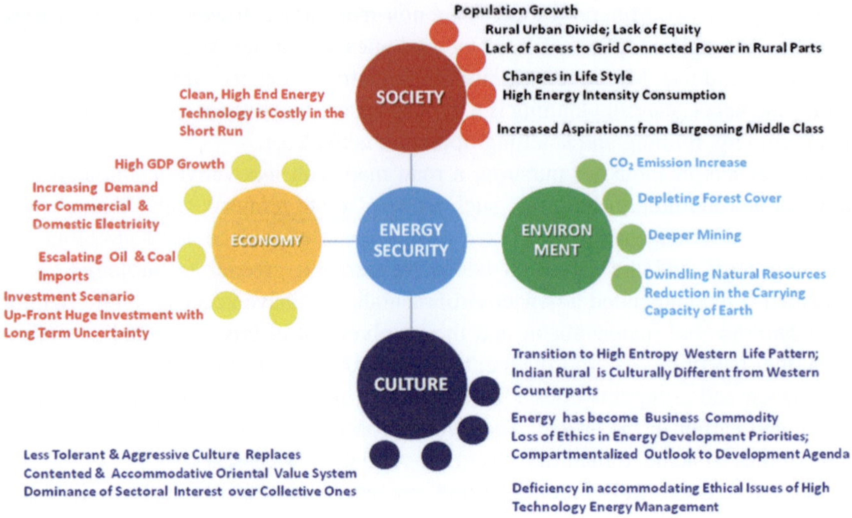

Fig. 2 Energy security: a multidimensional challenge

and risk management in new technologies combined with technology foresights are key to success in the energy demand facing India. Energy security is a multidimensional challenge (Fig. 2). It can be argued that without meeting the challenges of energy security with sustainability and affordability, Indian dream of being one among the first five countries in the world will not be realized. I am of the view that current models for forecasting energy needs assumptions in balance of energy resources and technologies, forecasts and foresights of individual technologies, essential availability of human resources, dynamism in sustained policies, etc. need engagement of appropriate experts drawn from fossil, nuclear, renewable, hydro, social sciences, climate change, specialized bureaucrats and politicians. We need an empowered mechanism for India to meet the challenge for the vital need of India, i.e. ensuring energy security, affordability and sustainability. India, by achieving success in this challenge, can be a pioneer nation for other nations to emulate and can simultaneously grow huge business (internal and external) for our growing economy. The challenge also provides a fertile area for national and international collaborations in science and technology at the cutting-edge frontier. Moreover, achieving this challenge in a cohesive manner shall enable establishing a robust mechanism for meeting other challenges, say in water, health care, infrastructure, etc. Currently, our R&D intensity, concepts to maturity pipelines and strategies are immensely inadequate in comparison with the challenge.

I wish to utilize this opportunity to dwell specifically on roles and concerns about nuclear power for energy security. The assessment of the risks of nuclear power generation should be done in the context of climate change and change the use and production of energy towards meeting large unsatisfied needs of the country now and in the coming decades, of our growing economy and the citizens of India for better quality of life. From the analysis, it can be argued that nuclear energy has to be utilized much more intensely in the decades ahead in the Indian context. India has done well to make significant advances in science and technology of water reactors, sodium-cooled fast reactors, thorium technologies, etc. for energy and in developing mature expertise in interrelated comprehensive range of science and technologies spanning almost all domains of science (including mathematics and life sciences) and technologies such as food, health care and water. India is among the world leaders in fast reactor programme with closed fuel cycle, a sustainable energy source, as assessed by International Atomic Energy Agency and leading countries of the world (Gen IV). This energy system is sustainable over centuries and can be made competitive in cost, once developed, with relevant science and technology, and operated in a business model based on a large-scale exploitation of this system. India is also a clear leader and pioneer in thorium technologies.

The concern for nuclear energy is the probability of access of fissile materials (by wrong hands) safety under extreme conditions of accidents and management of high-level wastes. Indian nuclear programme, encompassing three stages and doctrine of closed fuel cycles, has a clear possibility of addressing these concerns with significant improvements. The current challenges include scaling the quantum of nuclear energy to be multiplied expeditiously and public acceptance on the merit of informed opinion and not myths. The country and the Department of Atomic Energy need to initiate a well-conceived paradigm changes in the strategy to meet these vital future challenges to enable nuclear power to meet the energy demands when it is needed most that is now, in the next decade and a decade later.

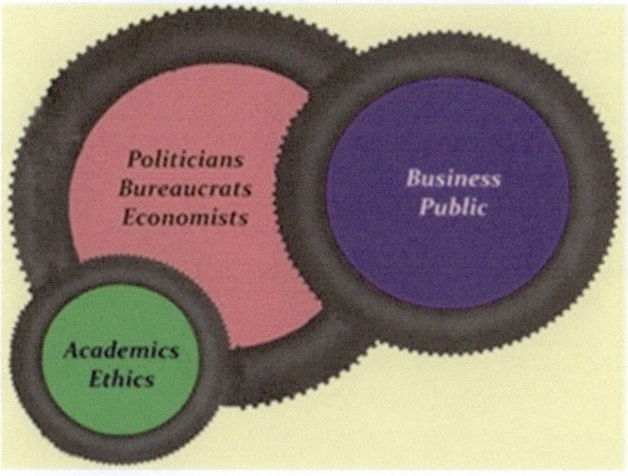

Fig. 3 This scheme shows overindulgence of business, politics, bureaucracy and economics combined with out-of-proportion interventions by the public. The steering wheel is essential in the hands of academicians and outstanding professionals in all domains with the clear mandate for balancing growth, equity and sustainability

In the twenty-first century, it is an inescapable context and trend for humans to seek a balanced mix of nature wealth—economy and societal drivers for attaining "sustainability" in the practice of materials, technologies and energy while continuing to improve the quality of life on this planet with higher ethics and equity. Currently, a disturbing feature amidst population of 7 billion human beings is vivid signatures of fatigue and even failures which need to be corrected with clear results to generate confidence and pave clear pathways towards sustainability. These requirements demand new approaches, innovations, materials and technologies. I have depicted in Fig. 3 that academicians and intellectuals in all domains should take the steering wheel to guide the society for balancing the progress with ethics and sustainability.

References

1. Xu JP, Li ZM. A review on ecological engineering based engineering management. Omega. 2012;40:368–78.
2. Ma SJ, Wang RS. Social-economic-natural complex ecosystem. Acta Ecol Sinica. 1984;4 (1):1–9.
3. Kyoto Protocol: Status of Ratification. United Nations Framework Convention on Members of the Commission. Our Common Future. In: World Commission on Environment and Development (WCED). Oxford: Oxford University Press; 1987.
4. Fiksel J. A framework for sustainable materials management. JOM. 2006;58(8):15–22.
5. Jacobs JA, Kilduff TF. Engineering materials technology: structures, processing, properties, and selection. 2004; 67:2.
6. Kumar M. Future prospects and challenges in aluminium industry in India. Metal News. 2011;14(4):14–9.
7. Khanna KK. Environment policy—a global perspective. IIM Metal News. 2011;14(5): 9–13.
8. Rao JR. Cleaner technology paradigm: a case of leather making. Central Leather ResearchInstitute, India (Personal communication).
9. Raj B, Rajan M. India's fast reactor programme in the context of environment sustainability. Int J Environ Stud. 2007;64(6):729–47.
10. Chichilnisky G. What is sustainable development? Land Econ. 1997;73(4):467–91.
11. Raj B, Raju S. Materials for energy systems. Kolkata: Publication of Indian Institute of Metals; 2013.
12. Raj B. NP Gandhi memorial lecture, energy, ethics and equity. IIM Metal News. 2011;14 (6):15–25.
13. Walter AE, Todd DR, Pavel V, Vetkone TS. editors. Fast spectrum reactors. In: Raj B, editors. Core materials. Berlin: Springer; 2012.

Treating Challenges as Opportunities

B.N. Suresh

As we move along in our life or in our career, we face challenges, which are natural. When we come across such challenges, there are two ways of treating them, either from a perspective of helplessness or from a standpoint of one's own belief. But when we choose the latter, we are consciously optimistic and these challenges open up a vista of opportunities. Once we accept a challenge, no doubt we step into the world of the unknown and it has a potential to change us forever.

With the right mental attitude, one can certainly reframe the way one treats the challenges. To look back and view such challenges I have faced in my long career of four decades and more at Indian Space Research Organisation (ISRO) and Indian Institute of Space Science and Technology (IIST) at Thiruvananthapuram, are some of the most valuable life experiences for me. Throughout my career, I was quite fortunate to come across a number of challenges, and knowingly or unknowingly, in all such cases, I have taken decisions to accept them. Each such challenge has transformed into an opportunity and has helped to learn something new and very valuable in life. As a consequence, it has made me more confident and not to shy away from facing challenges as I go along. I want to share some of the interesting challenges I have faced during my long career at ISRO and narrate to readers how it helped to open up gainful opportunities which have benefitted me immensely.

My entry into ISRO itself was not by design, but it happened. We were ten students, studying postgraduation in mechanical/machine tool design at IIT Madras, Chennai, during 1967–1969. In the middle of 1968, a few of us in the final phase of our project work bounced on a small advertisement in one of the national papers, about a few openings for engineers with postgraduate qualifications at Space Research Organisation at Thumba. The very fact, the advertisement mentioned that they would pay all expenses for the candidates called for interview; it attracted us, as we felt it is a good opportunity to visit Kanyakumari which is only 2 h by bus from Thiruvananthapuram. We did not even remember about our application, till four of us got interview calls on one fine day in November 1968. We presented

B.N. Suresh (✉)
ISRO Headquarters, Antariksh Bhawan, New BEL Road, Bangalore 560023, India
e-mail: byranasuresh@gmail.com

P. Ghosh and B. Raj (eds.), *The Mind of an Engineer*,
DOI 10.1007/978-981-10-0119-2_3

ourselves before a small selection board, and the interview was smooth. We returned to Chennai after fulfilling our main aim of visiting Kanyakumari. It was only in March 1969 two of us received the news of our selection, but it did not excite us, as we felt that this posting is only a springboard for us to look elsewhere shortly.

Nevertheless, myself and my friend who got the offers joined Space Science and Technology Centre (SSTC) in the month of July 1969. Later on, in 1972, it was named as Vikram Sarabhai Space Centre (VSSC) after the founding father of space programme Dr. Vikram Sarabhai. Upon joining, I was instructed to report to Dr. SC Gupta, the Head of Control, Guidance and Instrumentation Division (CGD), who subsequently went on to become Director of VSSC and steered the launch vehicle developments very efficiently. I met him the same day and he spent about half an hour explaining the various tasks of the Division and also broadly the goals set for the Organisation. He mentioned that he has a particular assignment waiting for me and stated that it is the development of a control system for the first stage of the proposed SLV-3 launch vehicle. He also gave a broad outline of the development involved. Obviously, I did not feel very comfortable with the proposed activity since the task assigned to me is in the control system, whereas all my studies were in mechanical engineering and design. To accept it was certainly the first challenge of my career. I spent two days thinking over the same, again and again, discussing with my seniors who were already there and spending time in library. At the end of two days, I had a great fear of failing in my first assignment, but I decided to accept the challenge as it would open an opportunity to learn a new discipline right in the beginning of my new career.

The difference between success and failure probabilities when we face a challenge is only in our perception. How we represent things to ourselves determines our response to that situation. Our positive response surely gives us the right solution. I spent a lot of my time in the next couple of months, studying the new subject intensely in library and discussing with the control experts in the Division. Deep and deep I started exploring the new subject, and it roused my interest and became more and more fascinating. This intense involvement and urge to learn something new led me to build a proto-unit of the control system successfully within a span of one and half years.

Dr. Vikram Sarabhai who was guiding the space activities in the very initial period of space programme used to visit SSTC frequently those days. During all such visits, he used to find time to visit some of the laboratories where interesting development works are being carried out. On one such visit in the last quarter of 1970, our laboratory was included in his itinerary. I was thrilled to demonstrate the successful functioning of the proto-unit of the first-stage control system to him. He spent fairly a long time showing keen interest in the system, asking a number of questions and probing my understanding in the area. I still vividly remember his open appreciation of this development work in the presence of several senior officials and visitors. We ultimately flight-tested the control system successfully in all our SLV-3 flights.

This experience helped me to expand my expertise in the new area of aerospace control system. During 1975, when I applied for the commonwealth scholarship for doctoral studies at UK, they were looking for scientists working in the area of cutting-edge technologies in India. Certainly, my experience and learning in this specialised field of aerospace control systems for five years facilitated my selection. I am happy that the challenge accepted in the very beginning of my career helped me to open up an opportunity to do doctorate in my area of interest in one of the premier institutions of UK. Also, my experience of building hardware and software in control systems at VSSC greatly assisted to complete my doctorate in just two and half years. After my return from studies, I continued to work in the same area of navigation, guidance and control, and my long experience in the control and avionics areas enabled me to head Avionics Group at VSSC, although I had my education all along only in mechanical engineering. Even now many of my colleagues do not know that I did specialisation only in mechanical area during all my studies. Learning and mastering control and avionics in my career coupled with my core area of mechanical engineering during studies helped me immensely to understand, develop and manage many of the complex aerospace systems needed for our launch vehicles.

Another interesting experience of facing a challenge was in the late 1990s. While I was at VSSC, I got an opportunity to participate in the reviews of a major National Project of Defence being developed at Bangalore. I was inducted as one of the review members from ISRO for the test readiness review board to clear the vehicle for flight. Several meetings of the board were conducted to review various aspects of the vehicle. During the course of this review period, the country carried out the Pokhran nuclear test. As a consequence, sudden embargo was imposed for very many critical elements. This resulted in denial of many high-end technology elements for some of the national programmes. This National Project also came under embargo and the result is no doubt inordinate delays in its development. In one of the test readiness review meetings, this aspect was discussed and some of us felt that we should undertake the indigenous development of all denied items in the country and replace all imported elements one by one in a progressive manner. We firmly believed that we have the needed capabilities in the country.

One of the critical elements denied for this programme was quadruple redundant electrohydraulic actuator. At that time, there was only one company in the globe, located at USA, who had the capability to supply this item to our specifications. Without understanding the full complexity of this subsystem, I took up the challenge of undertaking the development of this critical system at VSSC, based on our experience of successfully developing a range of actuators for our launch vehicle programmes. Although there was some reluctance initially, to accept this development at ISRO, we got the needed approval from our management. The interesting part is that some of the senior officials who have been closely associated with the National Project called me and cautioned me that I am making a grave mistake in accepting this challenge. Their perception is that it is impossible to develop the same successfully in the country due to the sheer complexity of several subsystems in the actuator and also due to non-availability of sophisticated facilities for their testing.

In all our careers, challenges are the real stuff of life. They certainly create an opportunity, make us stronger and smarter and provide us the way forward. The task of actuator development involved, among other things, networking with several agencies in the country, building a number of facilities at different organisations, facing a few failures as we move along with development and finding the right solutions. I knew that it was a very rough road, and the development took more than 14 years before the system was accepted by the flight certification agencies. Now, the system is ready for flight trials. Self-discovery is the outcome of this challenge, and today, we ourselves wonder how we could accomplish such a complex job, after travelling through the very rough terrain of development with several setbacks. It gave me and also to the entire development team tremendous confidence to undertake such complex jobs in future and do them successfully. It would not be wrong that if I mention that we became the second country to master this technology, next only to the country which supplied earlier. Although this development was for a National Project, it created an excellent opportunity for us at ISRO to master some of the complex technologies needed for the human space programme much ahead of our requirement. Many of the critical facilities have already been established in the country and are available. No doubt we at ISRO are immensely benefitted by this technology development since some of the new technology elements are already being implemented in the ongoing programmes of ISRO for improving the performance and reliability of the existing systems.

Important lesson I learnt from undertaking this challenge is that until we jump into challenges, we will not have any opportunities. It is one of the most empowering things we can do for ourselves. We chalked out well-planned development route looking into all aspects of development and testing. We started controlling our steps from the very first step, systematically and cautiously, so that we are in the right direction. This kind of planning and execution made it much easier for us to maintain the right direction and reach the set goals. In my opinion, it not only paved the way for success but also provided an excellent opportunity to learn many new and exciting technologies.

Establishing the Indian Institute of Space Science and Technology (IIST) under Department of Space at Thiruvananthapuram in a very short span of time was a very exciting experience. This is the period I faced series of challenges which were quite demanding, at times impossible targets to achieve and in some instances most frustrating too. The genesis for starting an educational institute under ISRO itself was very interesting. We at ISRO had the practice of briefing the Prime Minister of India, invariably after the successful launch of the major mission like PSLV or GSLV. The presentation team invariably consisted of 10 or 12 members, mostly key functionaries of ISRO, who were directly responsible for the execution of the project, and the team was led by the Chairman of ISRO. In the middle of 2005, after one of our successful launches of PSLV, some of us had an opportunity to meet the Prime Minister and brief on the salient aspects of the launch and also on the satellite. Immediately after the meeting, tea was arranged and the PM was gracious enough to join us, in spite of his very busy schedules, and spent some time. During the informal discussions, he raised a pertinent question on the availability of bright and young scientists/engineers for our future ambitious programmes. In response to our answer that we are having difficulty in getting the bright candidates as they are lured by the attractive salaries being offered by IT and other private companies, he seeded the idea of initiating a well-structured system or mechanism within the Department to overcome this problem.

This was the genesis for us to initiate our proposal for starting a Space Institute of our own. I was asked to prepare a project proposal by the then Chairman of ISRO, and the same was prepared in consultation with some of our colleagues at VSSC and ISRO HQ and submitted the same after all internal reviews. The name for the Institute was also decided as Indian Institute of Space Science and Technology (IIST) and be located at Thiruvananthapuram, under the control of Department of Space. We submitted the proposal to the Government in January 2007 and got the approval in April 2007 within a short span of four months. In the month of May, the then Chairman of ISRO called me and informed me that I should take over as Director of IIST and establish the Institute within a short time. I was quite surprised by this proposal. He further stated that we should start the first batch in the year 2007 itself. It is just four months to put in everything needed for the Institute, whereas we had nothing with us on that date. I was very shocked, as it is an impossible task to set up an Institute within such a short span just four months.

It was indeed a very big challenge, because as already stated, we did not have anything with us, like the building or space for Institute, laboratories, faculty, syllabus, how to source the students and so on and so forth. This challenge was certainly a frightening experience for me, but I felt that it is also an opportunity to test my abilities to meet such a daunting task. This feeling suddenly brought a positive spin in my thinking, and with that positive energy, I started working out the ways and means of meeting such demands. I was fortunate enough to have good colleagues at VSSC and important functionaries at ISRO headquarters including the Chairman of ISRO, who extended the full support in providing the solutions to many of our complex problems.

Identification of the temporary building was an important decision. We decided that we would utilise the new training facilities being built for the training programme of VSSC as temporary venue for IIST. Since I was also holding the post of Director of VSSC, it was easy to discuss with the Head training programme and convince her to spare the new facilities to IIST and continue her training activities in the buildings already existing at VSSC for some more time. We made some alterations to the new training complex building and created three nice classrooms to cater to three streams which we were planning to start. We had also constructed a community hall in the same complex which became very handy for us to house the IIST library and a few indoor sports facilities. We also got permission to have access to our faculty and students at VSSC library, which is one of the excellent libraries.

We decided that in the Institute, we will run three undergraduate courses in aerospace engineering, avionics and physical sciences, needed for space programme with more emphasis to the space activities. Framing of syllabus for all three streams selected in such a short span was indeed a very daunting task. We took a quick decision to constitute a National-Level Committee with members drawn from all prominent institutions in the country including IISc and IIT, R&D laboratories and experts from ISRO. Prof. R. Natarajan, an eminent academician, the former Director of IIT Madras and the Chairman of AICTE, accepted our request to Chair this Committee. After a number of sittings, reviews and discussions, we were able to put in place a very good syllabus for all three streams by August 2007, with a good option for electives.

With all these activities, we had already stepped into the month of May 2007. We need to initiate the classes for this academic year in the month of August or latest by September 2007. Neither we had any faculty members to start the classes with us nor we had finalised the mechanism to source the bright students for the Institute.

We finalised the minimum number of faculty members required for running the first year for all three streams and took a quick decision to release the advertisement with full schedule of interviews and joining date. The advertisement was released in June 2007. The interview dates were fixed during the first and third weeks of August 2007 and decided to organise the interviews at Bangalore. Fortunately, we had a good response for the interviews, and during the course of the interview, one of the external examiners from a premier academic institution asked me casually when we are going to start the courses. I mentioned that we are planning to start the classes by 20th September, just a month away from the date of interviews. He stared at me and reacted sharply, stating that he asked a serious question and not to cut jokes. It took lot of my energy to convince him that I was not joking.

In order to meet the very tight targets, I took an important decision of releasing the appointment order, of course with the necessary approvals from the Department, to all selected candidates, at the end of interviews every day with an earnest request to join the Institute on or before 10th September positively so that we can start the classes on 20th September. We also requested all the candidates appeared to stay till the end of the day to know the results and take their appointment letters same day in the event of their selection. No doubt that under such challenging situations, making this kind of crucial decision was very vital. This one decision enabled us to position the entire needed faculty at IIST by 10th September.

Since the whole process got delayed and also we were very firm to start the classes within a few months, an onerous task, we were in dilemma about the method of sourcing the bright students for IIST. We decided that we will have an intake of about 150 for all three streams. We also took a decision to join the IIT entrance process and consulted the concerned IIT authorities to allow us to source the students from their list. With great reluctance, they agreed, but with a condition that we can do it only after their selection process is fully completed. We did not have any choice but to accept it. We inserted a separate advertisement inviting the students who had appeared for the IIT entrance to apply separately for IIST. We had an overwhelming response, running into thousands of applications, and we were quite pleased.

Meanwhile, we worked on the development of entire procedure for counselling and selection in a professional manner within the short time available and declared that we would conduct the counselling in one of the prominent hotels in Bangalore. Since IIST is a new Institute and we did not have any infrastructure with us at that point of time, we were very nervous about the response on the day of counselling. But by 10.00 h, we were quite thrilled to see a big rush of students with their parents seeking the admission to the Institute. I presume that this overwhelming response was entirely due to the excellent performance of ISRO as an Organisation. We arranged an open session to explain about our grandiose plan for setting up all activities including the infrastructure for the Institute and also on ISRO's activities. Excellent presentations were arranged, followed by the long session of questions and answers clarifying all their doubts. I was very nervous that some of the parents might demand to show our facilities at Thiruvananthapuram since at that time there was none, although we had our own plans to set up the needed facilities in a short time. Fortunately, none of them insisted on this and it could be that all of them believed in the capability of ISRO in setting up the best facilities based on our earlier track records.

By 16.00 h, entire counselling process was over and all of us were filled with joy because not only we could fill up all 156 seats but also we had a long list of wait-listed candidates. The overall ranking of the students who opted for IIST was reasonably good, considering that we are a new Institution and yet to test the waters. Finally, 136 students stayed in the first batch and I am happy to mention that all these students did very well not only in academics but also in all other extracurricular activities. Most of them are presently with ISRO, making excellent contributions. Even today I admire the parents of students for their bold decision to admit their wards in this new Institute where we did not have anything at that time. I guess that their decision must be purely based on the ISRO reputation, and now, I have full satisfaction that we did not let them down and instead created innumerable opportunities for their all-round growth.

We were under tremendous pressure to declare the date for starting the session. But we did not have any accommodation for the students in the campus. We had no clue as to how to organise the same. I requested our VSSC administration headed by Sri K.M. Nair who was our efficient Controller to hunt for suitable accommodation in the city and hire. After hectic search for couple of days, we were able to identify two or three places for the boy's hostel. One of the buildings was well suited for the purpose, and the remaining accommodation was carved out of the regular houses. We had a few girls in the first batch, and we finally took a decision to accommodate them in our guest house located in the city, since it has all facilities and importantly the needed security. Once these things are finalised, we announced the date of joining as 20th September. We had the inauguration in the morning of the same day, and the classes were conducted in the afternoon. After a long time, I was able to breathe easily and go home with some sort of satisfaction of overcoming the almost impossible challenges.

While these things were being solved with hectic follow-ups, we had yet another problem of providing the laboratory facilities. We did not want to cut the corners here since it would make the students suffer. I set up a team of experts headed by Dr. K.N. Ninan, the then Deputy Director, Chemicals and Propellants systems at VSSC to survey in around Thiruvananthapuram and enter into memorandum of understanding with the Institutions which are having the best of the facilities. The team identified Government Engineering College for engineering and Mar Ivanios College, Thiruvananthapuram, for science. But to convince them to support us took a lot of our efforts because they had their own genuine difficulties to accommodate our students suiting the requirements of their students and also our students. With all these efforts which we thought one time impossible, we could conduct the practical classes for all our students very efficiently, although most of the times our students attended the practical classes late in the evening.

The greatest challenge we faced was to identify a suitable location and land for the new campus in and around the city of Thiruvananthapuram. We had estimated that we require about 100 acres of land to establish the Institute. Initially, we requested the Government of Kerala to provide the needed land, but they expressed their inability to do so since they have no vacant Government land of 100 acres in the vicinity of the city. Immediately, we took a decision to float a public tender for the same, and the advertisement was released in all Kerala newspapers. We had a good response, and after the due process of evaluation, we bought the land in Ponmudi from one of the bidders which was the lowest in cost and more suitable for the Institute. In order to complete the civil constructions in time, we had parallel floated a separate tender giving all our requirements seeking the proposals for undertaking the turn key contract for realising the entire infrastructure including the architectural features. Here too, we had a good response from the reputed builders all over India. Soon after the land was bought, we requested them to provide the full design tailoring the same to the land we had bought. We set up an expert team to evaluate the bids and recommend a suitable party which met all our technical and financial requirements.

While we were quite happy with the overall progress, we were shocked to hear one day that the Government of Kerala notified that the land we had bought is essentially ecologically fragile forest land and the matter went to court. All our efforts carried out in a hectic manner came to nought. I was terribly disturbed and had no clue on how to meet the demand for the new campus. After a great deal of deliberations on the matter and search for alternate land, we did not see any sign of getting one in the near future. This realisation brought us many sleepless nights as it was mandatory to create a new campus as quickly as possible; otherwise, it was impossible to manage the Institute in the temporary campus for long. The time was running out, and new batch of students were joining every year. We needed many of other infrastructure facilities in place to manage the efficient running of the Institute. When we are working towards an important goal, the last thing one wants is to face big and unexpected challenge or obstacle. When that happens, it means a major delay, mind stops working and that brings in unimaginable complications. This is precisely what happened to us with this unexpected turn of events.

This situation created a kind of panic in us, and both myself and the then Chairman of ISRO who was also Chairman, Board of Management, IIST, were forced to do a lot of brainstorming with all concerned officials at ISRO. Finally, we had to take a very difficult and crucial decision of carving out 50 acres space in the ISRO facilities already established in Valiamala. It also became necessary to convince our contractor to remap the entire architectural design and buildings they had worked out for Ponmudi to this new geographical location. Although they had a lot of reservation and reluctance to do so, I was able to persuade and make them accept our proposal. I started spending a lot of my time with the contractor and the architects on a continuous basis. In the process, it would not be wrong if I say that I almost started functioning like a civil engineer. Finally, the foundation stone for the entire civil works was laid in December 2008, and with our continuous concerted efforts, we completed the one large academic building, five hostels, two canteens, basic sports facilities and all other civil and electrical amenities associated with it in a record time of 20 months after the laying of the foundation stone. We shifted the entire Institute to the new campus in the middle of August 2010. On the Independence Day, I hoisted the flag in the campus, and during the address, I stated that my feeling is almost similar to Neil Armstrong when he landed on the moon and hoisted his national flag.

The question that comes to my mind is what is the takeaway for me from all these challenges I experienced. There were instances which made me very nervous during the setting up of the Institute, since if I fail in any of these ventures, it was beyond one's imagination to guess the consequences. In many instances, we would have ended up with the point of no return, but in the midst of difficulties, I learnt to respond with the right mental attitude and to completely reframe my way to overcome the challenges. All the challenges are definitely my most valuable life experiences. Each of these experiences helped me to see most challenges as opportunities and to harness my own personal abilities to an even greater degree. One important lesson I learnt in the whole process is to treat the challenge as a game and enjoy it as a fun. If we fear in taking a tough decision, it prevents us from facing the challenge and seeing it as an opportunity. It is always necessary to step out of our comfort zone, and all actions we undertake need much more attention than any one normally thinks. Debate and discussions among our colleagues to arrive at possible solutions are highly beneficial. Before I conclude, I am reminded of the two important quotes on this topic, one by Albert Einstein: "In the middle of difficulty lies opportunities" and another by Winston Churchill: "The pessimist sees difficulty in every opportunity, the optimist sees opportunity in every difficulty". How true they are!

Mind to Chemical Markets: Engineering and Managing the Complex Journey

K.V. Raghavan

The chemical markets in India and China are growing at 15–20 % per annum. Their chemical industry is gradually going through a period of transition from a follower to a leadership position in product and process discovery and development. The current trend in technology upgradation is somewhat incremental in nature. The chemical innovations look to be moving away from major breakthroughs which were hallmarks of twentieth century in Europe.

The chemical industry worldwide provides 50,000+ chemicals, materials and technologies for efficient use of consumer products, energy and natural resources in a way that improves the quality of life of people, their health and well-being. The chemical technologies are also potential sources of unwanted emissions and waste materials (gaseous, liquid and solid). However, the high-volume supply chains within the chemical industry, which are market driven, provide tremendous opportunities to chemical engineers to redesign material sourcing, process efficiency, energy and waste reduction and side product utilization to meet new regulations. The long-term sustainability of chemical products is often interlinked with environmental compatibility for chemical engineers to think about developing more novel products and processes for the global and national economies. New ideas have to be taken from the "mind to chemical markets" through multiple research and development initiatives, stakeholders and market environments.

The Mind to Market Journey

The journey is arduous with number of stages and stakeholders associated with the process. Figures 1 and 2 highlight them. Both top-down and bottom-up approaches are employed. The concept development stage demands creative design of a product or process that allows distinct improvements over those existing in the market. The

K.V. Raghavan (✉)
Reaction Engineering Lab, Indian Institute of Chemical Technology (IICT),
Uppal Road, Hyderabad 500 007, India
e-mail: kondapuramiict@gmail.com

P. Ghosh and B. Raj (eds.), *The Mind of an Engineer*,
DOI 10.1007/978-981-10-0119-2_4

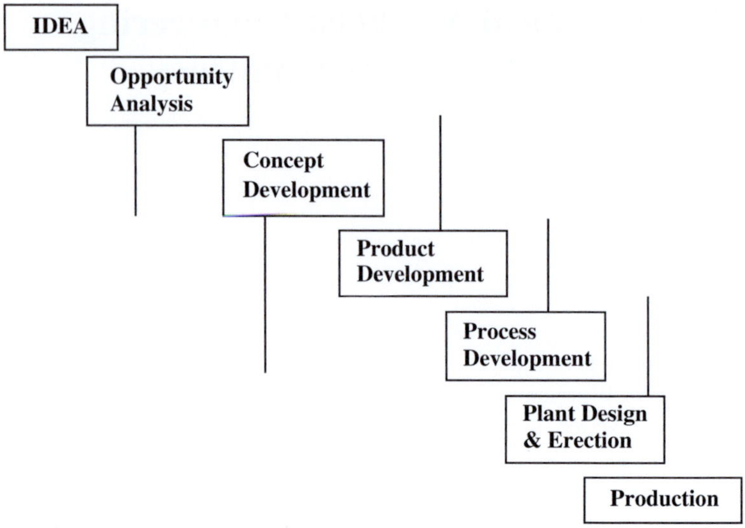

Fig. 1 The mind to market journey

Fig. 2 R&D and stakeholding areas in the mind to market journey

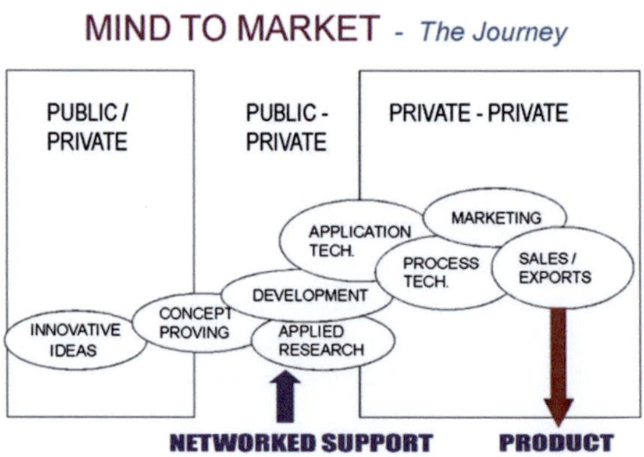

product and process technologies need systematic generation of new alternatives and better options that can enthuse the chemical markets. Modern-day technologies need good understanding of their life cycle issues (Fig. 3) to create products which can be manufactured on sustainable basis by adopting cradle to grave approach.

The stakeholding of various activities during mind to market journey provides the real challenge. Normally, public funded research institutions generate innovative ideas, prove their concepts through experimental validation and develop new products or processes on laboratory or bench scale. The public–private partnership then becomes essential for scale up through pilot-scale development and user trials of the developed chemical products. Commercial funded public–private or private–private partnerships have to take over for semi-commercial product or process evaluation, techno-economic feasibility analysis and test marketing.

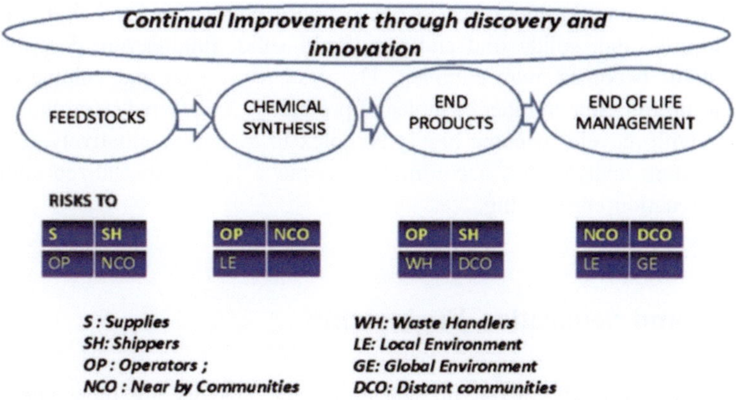

Fig. 3 Life cycle approach for analysing a chemical process system

Grey Engineering Terrains Created by System Complexities

During the 1950s and 1960s, chemical engineers adopted black box approaches for chemical process analysis. They looked at chemical unit operations from outside, viz. assessing what goes inside the concerned equipments and what comes out of them. The system properties were extrapolated with limited mass, heat and reaction engineering knowledge. Mathematically, the mass and energy conservation equations were employed for simple system description. Many chemical systems involve multiple material and energy inputs and outputs to make chemical engineering analysis more complex since they have to account for concentration of each chemical constituent for writing material and energy balance equations. In vast majority of chemical processes, a set of raw materials are processed to yield a set of desirable and undesirable end products by employing more than one unit process or unit operation through which the intermediates and products pass. The process engineers prepare process flowsheets linking all the relevant processes and operations. The design engineers super-impose material and energy flows and control systems on the flowsheet to produce the desired end product.

The chemical process systems become more complicated when recycle loops are introduced into the process flow schemes. The process streams are split and then recombined at specific points. System properties (densities, viscosities, solid content, etc.) change before and after these points. Grey engineering terrains become more evident when chemical reactions become more complex with some of them taking place in series or parallel or series–parallel combination modes. Chemical reactions introduce stoichiometry, kinetic of rate processes and thermodynamic equilibrium limitations. It becomes necessary to use terminology involving gram moles, molecularity, process conversion, selectivity towards desired product formation and reaction velocity constants to describe a chemical process system.

Chemical reactions do take place in one or more phases, viz. gas–liquid, liquid–liquid, liquid–solid, solid–solid and gas–liquid–solid. Engineering description of such systems becomes more complex. The system pressure and volume changes can occur unless same number of moles of product(s) are created or provided. Such reactions with recycles become highly complex to define quantitatively. The multiphase system analysis has accordingly become a highly specialized subject in chemical reaction engineering.

Reactor and Separation Engineering

Theoretically, perfectly mixed or near-plug flow reactors are most effective in achieving best conversions and yields depending upon the extent of backmixing needed for a process. Traditionally, chemical engineers have ensured best possible conversions, selectivity and faster reaction rates by employing batch, semi-batch, continuous and semi-continuous reactors. They operate them at steady as well as transient states. They designed reactors based on their mode of operation, nature of phases present and the vessel geometry. Stirred tank (batch and continuous) and tubular (empty and packed/fluid beds) reactors are employed extensively in many commercial processes.

Increasing demand for super-pure chemicals with impurity levels in ppb range as required by the chemical markets and emergence of high-performance chemicals and materials are driving the chemical engineers to look for more novel reactor configurations and separation systems. They found that interfacial phenomena control not only the underlying chemical processes but also the activity of the products. Novel reactor configurations employing high gravity, photocatalysis, membrane separation and circulating fluidization as major driving forces are being employed. They include packed bed, membrane, photo, fluid bed riser, reverse flow, monolith and other reactor configurations.

A large number of separation processes involving mass and heat transfer, viz. distillation, evaporation, extraction, absorption and separation, leaching, drying and crystallization and those involving mechanical operations such as crushing, grinding, filtration and centrifugation, are employed for intermediate recovery and final purification processes. The recent successes in high-efficiency separations have been mainly driven by the chemical engineering advances in heat and mass transfer. A new breed of separation processes involving chromatography, membranes, electrodialysis, enantiomeric affinity, etc., have emerged in recent years with high application potential in performance and bioactive speciality chemicals. In spite of these advances, there are several unconquered engineering terrains in these areas. Added to these, high-potential applications in gas processing platforms for energy sector requiring cryogenic distillation of air, condensation of minute quantities of organic vapours and membrane separation of gas mixtures have brought in new engineering challenges. Integration of chemical reaction and physical separation in a single process vessel as a combo operation is bringing large dividends in reaction and separation efficiencies.

Kinetics, Modelling and Simulation of Reacting Systems

A chemical reaction's equilibrium position defines the extent to which it can be processed. A study of reaction kinetics begins with the measurement of reaction rate through experimentation. The rate laws are employed to determine the reaction order and the reaction velocity constants. The activation energy of the reaction is evaluated by Arrhenius plots. The kinetic parameters of a reaction are employed in the modelling studies which consider the variations of process parameters in both time and space domains as dictated by the process. One-, two- and three-dimensional models are employed by the engineers by considering the radial and axial variations of reactant concentration and temperature. The model equations are formulated with the help of equations of continuity and motion. The fluid flow in catalytic reactions is modelled by using Darcy's Law and by considering the conductive and convective heat transfer mechanisms. Three basic reactor models, viz. batch, continuous stirred tank and plug flow, are employed. The model equations are solved by analytical or numerical means. They require physical, transport and thermodynamic properties of reacting system for their solution. Computational fluid dynamic (CFD) models are employed for simulating the performance of multiphase reactors with complex hydrodynamics.

The models for simulation of reactor performance are used at different stages of chemical process development. In the initial stages, engineers employ them to understand the less-known aspects of system behaviour; at later stages, they are used to optimize and control process variables and parameters through virtual experimentation; and in case of situations such as thermal runaway and other phenomena, they are used extensively since they are difficult to investigate experimentally. In recent years, sophisticated tools such as Monte Carlo or molecular dynamic simulations and molecular recognition are employed to follow the reaction intermediate formation in the porous structure of the catalysts and the formation of bilayered vehicles in the microencapsulation processes.

Modelling and simulation will assume a more central role in developing novel strategies that can effectively deal with the concerns of future societies in 3Es, viz. energy, environment and economics. Optimization, intensification and scale up of complex chemical processes will bring more engineering challenges to the fore. The CFD models currently employed for multiphase reactors are insufficient in terms of simulating the backmixing, turbulent phenomena and in describing the microscopic material and thermal processes occurring at interphases. They also have to look at the interstitial phenomena in packed and fluid beds in terms of particle boundary layer formation. Grey engineering terrains are evident in simulating the reaction and transport processes occurring on different timescales and space domains, turbulence around gas bubbles and boundary layer movement. They will place more demands on measurement technology to employ tools such as particle image velocimetry and intelligent laser applications.

Process Intensification Opens New Engineering Terrains

Process intensification, which is hardly a decade old, is closely linked with green chemistry. The basic aim is to optimize capital, energy, environmental and safety benefits by radical reduction in physical size of process plants. Several novel engineering options have been reported under alternative methodologies and equipments as highlighted in Fig. 4. Many alternative equipments are of type never known before such as rotating packed beds and oscillating flows in reactors. Structured packings, static mixers, ultrasound and microwaves to enhance the efficiencies of reactions and separations.

The reactive separations in which reaction and mass transfer operations such as distillation, extraction, absorption, adsorption, membrane separation, chromatography, crystallization and precipitation occur simultaneously offer special challenges to engineers in experimentation as well as in their design. The additional degree of freedom offered by them enables the concentration profiles to be tailored and better performance achieved in terms of enhanced reaction rate, conversion,

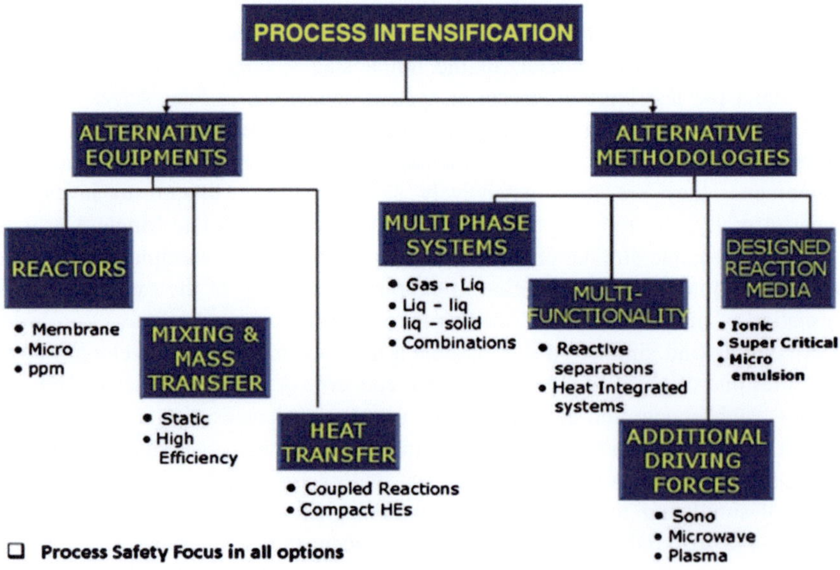

Fig. 4 Various process intensifications

selectivity and yield. The unconquered engineering terrains in reactive separation are broadening their windows of operation and coupling of two or more reactions in a single unit. A typical example is the dimerization of isobutene to isooctane with subsequent hydrogenation in the same distillation setup. The biorefinery operations will greatly benefit from future advances in reactive separations.

Management Issues in Mind to Market Journey

Mind to market journey requires several independent scientific and engineering efforts executed under different institutional support systems with different stakeholders and different targets. They have to be developed adequately at each stage addressing the unexplored facets of long-term product or process stewardship. The research laboratory providing the basic concept is often under pressure to prove its implied technological feasibility. This being the idea generation stage, a feasibility analysis of the product or process being developed is less expensive than at pilot scale. It is an effective way to prioritize various development options. It is important at this stage to keep in mind the availability of limited database on market needs and to depend more on opinion of market leaders.

Market Orientation

Market orientation has to be the key driver of R&D activities at laboratory, bench and pilot scales till their eventual commercialization. The actual market introduction of a new product or process determines the efficacy of a R&D programme. It was demonstrated, on several occasions in the past, that the proper alignment of all R&D efforts from idea or concept generation to the launch of a product or process holds the key to commercial success. In the case of chemical industry, technological change has been found to have a decisive impact on the competitive markets.

Knowledge Management

Non-technological innovations are necessary in knowledge management during the mind to market journey. The focus should be on the sustainability of knowledge generated from strategic, tactical and operational angles. Good management helps to spark technological innovations that eventually improve the product and process utility in competitive markets. New engineering terrains have to be crossed for effecting significant product and process improvements as dictated by the market forces. Technology portfolio management is an important component of knowledge management during the mind to market journey. It has to be integrated to the market needs. The patent portfolio has to be approximately designed to protect the technological inventions and at the same time enhancing their attractiveness to the market.

Stakeholding

It is to be recognized that a very small percentage of new ideas or concepts will find commercial attractiveness in chemical industry. They alone will receive funding commitments to proceed further. An effective methodology has to be found to identify appropriate stakeholders, viz. researchers, peer review groups, technology funders, risk investment financers, user industry consultants and investers, market forecasters, and pilot plant manufacturers, at appropriate stages to reach the final end user. Clear-cut management criteria are needed to monitor the overall R&D project progress vis a vis the market friendly milestones, to assess the probable techno-economic viability of a product or process option and to decide on the involvement of concerned stakeholders to give a go forward, stop or hold and return signals. Their inputs are vital for resolving interdependent scientific and engineering issues related to 3Es, viz. energy, environment and economics.

Summary

The mind to chemical market journey is arduous since new ideas or concepts have to be taken through multiple institutions, stakeholders and industrial environments. Chemical process systems have created several grey engineering terrains due to the system complexities. They need practical solutions. The main focus areas are chemical reaction and component separation areas. Advanced knowledge is vital in process intensification, modelling and simulation and scale up. There are several management issues related to market orientation, knowledge management and stakeholding which need attention during mind to chemical market journey.

Acknowledgement Thanks are due to Sri Yadu Krishna of IICT, Hyderabad, for word processing assistance.

The Search for Meaningful Purpose: Reminiscences of an Indian Engineer

Sanak Mishra

As a school student in Cuttack in the State of Odisha, I read about the great visionary Jamsetji Nusserwanji Tata and his magnificent contributions to science, industry and society. Tata Steel then, as now, was a household name in Odisha and one of my elder brothers, a Chemical Engineer from IIT Kharagpur, who had taken summer training there, had mentioned that Tata Steel had a number of brilliant metallurgists doing exciting work, mastering the flow of liquid iron that had the colour of ripe orange and imparting shape, strength and toughness tored hot solid steel.

This was the background in which I went to the Indian Institute of Science (IISc) in Bangalore in 1965, to study the Bachelor of Engineering degree in Metallurgy, after receiving a B.Sc. Honours degree in Physics from the Ravenshaw College. Back then, very few people in Bangalore City would call it as IISc; it was far better known as Tata Institute, having also been founded by none other than Jamsetji Tata. In fact, arriving at the Bangalore Railway Station, I had some confusing moments as the auto-rickshaw drivers told me that the only Institute they knew was the Tata Institute. The world-famous institution that it was, the IIScal so embodied the value system of the House of Tatas and that touched me profoundly. *Years later, it was a proud moment for me to come back to the Indian Institute of Science to receive one of the Distinguished Alumni Awards presented on the occasion of its Centenary Year in 2008*. This time around the young driver of the taxi, who took me to the IISc campus, knew exactly where I wanted him to take me. He even commented that he had heard from his grandfather that in his time the whole of Bangalore was like the IISc campus, an Oasis City of gardens and flowers.

It is at the Indian Institute of Science that I learnt much about the charismatic Jehangir Ratanji Dadabhoy Tata, or, simply JRD Tata. In fact, I also had the good fortune to shake his hands and speak with him, on one of his visits to IISc. Needless to say, in my generation, he was an iconic role model for all of us. His urbanity, humanity, integrity and his love for the Indian Nation left indelible impressions in our young minds. *It would be my great privilege to receive the Fifth JRD Tata*

S. Mishra (✉)
SAIL-CGHS, 341 Green Heavens, Plot 35, Sector 4, Dwarka, New Delhi 110 078, India
e-mail: mishra.sanak@gmail.com

P. Ghosh and B. Raj (eds.), *The Mind of an Engineer*,
DOI 10.1007/978-981-10-0119-2_5

Award for "Excellence in Corporate Leadership in Metallurgical Industries" for the year 2012, from the Indian Institute of Metals.

From Bangalore, my metallurgical ambitions, and a full scholarship, took me to the University of Illinois at Urbana-Champaign (UIUC) to pursue the M.S. and Ph.D. degrees in engineering. It was a stroke of good luck that I got to work for five years as a Research Assistant to Prof. Paul Beck, a pioneer of modern Physical Metallurgy and one of the founding fathers of the new interdisciplinary field of Materials Science and Engineering. My experiments went down to the temperature of liquid helium, and on some occasions even below that. This by itself was a magnificent challenge. Our research resulted in the discovery of *super-paramagnetism* in dilute alloys and also significantly advanced existing knowledge base on the phenomena of atomic short range order and that of magnetism in metallic systems. Subsequently *in 2010, I was delighted to be conferred with the Distinguished Merit Alumni Award from the Department of Materials Science & Engineering at UIUC.*

I returned to India in 1973, having been directly recruited from USA by the then Hindustan Steel Limited, now the Steel Authority of India Limited, to join as a "nucleus member" of the just-founded Research & Development Centre for Iron & Steel at Ranchi (RDCIS), thereby began my 32-year-long career at SAIL, including a 2-year sabbatical that I took at the Technical University at Aachen in West Germany, on receiving the *Alexander von Humboldt Fellowship*—till then very rarely awarded to persons like me engaged in industry. In Aachen, I collaborated with Prof. Kurt Luecke, the highly respected Director of the Institute of Physical Metallurgy and Metal Physics. Interestingly, this internationally renowned Institute had never worked on body-centred cubic metals such as steels and he encouraged me to establish a strong programme in ferrous research, which I was able to do in less than two years. Further, my work as Humboldt Fellow resulted in arriving at a far better understanding of the critical factors in the production of automotive steels and grain-oriented electrical steels and introduced new concepts such as the *Texture Memory Effect* and the *Inheritance Model.*

Coming back to SAIL, I will always cherish my long innings in this August Company. In the initial years, it gave me the opportunity to build, along with my peers, a world-class laboratory complex, formulate long-term R&D plans and along the way lead teams of very competent technologists to culminate in a plethora of new products and Patents, such as steels for domestic gas cylinders, high strength steels, corrosion resistant steels and specialty alloys. Then, in the second half of the 1980s, SAIL went through a significant make-over, through a multidimensional and comprehensive intervention called "*Priorities for Action*" and a few of us were hand-picked as facilitators, drawing us deep into a unique and intense engagement in corporate strategy and planning. It was very exciting for me. In later years, I was actually called upon to head the Corporate Planning Directorate at SAIL's headquarters in New Delhi. I might add that it was during this period that an opportunity to work with Tata Steel also came up. In fact, as the Chairperson of SAIL's Task Force on e-commerce, I was able to play a key role in creating, jointly with TATA Steel, one of India's first digital market places, namely *metaljunction.com.* It is a matter of satisfaction that re-engineered later as *mjunction*, and it has grown rapidly to emerge as the *world's largest* e-Market place *for steel.*

Towards the end of the 1990s, globally the steel sector went through a major downturn. SAIL, with the baggage of several internal weaknesses, was badly affected and its bottom line eroded rapidly. Just before that SAIL had been given the status of a Navaratna Company by the Government of India and naturally the Steel Ministry was deeply concerned that the Company must arrest the slide and bounce back at the earliest. SAIL chose, rightly, to undertake an ambitious and comprehensive revamp. I found myself roped in as the Head of the Office of Business Restructuring, specially created to drive the revamp process. Working closely with the Board of Directors and the Ministry of Steel, my team was able to contribute strongly towards SAIL achieving a major turnaround, and overall rendering the company to be far more efficient and sustainable. This eventually helped SAIL to achieve the Maharatna status from the Central Government. In recognition of my contributions in SAIL, I was conferred the National Metallurgist Award by the Ministry of Steel in 2003.

In 2002, I was drafted as the Managing Director of the heavily loss-making Rourkela Steel Plant with the unenviable task of recovering it from the brink of possible closure. I was able to establish and institutionalise a new work culture, by resetting the minds of its 26,000-odd employees. A principal vehicle for the remarkable transition was a massive human resource intervention, named "SAMSKAR", backed by a sustained, large-scale, interactive campaign called the "Mass Contact Exercise", in which sets of 500 different employees participated every week, thereby covering each employee once every year. Prof. Paul Argenti of the Tuck School of Business at Dartmouth, in his book "*Strategic Corporate Communication*", has referred to this exercise as one "*that had no precedence in Indian corporate history*". More details of these efforts can be seen as a full chapter in "*Break Free*", a book on leadership authored by Prof. D. Chatterjee, Director, IIM Kozhikode, who had taught at Harvard University.

By design, the "SAMSKAR" movement was based on Vedic principles and focussed on reducing the distance between minds. It centred on a unique code of leadership practice that ran through all levels of management, becoming in the process a model of good governance. The momentum of the "SAMSKAR" movement, together with several bold asset restructuring measures and major technological upgradations, which included the total reconstruction of its largest blast furnace, resulted in rehabilitating the steel plant as one of the most consistently profit-making industrial enterprises of India. In fact, *from a level of making a loss of Rs. 3 crores a day in the financial year 2001–2002, the steel plant started making a profit of Rs. 3 crores a day by the middle of 2004*. The high level of profit-making regime has been sustained to this date.

In more ways than one, the transformation of the Rourkela Steel Plant was quite dramatic. At the same time, I was very conscious of the Steel Plant's responsibility to the Society at large and to the environment. We took up a massive rebuilding and rejuvenation of Rourkela Steel City, involving infrastructural improvements and the promotion of education, sanitation, ecology, health, sports and culture. In the process, it also became one of the greenest and cleanest towns in the country. It received the prestigious Indira Priyadarshini Brikshamitra Puraskar from the Ministry of Environment & Forests in 2005, a rare recognition for an industrial township. On an equally strong footing, we set into motion a major intervention in the societal development in the areas beyond Rourkela, which comprised mainly of tribals. In the first step, we prepared, in association with the Ranchi-based Society for Rural Industrialisation, *SRI*, a long-term master-plan called "Shankhadhwani" for peripheral development. Next, we set up an Institute for peripheral development to carry out the identified tasks, guided by a Memorandum of Understanding with the Basic Agro-industries Foundation, BAIF, Pune.

A bright memory that I have my days at Rourkela Steel Plant was the visit in May 2003 of the then President of India, Dr. Abdul Kalam, who, amongst many other interests, was very curious to see what a blast furnace looked like. His child-like enthusiasm, to witness for himself how liquid slag and liquid iron coming out of the blast furnace were physically separated, was inspiring, to say the least.

Dr. Sanak Mishra with the President of India Dr. Abdul Kalam (May 2003, Rourkela)

In January 2006, I was hand-picked by the then Mittal Steel Company, as its Chief Executive Officer in India, to build up from scratch its Indian subsidiary. I also helped to install in Kolkata a global group facility, namely the Arcelor Mittal Design & Engineering Centre (AMDEC).

It is not possible to leave out INAE from any professional discussion. I owe much to it for what I am today. It is an institution which embraced me with open arms as a Fellow in 1997 and has nurtured me and supported me ever since. I was fortunate to be chosen as its Vice-President for the years 2007–2008. *I received from INAE in 2010, the Jai Krishna Memorial Award*. I also feel a sense of pride to have served as the President of the Indian Institute of Metals for the year 2009–2010. Like the INAE, IIM is a body of eminent persons of high distinctions who have contributed immensely to national cause in the metals and materials fields. I was privileged to receive its Honorary Membership in 2005 and the Platinum Medal in 2011. My association with IIM led to my election in 2010 as the

Chairperson of IOMMMS (International Organisation of Materials, Metals and Minerals Societies), an Apex body headquartered in the USA.

It has been a magnificent lifetime opportunity to work in India and for the prosperity of India. Looking back, I am glad that I chose the Engineering profession.

I must acknowledge that all through my professional career I have had immense support from our great path-finders in Academia, Science & Technology and Industry and I am much thankful for that. I should confess that more often than not I have also received much inspiration from the young engineers. Many of them are brilliant and if their energy and passion can be harnessed purposefully and meaningfully, India would be in a commanding position. To them I can be honest in sharing what I myself have learnt that "Taking responsibility is the ultimate essence of personal courage". I would, therefore, urge the young to be responsible and courageous. I believe further that "True Leadership lies in taking up responsibility, in facilitating others, and in letting people make a significant contribution to a befitting cause- with full freedom, dignity and without fear". To our policy-makers and planners, I appeal that they provide that facilitation and that ambience to our young aspiring youth so that they can engineer what could be the "Brave New India". Indeed, the relentless pursuit of new knowledge, comprehension and innovation, the three building blocks of leadership in the new-age world, requires us to be bold and committed to the cause of the future.

An Engineer or a Scientist—A Perpetual Dilemma

Indranil Manna

Prelude

Mind of an engineer: who is an engineer? The one who produces, manufactures, constructs, builds, or repairs? The one who designs? What is his/her role? What does one get trained into be an engineer? Who needs him/her and for what? With three tertiary-level degrees from engineering faculty of three different institutions, am I an engineer, even if I am not engaged in production of engineering goods or build/construct engineering structures, machines, or components? And most importantly, what kind of a mind does an engineer carry or develop? Let me make an honest attempt to trace my own journey through the years and seek an answer to this dilemma, if any.

The Early Phase

As a child, my father was my hero who each one of us in the family dreaded and yet dearly adored. He could do many things which I could seldom do but always fancied to master. If anyone asked my name, I would promptly answer without wasting a second: Dr. Bonny (my pet name) G.K. Manna (my father's name). Before even realizing, a dream was born and a commitment was made: a dream to be a chip of the old block (like most of the kids who grow up worshiping dad or mom as a role model), and a commitment that I must earn a doctorate degree someday. While the latter did happen twenty five years later, the former remains a fervent wish even today.

I. Manna (✉)
Indian Institute of Technology Kanpur, Kalyanpur, Kanpur 208 016, India
e-mail: director@iitk.ac.in

P. Ghosh and B. Raj (eds.), *The Mind of an Engineer*,
DOI 10.1007/978-981-10-0119-2_6

I was born in Calcutta (not yet Kolkata) and my father moved to Kalyani to establish the Department of Zoology in the newly established University of Kalyani in that god forbidden place, only about 50 km north of Calcutta, where jackals and snakes, besides the innumerable cattle would easily outnumber the humans encroaching into their safe haven. And yet, Dr. Bidhan Chandra Roy, the visionary and architect of modern Bengal wished to create a knowledge hub precisely sixty years ago with the foundation of not one but two universities, in addition to hospitals, schools, industry, training institutes, roads, parks, bridges, residential area—all minutely detailed in a master plan that even today is smarter than many of the highly gloated 'smart cities' and has no parallel, not only in (West or East) Bengal, but over a vast majority area of the country simply because every inch on the ground was logically accounted for and every dimension above ground was filled with an imagination and vision that nobody has been able to surpass even today.

I grew up in Kalyani starting from nursery to higher secondary. There was only one mechanical workshop, two blacksmiths, three hair dressing salons, a few cobblers, a handful of cycle repair shops but no motor repairing garage and not even a dozen shops in this small locality where almost everyone in my childhood would have something or other to do with Kalyani University. As the township was new and in the making, construction and repair jobs were commonplace. The whole country, hardly fifteen years old from independence, was in development mode. A neighbor, who owned a factory, took me once to his works. The interplay of men and machines to create attractive toys and components fascinated me no ends. Even my father's department was a favorite stop for me as it had a museum of mounted animals in strange looking and smelling glass jars: insects, snakes, fishes, small mammals, and even a few models and skeletons carefully guarded in glass paneled showcases. Though the biological world fascinated me, the urge to create and understand the role of symmetry, function, and relationships in the world around us was slowly but steadily casting a spell on me.

The Journey Begins

My uncle was a metallurgist from the hallowed Bengal Engineering College who quit lucrative jobs and decided to pursue engineering in the true spirit and founded an engineering works in Calcutta, Dhanbad and a few other places. I always looked forward to visiting his factory in Behala and spending hours watching fabrication of complex steel structures and translation of design from drawing board into three dimensional products in a mystic atmosphere of dust, heat, shrill metallic noise, sudden flash of welding arc, and smell of sweat, paint and iron—all creating a web of a challenge to create something new. How I wanted to emulate the skills of one of those workers who would weld, cut, drill or form at ease and soon fabricate a structure, component or machine body out of sheets, rods, angles, and nuts and bolts!

What would you like to be? Without batting an eyelid, I replied to one of our family friends: *an engineer*. I am not sure if I disappointed my father with my prompt riposte, but he gracefully accepted, even if he wanted me to follow his footsteps in cytogenetics (genomics today) or at least biology or medicine. Actually, my other uncle, who was a doctor himself and who would occasionally pamper me more by taking me to movies like Ben Hur or Born Free, once thought that I would make a good doctor because I knew 'alcasol' was the antidote for fever and 'enteroquinol' for dysentery. However, I realized, having already seen my elder bother taking more than eight years to earn an MBBS without failing even for once (thanks to the tumultuous seventies in Calcutta and tortuous curriculum of medical education in Calcutta University), that I am born to be an engineer, a creator and nothing else.

Because of that metallurgist uncle, I had very little to decide once I had declared my intention to be an engineer. Since my father quietly passed the onus (to mentor me) on to his elder brother, this uncle in turn ordained without an iota of doubt that I must go to engineering college and study metallurgy. My feeble objection citing my inclination to mechanical engineering was blown away in a whiff faster than a bullet train as if I was destined to pursue metallurgy that too only in engineering college and nothing else.

To be honest, the deadly combination of long and boring drawing classes, mundane workshop practices and dreadful hours of differential or integral calculus was enough to make me wonder very soon—is this what I wanted to do? My proficiency in outdoor games, ability to blurt out something extempore in public, and knack of learning from faithful and studious friends or their notes soon offered me a much more enjoyable alternative: pursue cricket in college and club teams, turn into a overnight hero by public speaking and make up for the missed classes by learning from friends in the hostel rather than the teachers in the classroom. I managed to stay afloat but was perilously balancing between being declared non-collegiate due to poor attendance and barely managing enough in grades to survive.

Something dramatic happened in my third year. Professor Seal and Professor Das, who knew my metallurgist uncle well, forced me to sit in the first row in the class on the very first day and left no room for me to be, leave alone bunking, even being inattentive during the lecture or practical. By a strange divine intervention, the likes of professors Mitra, Banerjee, and Roychoudhury started talking about subjects that made sense to me. Without realizing, I stated liking the concepts of phase transformation, microstructure, or dislocation dynamics and found some interesting possibilities of tailoring material properties to suit the desired range and play god in designing materials for specific applications. By the time the year and even the course ended, I felt grossly embarrassed that I did not miss any class or lecture, did reasonably well in terms of grades and actually admitted that I had fallen in love with physical metallurgy—the science of correlating structure with properties of solids. Some *teachers make a huge difference to the taught*. In my early life, one who literally cast a spell and guided me during the school education was one Dr. Sudhanshu Dutta, a mathematics teacher, who single handedly produced generations of bright students from our small township with an unmatched sincerity and devotion that is worth adoring and emulating forever.

The job market was awful in the early 1980s, and the offer that I managed to get was surely not worth pursuing. Since I already started to wear glasses to correct my eyesight, I was ruled out for a job in steel plant or shop floor. I decided to move to IIT Kanpur to pursue master degree. Not only did my horizon broadened, soon my diffidence turned into sharp determination to master the subjects that were either considered dreadful or out of bounds in the undergraduate days. Though penchant for sports continued, the love for understanding the complexities of evolution and tailoring the microstructure of solids and correlating the same with their composition and performance for various applications was overwhelming.

On the Job

For a brief one-year period, I experienced the practical world of metallurgical profession in one of the finest industry of the country, Mishra Dhatu Nigam (MIDHANI) in Hyderabad. However, I soon realized that my grooming will remain incomplete if I continue only to follow and implement the production schedule of Forge Shop in MIDHANI and I shall never attain my dream of being a complete engineer.

Opportunity came in the form of a fellowship for Ph.D. abroad and a lecturership at a premier institution in India. My father convinced me for the second despite my strong inclination for the former. He maintained that his stand would be vindicated in future and I had very little to contradict at that stage. Subsequently life taught me enough to accept that *decisions are seldom taken correct rather proven correct.*

I joined IIT Kharagpur as a lecturer at the age of 24 without Ph.D. and did my Ph.D. just in 2 years despite taking my usual quota of teaching and other jobs that came along with the appointment. To make life more adventurous, I got married within a month making several of many of my senior but still bachelor colleagues raise eyebrows in utter disbelief! In research, I learned a few techniques, wasted my time in pursuing something intuitive that never got substantiated, but subsequently collected some new results and published and eventually submitted my thesis. I applied for and landed with a fellowship to spend a year at Max Planck Institute in Stuttgart to work on a particular type of transformation with an authority in the subject. I came across several giants in this area, made friends with a handful, handled sophisticated equipments myself, and gained confidence as a researcher when I published six papers in a year which made Professor Gust effusive in his recommendation letter for my assistant professorship that he had never had a more successful coworker like me before. I came back flushed with energy, dream and a purpose, and to top it up, was blessed with our first child, a daughter, to which my mother prophesied that it would bring 'good luck' to me.

On the Roll

The next twenty years passed in fast forward mode: students, teaching, publications, conferences, projects, results, awards, promotions, attempts, failure, success…the journey continued in frenzied fury. I taught courses and demonstrated experiments, some old and some new created by me, set up laboratory for research, and desperately pursued collaboration with elite institutions and industry for there were fewer facilities than what was needed to make a mark and foray into new domains and challenges. Research visits to a dozen countries, active cooperation with many, bringing in sponsored funds to support students and activities, and guiding theses and projects, establishing new laboratories and facilities, and above all, getting a foothold for myself as a researcher in India—all happened in a span of barely twenty years.

Research is all about establishing a new truth, creating new knowledge, not necessarily, conforming to common belief or anticipated trend. The difference between conducting a laboratory experiment and meaningful research is simply solving a problem with knowing the answer for the former and not knowing the same for the latter. I started using resistometry as a tool to study the mechanism and kinetics of a special type of solid-state transformation, called discontinuous precipitation, which necessarily occurs behind a moving boundary that acts as a short circuit part for diffusion (or solute transport) and differentiates between the transformed and untransformed regions of the solid matrix. An easy analogy is what happens when the wave invades the sea shore with a curved boundary and leaves a distinct pattern on the beach as the water fades out in the sand. Such moving boundary transformations were known to affect the mechanical properties of many important metallic alloys including steel (e.g., through pearlitic or eutectoid transformation). Despite long history of research, there were many unresolved issues related to nucleation (or starting), mechanism (path and process followed) and kinetics (rate). Such investigations are usually carried out by monitoring the microstructure under optical or electron microscope, mostly ex situ (postmortem analysis). I ventured to explore the feasibility of adopting an easy technique based on resistance changes with time under isothermal condition which would allow me to predict and examine the progress of the transformation in situ (continuously in real time). Except the Double Kelvin DC resistivity bridge, rest of the equipments were either fabricated or cannibalized from junks fit for a museum. Yet wonderful results were obtained that gave nice and new insights, some of which could be correlated with the prevailing theory, while the rest proposed as new features not known to date. I extended the work to some new directions and failed to obtain what I desired but got something more interesting that fetched a few more publications. The net result was that I concluded my doctoral study in flat two years, despite having full teaching load. I continued working in discontinuous precipitation, coarsening, and dissolution for another few years in India and abroad and slowly veered toward two new directions: laser-assisted surface engineering and nanometric solids synthesized by mechanical alloying.

Some of the typical degradation or failure of engineering components initiate or occur primarily at the surface with having very little to do with the interior or bulk. Hence, modifying or tailoring only the surface either by changing the microstructure or both microstructure and composition, without affecting the bulk, could be an economical and clever way of prevention or remediation. After all, is it not more common to apply a coat of paint on the car body to prevent or preempt tarnishing or rusting than changing the entire shell or body every time? Laser is a wonderful non-contact tool that allows very fast, convenient, and accurate heating (including melting or vaporizing) of an engineering component using a directed beam that delivers the required quantum of energy at the desired spot with an unmatched precision. Laser-assisted surface engineering allows simply heating/melting and also, alloying or cladding the surface with a new layer. In the process, one can significantly improve surface-dependent properties such as color, adhesion, roughness, wear, friction, oxidation, and corrosion. Suitable choice and optimization of process parameters can allow the same laser machine to do cutting, drilling, joining, cladding, or fabrication (direct manufacturing). Furthermore, the ultrafast heating/cooling (say up to 10^6 K/s) can develop a metastable aggregate that is not otherwise possibly using conventional processing routes. In the last two decades, my colleagues and students have achieved a rare distinction of having more number of published papers or citations, conducting more funded projects or supervising more students in this area of laser surface engineering than any other group in the country, and that too, without having a high power laser at our disposal. *When you pursue a difficult goal, no difficulty can be bigger than your own intention*!

Engineering solids are usually fabricated through elaborate processing routes including synthesis by melting, casting, deposition or chemical reaction involving change of state, say, vapor or liquid to solid. Instead if synthesis can be in the solid state, the advantages and economic benefit can be manifold. Mechanical alloying using high-energy planetary ball milling is one such easy, cheap, and convenient processing route that can convert elemental powders into alloyed product with nanometric crystallite size and even result into amorphous state. We utilized mechanical alloying to develop amorphous aluminum alloys with an intention to obtain strength comparable to high strength steel but retaining the lightness of aluminum. One must realize that equilibrium makes it impossible to convert molten aluminum into non-crystalline solid even by ultrafast cooling, leave alone in solid state at room temperature. Through sustained effort, we indeed synthesized aluminum-based ternary alloys in non-crystalline state by mechanical alloying. While synthesis of the desired alloy with amorphous microstructure was difficult, the bigger challenge was to consolidate or sinter the powders into solid component without destroying the metastable amorphous state that always tend to crystallize under slightest thermal activation (heating). We explored multiple routes of sintering and eventually achieved consolidation to sufficient density and integrity to conduct series of test to prove indeed the product met the desired level of mechanical properties except that the ductility or toughness was extremely poor. Emboldened by some level of success, the efforts continued further.

Through pain-staking and dedicated efforts sustained over two decades involving multiple students, scholars, colleagues and sponsored projects, it was possible to synthesize several nanostructured materials in different form/architecture (bulk alloys with in situ nanometric precipitate-phases, ex situ nanometric solid dispersed solid alloys and liquids (nanofluids), nanometric films/coatings, nanocomposites/hybrids, etc.). More interesting was to establish correlation between structure (length, size, volume or type of the crystallites, and their morphology, free volume, interface and crystal structure) and engineering properties of interest (mechanical strength, hardness, softening, thermal conductivity, superparamagnetism, surface wear/friction, sensing response, diffusivity, catalysis) in nanostructured particles/aggregates synthesized by various advanced material processing techniques (mechanical alloying/milling, vapor deposition, sputtering, laser surface engineering, plasma ion implantation, etc.). These investigations revealed and established some of the significant fundamental knowledge or understanding about material properties at small length scale like (a) size-dependent polymorphic change of Bravais lattice in early transition metals below a critical grain size due to structural instability caused by negative hydrostatic pressure, (b) enhanced diffusivity and boundary softening at room temperature due to nanocrystallization, (c) paramagnetic to superparamagnetic transition of Mn–Zn (perovskite) spinel ferrites in ultrafine regime, (d) super-high strength (1500–1800 MPa) of amorphous Al-alloys due to in situ nanometric intermetallic precipitation or ex situ nano-oxide-phase dispersion in glassy matrix, (e) high conductivity Cu(Cr, Ti) alloys with $\sim$55 % IACS and extremely good wear resistance due to nanometric oxide dispersion in supersaturated Cu-matrix, (f) 50–80 % improvement in thermal conductivity in nanofluids (stable colloidal dispersion of nanometric solids in conventional thermal fluids) in single-/two-phase heat transfer regime promising significant rise in heat transfer coefficient, (g) very attractive reversible magnetization behavior in ferrofluids, (h) highly sensitive (and selective) gas-sensing property of ZnO–SnO_2 nanometric thin films synthesized by pulsed laser or vapor–liquid–solid deposition with specific doping (Pd, Hf, Zr), (i) special FeCrTiAl ferritic alloys with nano-yttria dispersion to reach extremely high compressive strength (2012–3325 MPa), Young's modulus (230–295 GPa), fracture toughness (4.6–21.8 MPa$\sqrt{m}$), and hardness (15.5–19.7 GPa) with >10 % ductility, and (j) low-alloy high-carbon bainite + martensite duplex steels with over 2000 MPa strength due to nanometric thick ferritic sheaves with uniformly embedded 10–20-nm carbide particles. I may add that the activities on amorphous and nanocrystalline Al-alloys, nanofluid and laser/plasma-assisted surface engineering have evoked wide interest in the scientific community.

A New Chapter

Strangely enough, some respected peers in the country asked me in 2009 to lead a national research laboratory dedicated to glass and ceramics. I had scarce administrative experience, was never even a head of a department, and yet was given the responsibility to head an institution of over five hundred scientists, staff, and scholars with a mandate to develop and deliver crucial technology and products in glass and ceramics. As an academic reporting, new results and proving possibilities were enough. Here, the target was beyond: to develop actually deployable products or processes and not just demonstrate a prototype. This was the first time I realized the difference between *a technique and a technology* and learned the difficulties and challenges of engineering practices.

The Realization

In order to create and deliver new technology, does one simply try tinkering with an existing practice, borrow from past experience and depend on the skill sets of the technicians available? Is engineering an art of only producing objects based on acquired skills and societal demands? Is science of any relevance to engineering or technology or is science an esoteric exercise only meant for the gray haired or bald professors who would exclaim 'eureka' once in a blue moon after many years of secluded perseverance? Could there be a connection and relevance among science, engineering and technology, or there are disparate subjects and endeavors pursued in isolated silos? When a man flies an aircraft, does he care who makes it, what engineering branch is more important, what was the degree or curriculum of study for the guy who designed, fabricated, tested or assembled the aircraft? All one cares is to reach the destination in stipulated time with a comfortable and safe journey and encounter no uncertainty, leave alone any risk or danger on board. This is exactly what society demands from engineers—learn and define the problems/issues, meet the demand, serve and enable the stakeholders and create wealth for the nation, wealth both in terms of knowledge and opportunity. For that if aeronautical engineers are needed to build an aircraft and mechanical engineers for automobiles—so be it. But when it comes to making the structure lighter and more durable, extending the efficiency of the engine propulsion, creating safety interlocks more robust by new electronic circuits and devices, and improving communication and navigation systems to such an extent that high altitude flights can be controlled remotely or emergency landing can be maneuvered from ground stations in emergency—the engineers and engineering must borrow from each other, learn from fundamental theory and rules, and translate mutual knowledge and experience into the product of highest quality. The journey does not end there as continuous evolution and adaptation is needed to overcome, solve, improve, and invent with the passing of time, acquiring of new experience and addressing unforeseen issues and challenges. How do we connect then science to engineering to technology?

If science is driven by the primary urge to learn about nature and origin of certain truth or phenomenon, engineering or technology utilizes that fundamental body of knowledge to convert into a viable and useful product or process by pursuing the next logical course of action so that the fundamental knowledge and theory can be translated into products and processes that can eventually benefit the society or humanity. In simple words, if science is 'know-why?', engineering is 'know-how?' and technology is 'know-what (sells)?' Why certain objects conduct electricity and some do not is science as we do not immediately are interested to make use of a conductor, semiconductor, or insulator. This healthy curiosity about nature and natural phenomenon has always driven science and scientists to discover, invent, and innovate. When the flight of a bird inspired Leonardo da Vinci to design a flying machine, not meeting the ultimate success though, he was simply trying to convert scientific understanding into a reality to fulfill a long standing wish of common man—*to fly like a bird.* But it was not until 1903 when the Wright brothers first demonstrated the successful, sustained, and controlled heavier-than-air powered flight, based on the works of Sir George Cayley dating from 1799. But that was a mere demonstration. Sustained efforts over the next hundred years brought us into today's world where 3.6 billion passengers worldwide are likely to take air travel to various destinations only next year. This journey was a long and arduous one for development through the most difficult test of commercial viability, stiff competition, energy, and environmental issues, and above all, stringent safety and ethical regulations. This is how curiosity to application happens through time and translation of science to engineering to technology. This advancement is neither unidirectional nor sequential, it is an act of adaptation and evolution as situation and society demand.

Who invented electricity or current electricity? The simple answer is many over the years. In modern times, from English scientist William Gilbert (1600), who coined the term electricity (from Greek amber) to Otto von Guericke (1660), Galvani (1786), Volta (1800), Oersted (1819), Ampere (1826), Ohm (1827), Faraday (1831), Joule (1841), Kirchhoff (1845), Edison (1883)—the science of current electricity developed through individual and collective contributions of many genius. The electrical power stations were constructed and electrical machines and appliances were developed much later. The science and technology of electronics developed even later during the early twentieth century.

Thus technology does not arise overnight on its own; it is based on the fundamentals of science and virtuosity of engineering. This endeavor is continuous that never starts never ends and encompasses, again not sequentially or irreversibly, discovery (sudden and new, e.g., Raman effect), invention (gradual, e.g., Thomas A. Edison) and innovation (e.g., Internet, 1969). Humanity and society thrives and flourishes with valuable and continued inputs from science, engineering, and technology without any barrier among the domains or the custodians and champions of them.

Epilogue

I must sign off now with a humble request to the readers, learned or green: the views expressed are entirely my own, based on my own understanding and experience. While the dilemma (scientist or engineer?) persists, for simply there is no easy distinction. Yet if I could ask my father now: I wanted to be an engineer, with thirty years spent in academia and R&D, what am I today—a scientist or an engineer? He would have answered with that same benign smile: serve your folks with whatever you are good at and that is more important.

Is an Engineer's Mind Different?

P.S. Goel

I was born in a small village in UP and brought up in a large family. Being the youngest, I was on call for small works like bringing a glass of water for who soever feels thirsty. Seeing a bus for the first time at the age of eight was most exciting. The driver of the bus was the hero. Travelling to a big city by train, seeing a bulb and a ceiling fan running on electricity at 10, was wonderful.

The impressions on mind are vivid and large in number, be it mother's worship in the nearby temple, her milking a cow or seeing sister delivering a baby in cold pitch darkness with serious complications and no professional help. Ultimately, what drive you are some impressions, some incidents and some advices that happen in early life. Was this the excitement of seeing the first bus at 8 a cause for choosing science in 9th standard at Sir Pratap School, Jodhpur? Was the general science lesson in middle school the reason for opting engineering after higher secondary? Or being good at mathematics drove to opt for electrical engineering? Perhaps, I do not really know. Everywhere, I seemed to be following instinct, perhaps driven by those excitements of being exposed to bus or electricity at much later in childhood. A child born today would see them from birth, so no excitement from these. Perhaps a drone delivering milk at door step will excite a child at 10 in 2015.

Once admitted to college, doing engineering was more of routine and mind started to look into the society. Born around the freedom, stories of the freedom fighters and sacrifices by thousands of youth was another thought churning the mind. Then, suddenly, an old story gave a clear direction, "living for the nation is more difficult than dying for the nation", perhaps, Swamy Vivekananda guided the path.

While in the final year of engineering, towards end of the term, the Head of the Department called and said "Go to Indian Institute of Science, Bangalore and do masters, do not look for a job in Rajasthan Electricity Board".

The 2 and 1/2 days journey to Bangalore from Jodhpur with three changes was an experience. Coming out of north for the first time was full of curiosity and uncertainty. Luckily post cards used to reach faster, so only 5 days of agony for my

P.S. Goel (✉)
ISRO Research Centre Imarat (RCI), Vignyana Kancha, Hyderabad 500 069, India
e-mail: dr.psgoel@gmail.com

P. Ghosh and B. Raj (eds.), *The Mind of an Engineer*,
DOI 10.1007/978-981-10-0119-2_7

brother at Jodhpur. Five other college mates joined IISc and the 2-year stay turned out to be fun. All were applying for abroad, so I also applied. Got a couple of admissions also, but, then there was an offer from ISRO. The sentiment of living for the country prevailed. Had heard about rockets, not much about satellites, but, I was allotted to the Satellites Systems Division.

It was a bit puzzling, what do satellites do?

The quest of solving this puzzle became an engagement of a life time. It is not just about satellites, it is all about space, the environment including vacuum and radiations, the orbital dynamics and the dynamics of the bodies, inertial reference and sensing directions, controlling motion and working remotely. It has been wonderful to explore new challenges for every new mission. Finding ways to put space to the service of mankind (the vision of the founder of ISRO, Prof. Vikram Sarabhai) and of course giving some extra leverage to the soldiers at the borders became a purpose of life.

I was asked to work on spin stabilisation of RS-1, a 40-kg satellite to be launched on first Indian launch vehicle PSLV-3. Prof. U.R. Rao, then head of the Satellite Systems Division was a scientist, born with a mind of a scientist and create leaders, came to our house on a Sunday and asked for method of fine control of gas flow in a high-energy X-ray counter. Why head of a division has to come to the house of a very junior colleague?

A visit from Soviet Prime Minister and his meeting with Madam Indira Gandhi culminated in satellite project, ISSP (Indian Scientific Satellite Project) and was to be spin stabilised as well. None of us had seen a satellite and had to learn all aspects of satellite technology very quickly. This satellite, realised in just three years, named Aryabhata after launch, is a major mile stone in India's Engineering history.

What Aryabhata gave India was confidence to do complex technology, and to us to do something new through an engineering innovation in every subsequent mission. Bhaskara was converted to do remote sensing from a spinning satellite. Many new control logics, increased data rates, very new concept of vidicon camera and passive micro wave radiometers payloads were developed. Then came an opportunity to build an experimental communication satellite APPLE for free ride on newly developing launch vehicle of European Space Agency ARIANE. Development of 3-axis stabilisation using biased momentum, magnetic stabilisation, solid apogee boost motor, Earth sensor, monopropellant system, communication transponders in C band and many others were new technologies. APPLE was a major technology step and based on its success, we embarked on IRS-1. Zero momentum based 3-axis stabilisation, high data rate in X-band, complex thermal management of satellite and precise thermal control of optical payload, and a very new, CCD-based high-resolution optical payload development were new challenges. IRS-1 became a land mark project laying the foundation of most successful national remote sensing programme in the world.

TIFR wanted an X-ray payload to look at stars very precisely for day and months, and ISRO wanted a remote sensing payload looking at Earth. We introduced inertial pointing and anywhere to anywhere manoeuverability in IRS P-3, launched by PSLV. This developmental satellite remained in demand by both competing agencies for all its operational life of almost 10 years against designed life of 1 year. All subsequent missions are similar stories, but, Technology Experiment Satellite (TES) is different. Borne out of our desire to do our bit as fall out of Kargil War, we decided to develop high-resolution satellite with agility, to image area of interest at will. Thirteen new technologies were developed and it took just 24 months from concept to launch, a feat that perhaps has never been and can never be achieved anywhere in the world. Nationalism is still a most powerful driver, as it was before independence. Recovery satellite (SRE), IRS series, INSAT series are all links of this continuing chain as new generation of engineers are replenishing the outgoing ones.

At engineering college, we are taught how things (say machines) work, but, as practicing engineers we have to understand how things fail. Bringing reliability to a product is nothing but to understand failure mechanisms and generate methodologies to mitigate the failures.

How I became an engineer, a satellite technologist is perhaps a question to be analysed and understood, but, once you are an engineer, it is not a profession but a religion. Being an engineer at ISRO brings you into another cultural dimension, where your individual identity gets submerged into common goals, societal causes and nation first. Engineering becomes a means to solve problems, not just a profession.

But, if it is so, then why do we have so many problems? We produce more than 10 lakhs engineers a year in our country, but hardly any Indian product in the world market. All rural areas are crying for engineering solutions, but engineers do not reach there. Recently, we organised a study on engineering interventions to transform Sundarbans (West Bengal) into a developed region like Netherlands, but the study has not been read by any decision maker. The last engineering intervention to Sundarban was by a British engineer in 1942, to bring sweet water through a canal. Is a distributed solar power not a solution to provide electricity access to remotely located houses? Questions are millions and we also have tens of millions of engineers. Certainly, there is a disconnect. Most of our engineers are job seekers not creators (jobs or products).

Coming to the Department of Ocean Development was with a purpose, to create a new ministry, the Ministry of Earth Sciences, that will look into problems of Earth in holistic way. It was a new dimension of science, but the problems once again call for engineers to find solutions. Later, on becoming a member of National Security Advisory Board, I discovered the insecurity in which we Indians live. We all think that Israel is most insecure country in the world, not true. It is India and most Indians do not even realise this. My stay at DRDO for last six years in different capacities and my association with engineering community in a variety of forums have exposed me to only one truth, most of our engineers are doing jobs, not solving the problems of nation, security or otherwise.

The recent INAE initiative to organise "Engineers Conclave" is just a small step to find engineering solutions to the problems of society, through space, atomic energy, defence technologies, technologies for earth sciences, etc. Ultimately, the solution lies in creating engineers, with a "mind of an engineer".

Technology and the Global Community

N.R. Narayana Murthy

Technological advances have made life more prosperous, productive and comfortable for most of the people in certain parts of the world and for many people all over the world. Yet, we see enormous suffering everywhere. A large number of children die as infants. Many of those that escape from death are condemned to a life of malnutrition, illiteracy and hopelessness. Over a billion people in the world live below poverty line and eke out a living with barely a dollar a day. These people do not have access to basic facilities such as clean water, power, nutritious food, schooling, shelter and health care.

While access to basic necessities remains a challenge to a large segment of world population, we also see unsustainable consumption in certain parts of the world. Global warming is a real threat leading to climate change, melting of snow caps and rising of sea levels. The biggest challenge of this century is to ensure the well-being of 7 billion human beings without disrupting the balance of nature.

On the positive side, we, engineers, have technology, an extraordinary power, to make a difference to this world. Technology can achieve this in three important ways:

1. **Bridge the divide** between developed and developing countries and enable us to collaboratively solve our common problems to benefit the humanity as a whole.
2. **Build a sustainable planet** by enabling us to conquer the problems resulting from urbanization and unsustainable demand on natural resources.
3. **Solve problems of social inequality** resulting from the lack of access to housing, nutrition, education, sanitation and financial inclusion for over one billion people in the world.

Let me elaborate on each one of these.

N.R. Narayana Murthy (✉)
575, 21st Main, Jayanagar 4T Block, Bangalore 560 041, India
e-mail: nmurthy@catamaran.vc

P. Ghosh and B. Raj (eds.), *The Mind of an Engineer*,
DOI 10.1007/978-981-10-0119-2_8

Bridge the Divide

A study by the World Bank shows that technological progress has helped in reducing poverty in developing countries from 29 % in 1990 to 18 % in 2004. Globalization and rapid advances in technology have brought nations closer. The problems of the poor are more visible on global platforms today than ever before. There is much effort among the developed nations to help the developing countries in their problems.

Take the case of Malaria. Malaria is one of the three lethal diseases on earth. In Africa, a child dies from Malaria every minute. *The World Malaria Report 2011* shows that there are 216 million cases of Malaria and an estimated 655,000 deaths in 2010. Although there are drugs available today to cure Malaria, they are not effective in combating multi-drug-resistant mutant "superbugs". To find a cure to this drug-resistant form of Malaria, Scripps Research Institute (SRI) in the USA has a project to simulate molecular interactions to design a cure for Malaria. But, performing such complex simulations requires heavy computing power. The World Community Grid (WCG) has helped SRI by providing computing resources. WCG aims to create the world's largest public computing grid to provide computing resources to scientific research projects that benefit humanity. The basic idea behind this initiative is to use the unused computing resources of any registered person. Today, this initiative has over 602,000 registered participants. WCG has provided computing resources to researchers working on human genome, HIV, dengue, muscular dystrophy, cancer, influenza, rice crop yields and clean energy. It has, till date, accumulated over 652,683 cumulative years of donated computing time. The results computed on the WCG will be made available in the public domain for all scientists.

My next example is on cancer. A World Health Organization study projects that cancer is a leading cause of death worldwide and accounted for 7.6 million deaths (13 % of all deaths) in 2008. Without immediate action, the number of global deaths from cancer will increase by nearly 80 per cent by 2030. A key to cancer therapy is unravelling the genetic sequence of the patient. Using genomic methods, researchers are also trying to find ways to predict and cure cardiovascular disorders, depression, bipolar disorder, Alzheimer's, Parkinson's disease, attention deficit disorders and diabetes, to name a few. We have come a long way from the first genome project which cost several billion dollars. The search is on for a more cost-effective method. *1000 genomes* is a project launched to provide an impetus to this kind of research. 1000 genomes is led by a consortium of 75 companies and organizations to establish the most detailed catalogue of human genetic variation. By 2010, the project had completed its pilot phase. The 1000 genomes, in its current production phase, targets to sequence over 2000 people. It intends to make this compiled genome sequence data freely available for further research. A critical challenge is the quantum of data generated and making it accessible to global researchers. The genome sequence of a person has over 100 GB of data. The genomic sequence data for a million people would require hundreds of petabytes of data storage. Today, with cloud technology and big data analytics, such vast amounts of data can easily be stored, retrieved and analysed. A gigabase is available at USD 10 for an academic. A gigabase is a billion letters of DNA, and for that price storage, tools, and visualization, are bundled into one package. With advances such as these, it will not be too far before personal genomics becomes a reality.

Build a Sustainable Planet

According to a UN report on global sustainability, the world population will reach 9 billion by 2040. The demand for resources will rise exponentially. By 2030, the world will need at least 50 % more food, 45 per cent more energy and 30 % more water. The technological advances in combating global warming are well known. Therefore, I will just take two unusual examples.

The electricity consumption in the USA reached 3856 billion Kilowatt hours (kWh) in 2011 which is 13 times greater than electricity used there in 1950. Stan Cox, a scientist and author of '*Losing our Cool: Uncomfortable truths about our air conditioned world*', points out that the USA uses more electricity for cooling than the total electricity consumption in Africa. Global consumption for cooling is projected to grow to 10 trillion kWh per year. This is half of the world's entire electricity supply today.

To overcome such an unsustainable consumption pattern, energy-efficient systems are the need of the hour. Advancements in sensor and grid technologies reduce energy consumption dramatically. Burcin Becerik-Gerber, an innovator featured in MIT technology review, has developed a system which uses smart sensors. The application uses occupant's cell phone to negotiate energy-efficient settings in office buildings. The system asks people how satisfied they are with the temperature, lighting, air quality and even noise level at the workplace. The system then works with the building's agents to find the most energy-efficient way of adjusting the settings to make the maximum number of people in that workplace happy. System simulations indicate that it is possible to satisfy 70 % of occupants while reducing overall energy consumption by more than 30 %.

A consequence of rapid urbanization is the increasing number of cars on the roads. As car ownership in developing countries becomes more pervasive, the number of vehicles on road will increase from the current one billion. Based on current trends, the World Health Organization predicts the death toll in road accidents to exceed 150,000 deaths a month by 2020. Around 90 % of accidents are caused by human error. Driverless cars are a great solution for passenger safety. These cars can also reduce congestion by coordinating their routes, travelling in close formation and enhancing the capacity of road networks. With an array of sensors, ultrasonic detectors, gyroscopes, accelerometers and altimeters, a car can become more aware of its surroundings. Driverless vehicles developed by Google is a good example of transforming car design, redefine car ownership, affect urban planning and provide means of transport for people with disabilities.

Solve Problems of Social Inequality

Today, 22 % of the developing world's population or 1 billion plus people live below $1.25 per day. A recent report by the United Nations Food and Agriculture Organization estimates that 925 million people remain undernourished. UNESCO points out that 796 million people lack basic literacy skills in the world.[1]

Education is a powerful tool for national development in developing countries as it provides a route to economic prosperity for both individuals and the nation. In many of these countries, government infrastructure for providing education is just not enough to meet the demand. Online educational solutions like Khan Academy rely on basic Internet and computers. The inherent problem in developing economies is the lack of such infrastructure. This impedes the access to education in such economies.

To overcome this difficulty, *Teach a Class* was founded by Neil Dsouza and Leila Al-Muthashib. The key motivation behind this project was to facilitate education in orphanages. They use technology for "recycling education" by harnessing the power of existing open-source courseware and virtual classrooms. *Teach a Class* uses 'Education Hotspots', a self-contained mini wireless network that contains built-in educational content. The organization also collaborates with local educators, partner organizations, community stakeholders and volunteers around the world to enhance the way children learn. Starting off with a pilot project in Indonesia, this initiative has spread to villages of India and Mongolia.

According to WHO, about 39 million people are blind in the world. 90 % of visually impaired people live in developing countries. Blindness and poor vision have a tremendous impact on quality of life, particularly for those living in poverty. Yet, the tools to provide mobility to the blind have changed very little since the 1920s.[2] Anirudh Sharma tackles this problem with a cost-effective and simple solution. *Le chal* is a USD 20 dollar, cheap and unobtrusive navigation aid for the visually impaired. *Le chal* means 'take me along' in Hindi. The fundamental technology behind these shoes is haptics, based on the sense of touch. *Le chal* shoes provide haptic feedback by guiding users towards their destination through vibration. By pairing it with a GPS-enabled Android smartphone, the user can speak a destination into Google maps and then move around using Google's navigation. Four small vibration motors embedded in the front, left, right and back of the shoe mildly vibrate to indicate the correct direction. The vibrations get stronger to indicate that the destination is nearing. In addition, a built-in proximity sensor in the shoe gives feedback to the users on their immediate surroundings.

These examples are just a few examples of the power of technology in solving societal problems. Technology has the potential to vastly improve the quality of

---

[1] http://uil.unesco.org/home/programme-areas/literacy-and-basic-skills/news-target/literacy/fe7ac3ca6636388c8a10f7d43a2b6a7b/.

[2] http://www.economist.com/blogs/babbage/2012/07/footwear-blind—The Economist article on *Le chal*.

lives in developing countries. These countries are tackling their social challenges by innovative use of technologies. China and India rank among the top two in INSEAD's Global Innovation Efficiency Index.

When I see a lot of extraordinary innovations that are taking place around the globe, I feel enthusiastic and confident of the success in applying technology for the benefit of the global community. And I hope every one of these researchers will reflect on this challenge and provide a social purpose to their areas of research.

The Anatomy of Change

Tarun K. Ghose

Over the years, the education system all over the world has gone through major changes. The static body of 'necessary' knowledge, considered essential for preparing students to function for a lifetime in rapidly changing professions, has also rapidly shifted. In engineering, there has been an increasing decline of emphasis on handbook design, and more and more educators have realized that they must turn out engineers whose minds will be trained to enable them to identify and solve problems not yet visualized. 'It is not sufficient to educate our students to solve problems others will define. It is essential that we cultivate in our students a curiosity and vision such that, on their own, they can discuss new, real and important problems', said one of the former Deans of MIT. Certainly, expectations from the present-day engineering graduates are much different from what it was a decade or two ago.

Changing Boundaries Between Physical and Biological Sciences

It seems generally accepted that in addition to new computer-oriented design courses, the engineering students should have a solid foundation in the principles of modern sciences and mathematics. This enables them to appreciate and approach engineering problems originating from complex biological and other self-adjusting systems. In the engineering analyses of biological processes such as the biosynthesis of primary and secondary metabolites, catalysis in solid substrates, separation processes, and growth/respiration/product-formation models, the use of classical mathematical tools helps the understanding. The application of computer methods to this highly complicated area of study is more suitable for the solution of problems than for the understanding of them. Nevertheless, much of the basic micro-

T.K. Ghose (✉)
C-518, IInd Floor, C.R. Park, New Delhi 110 019, India
e-mail: tarunghose@comcast.net

P. Ghosh and B. Raj (eds.), *The Mind of an Engineer*,
DOI 10.1007/978-981-10-0119-2_9

biological information required, for example, for scale-up, can be obtained from computers rather than from shake flask experiments.

Modern engineering curriculum is needed to be designed to develop an understanding of the laws of nature, an understanding provided by the basic sciences, including the biological sciences, not merely by abstract mathematical formulation, but also by determining the range of its validity and the limitations of its applicability with properly designed and conducted experiments.

Investigation into the sensory routes through which animals gather information has been possible through a deep interaction between biology and electronics. The need to resolve the physical and mechanical limitations of machines promoted the study of living organisms as prototypes for man-made devices. Only through a study of biological systems displayed by single living cells was it possible to discover how to simplify information circuitry, how to eliminate irrelevant information from data-processing operations, and how to construct miniature machines that may evaluate concepts.

Studies on decision-making circuits in many lower animals show promise of clarifying many extremely complex problems in space projects, for example, in weather satellites, computers, and guidance systems. The frog's eye is an example of a highly sophisticated natural computers of precision not well understood. How do its eyes draw in the surrounding scene, filter off the unnecessary background information, transmit to its simple brain only what is relevant and, simultaneously, receive the decision on what to do about the insect that is hanging about its nose? Several living species such as the bee, the beetle, the dolphin, the cod, the bat, and the homing pigeons have and are going to contribute to our scholastic pursuit in the development of data-obtaining and data-processing machines.

The understanding of the physical state of water in biological systems is limited. According to the membrane theory, living cells are sacs of dilute aqueous solutions surrounded by submicroscopic membranes. In this physical understanding of the biological systems, the water within the cell is not different from that in a dilute solution. But there is mounting evidence against this concept. Because equilibrium concentrations of permanent solutes are not found across the membranes, asymmetry in the distribution of ions and of other solutes between the intra- and extra-cellular phases is the rule.

The only conceivable way to modify the theory is by introducing the pump concept. This certainly explains the existence of a low concentration of $Na+$ in the cell water because it is constantly rejected by the pumps situated in the cell membrane. If the energy balance problems of the $Na+$ pump is considered and all the pumps essential to maintain the existing concentration of solutes inside the cells are taken into account, a perplexing situation arises which appears to violate one of the most fundamental laws of nature, the second law of thermodynamics. Theoretical calculation and experimental results appear to show that the total energy needed by such membrane pumps far exceeds the energy available to the living cells, even when the performance is accepted as 100 % efficient. The association-induction hypothesis postulates that the water content of the living cells exists in polarized multi-layers. It is this state which allows the cellular water to accept a much lower concentration of solutes within the cell and away from the membrane.

Studies indicate that in living cells, protein and water in close association constitute a cooperative assembly. It is this property which renders the hydration and dehydration of fresh foods or biologicals a much more intricate process than the hydration and dehydration cycles of simple molecules or substances, which may be either diffusional or surface evaporation controlled. The engineer's understanding of this phenomenon leaves much to be desired, but the problem has a lot to do with the kind of mass transport taking place in biological systems, both in the normal and low-temperature regions.

Another paradoxical situation existing in the living systems shows how an engineering approach has made use of both aspects of the phenomenon. Freezing is universally known to be lethal to most living systems, yet it can also prevent cells and their constituents from destruction. It may also permit long-term storage of whole, viable human organs or micro-organisms. Freezing can slow down or even stop some biochemical reactions, while it accelerates others. It is used, in fact, both to preserve and disrupt the fine structure of cells. Present knowledge is not adequate to explain why this is so, but engineers have developed hardware and systems to make full use of both the destructive and perspective powers of freezing in their manipulation of living systems.

Integration of Biological Systems with Chemical Engineering Sciences

In the context of the examples of the exciting interaction between the biological and physical sciences, it is pertinent to ask how chemical engineering has reacted to the impact of this interaction. Leadership in engineering assumes the necessary motivation and the capacity to enter into new areas rather than to extend the existing ones. The interactions of the biological, physical, and chemical engineering sciences have introduced a great quantum of knowledge to the main stream of chemical engineering. This interaction is a unique engineering activity with which the curriculum changes are closely related. Indeed, the rapid strides that have taken place in this development have been promoted by increased understanding of biological systems.

The origin of biochemical engineering can be traced back to the First World War when Chaim Weizmann almost single-handedly rescued Britain from a serious ammunition shortage. Acetone, produced from maize starch by the tetanus bacterium, *Clostridium botulinum*, helped to keep up the production of cordite. After the war, in 1923, the first citric acid plant went into production, and within a few years, this microbiological venture quantitatively replaced the established source of supply from citrus fruits from all over the world. Until now, no chemical means of producing this acid have been found to replace the microbiological process.

Between 1928, when Alexander Fleming discovered penicillin, and 1944, when Howard Flory and Ernst Chain (both Nobel laureates) made known the therapeutic use of this 'wonder drug', attention was directed, mostly in the USA, to solve the complex problem of producing penicillin on a large scale. The demand was so great that it soon exceeded that country's capacity to produce milk bottles which were serving as principal fermenters for the production of penicillin. The challenge was met by a combined effort of several federal agencies, top US chemical industries, and a few of the finest university laboratories. The chemical engineer was never before confronted with problems presented by large-scale aseptic work like the continuous daily supply of 50–100 million litres of sterile air (>99.99 % collection efficiency of microbial cells of >1.0 μ size) in one single plant, and the transport of O_2 from the gas phase into the active sites of the cells in a highly viscous non-Newtonian suspension. New separation techniques for the recovery of biologically active penicillin from the fermentation broth were created. The development of processes for the production of other antibiotics followed in rapid succession.

One of the most notable contributions the biochemical engineer made was the advancement of the contamination-proof philosophy in the design and operation of fermentation system. Almost every classical approach for vessel fabrication, valve design, piping layout, and even gasketting underwent rapid and profound changes long before they could be analysed and reproduced as textbook materials. New kinds of probes, monitors, and control systems, based on their sterilizability and fast-response reversibility, for the proper record of variables and biological systems, took a considerable time to develop. The greatest impact that this activity made on the engineering profession, as a whole, lays in the continuous achievement of new products derived from microbial and biochemical systems. It surpassed all other production processes in reducing the costs of the products.

System instability imposes a prolonged transient state to a culture, which is analogous to a batch situation and is, in effect, a deterrent to an induced mutation of the concerned organism. However, unlike in a continuous reaction in a pure chemical system, it has not yet been possible to maintain over a prolonged period a truly steady state and a stable, continuous, and pure culture microbial system. This has resulted, in several instances, in the temporary adaptation of the culture. It has been shown in a semi-continuous mixed culture that substrate tolerance of methane formers can be increased eightfold. In this area, biochemical engineers have been feeding back valuable information to microbiologists, providing new tools for mutation studies. Similarly, the space velocity concept of continuous chemical reactions in back-mix or plug-flow systems is not directly applicable to microbial reactors. Here, one is confronted with a situation of growing catalysts, which is not the case in a chemical system. Again, the physiological stages of growth and biosynthesis are all nonlinear sequences, and each phase appears to dictate what the next should do.

The biological deterioration of organic substances is perhaps the biggest mass hydrolytic reaction occurring in nature, contributing every minute by photosynthesis over 5.7×10^5 tonnes of carbon to the generation of new materials in nature's continuous cycle. About one-third of this quantity is in the form of cellulose. Excluding air and water, no defined chemical substance is more abundant on earth than cellulose. The knowledge on how the intra-cellular enzyme system is capable of depolymerizing a large molecule like cellulose has been resolved. One of the biochemical engineer's principal concerns has been the search for new sources of cheap and abundant substrate for future biological processes as existing sources have either become scarce or uneconomic.

On the one hand, in conventional agriculture, the differences between the best and the poorest yields are rarely more than a factor of ten and very frequently less. On the other hand, the growth rate of a single microbial cell is between 100–1000 times faster than the growth rate of plants and animals and, therefore, places this method of reproduction in a category wholly different from that of conventional agriculture. Given unlimited space, optimal O_2 supply, nutrients, and the right environment, a single microbial cell, dividing every twenty minutes, could produce 2.2×10^{43} cells in two days. Even though the weight of a single cell is only 10^{-12} g, the total weight of this microbial mass would be 2.2×10^{25} tonnes roughly 4000 times the weight of this planet. If C_6 sugars are used as substrate for such a conversion, the amount of O_2 required would be 2.34×10^{29} m^3 at NTP. However, this computer manipulation is not applicable in practice because of the obvious limitations of space, nutrients, and O_2 supply, but it does show the potential reproductive force of microbial cells which can be most profitably harnessed for man's greatest need, namely food. It is predicted that one of the biochemical engineer's most serious and meaningful occupations would be in the development of innocuous, cheap, and abundantly available protein foods from an unconventional source.

Environmental control is another extremely important area where biochemical engineer's contribution has been and is profound. It is no longer possible to solve problems of complex industrial effluents on the basis of the so-called more BOD, more air concept, as certain complex molecules, including some fairly large ones, cannot be synthesized by means of other than microbiological route. It is also not yet known how to provide a better and more economical way of disintegrating some other kinds of molecules to a hazard-free state. In India, huge quantity of waste urea as effluent is discharged daily into rivers from the production plants. It is possible to provide a microbiological solution to this. A right approach would enable back-feeding of the NH_3 in the synthesis cycle through a biochemical breakdown of the waste urea, thus discharging N_2-free effluent into the rivers.

In industrial biological activities such as food processing, enzyme technology, waste treatment, and also in the large-scale desalination of water, the economic use of membranes has been established, and the application of membrane separation techniques is undoubtedly important in separation processes. The limit to the improvement of mechanical designs of transport processes such as distillation, absorption, extraction, and evaporation has been reached. The use of polymeric membranes as a new tool for separation processes has made remarkable headway. The ability to tailor-controlled permeability and selectivity into a polymer film by rather simple manipulative techniques is regarded as an important step forward in solving specific industrial separation problems. The separation of two molecular species, despite equality of driving potentials, is made possible by the alteration and improvement of their selectivity and permeability. It is this unique characteristic of membrane separation, which differentiates the process from most other common transport operations. Because this process requires no change in state, it offers the ultimate capability for low-cost operations.

The extraordinary capacity for information storage in microbial cells may help evolve new concepts of large memory computation. It is possible to build up about 10^{12} *E. coli* cells per ml of a culture medium. Each of these cells has 2000–3000 different kinds of enzymes. These enzyme systems are coded by the somewhat redundant purine pyrimidine base DNA materials. Each of the three bases code specific amino acids of which there are, on an average, between 100 and 1000 in an enzyme system. Each cell has a capability of storing around 10^6 different items of information. Thus, a ml of the medium can be considered as a memory storage system capable of handling 10^{17}–10^{19} items of information. This is a challenge to the understanding, but it is one of the many examples of challenge that biological systems present in an effort to utilize their principles for the benefit of mankind.

Looking Ahead

It is clear that historic development as well as the insistent demands of modern engineering have created an intellectual ferment, an overlapping of interest and a consistent federation of thought across the traditional boundaries of the physical and biological sciences, electronics, and chemical engineering. The biological sciences now provide sound trans-disciplinary bridges over a wide range of engineering activities. As biological sciences developed the application of technology certainly moved far beyond the established concepts in engineering education and produced a 'generation of greatness'. The quantum of chemical engineering knowledge itself has expanded enormously, particularly in the areas of systems dynamics, kinetics and catalysis, transport processes, and in new operations. It is for the reasons that a good case for sophistication can be made for a strong and viable biological science's activity in modern engineering schools committed to breaking departmental boundaries.

Graduate schools in life and biological sciences in engineering campuses are commonplace today. Many forward-looking engineering schools have started full-fledged undergraduate programmes in life sciences. There is hardly any alternative in the matter, nor any question of possibility or priority for a country like India whose food, health, sanitation, and population problems are in the ultimate analysis biological in nature are acute, if not explosive. We should not underestimate the power of the biological processes.

Because the future of our industry and its progress depends principally on superior talent, some soul-searching by the industrialists is called for to ascertain why our engineering educational system does not provide the biological input to meet industry's challenges. Those industrial companies who take steps to make up their deficiencies in this respect will undoubtedly reap the rewards of leadership and enhanced profit. Furthermore, discoveries in microbiology and biochemical engineering sophistication in turn would lead to more discoveries and greater sophistication.

In his aphoristic words written over one hundred years ago, the master microbiologist of all time, Louis Pasteur, wrote "There are no applied sciences, there is only the application of sciences. The study of the application of sciences is easy to anyone who is master of the theory of it". Earlier he warned, "Without theory practice is but routine, born of habit. Theory alone can bring forth and develop the spirit of invention".

When an engineer or a scientist finds that he is a part of an organization that is continuously endeavouring to be ever more useful to the society; when he feels that the management is conscious of the implications of new knowledge in science and technology and is ready to look beyond traditional concepts; and when he sees that his views constitute an effective part of the decision-making process; when he realizes the constraints of the system, but there exists an honest recognition of the difference between paper work and creativity, in such circumstances he will throw his full enthusiasm, talent, and effort into his work. Such a system will create new concepts, attract more talent, and enjoy society's support.

India's Tryst with Destiny Establishment of the Highest Level of Engineering and Technology

Mahesh C. Chaturvedi

Introduction

The national commitment to build the new India was eloquently expressed by Jawaharlal Nehru in the first breath, as India achieved independence at the midnight, 15 August 1947: 'Long years ago, we made a tryst with destiny, and now the time comes when we shall redeem our pledge… We have to build the noble mansion of free India…' India has very far to go. Engineering and technology will be the central activity in building this 'noble mansion'. In view of the vast challenges and some severe constraints, determined efforts with vision and discretion are needed.

I got involved in this challenge by several chance events, and my experiences are presented to contribute to the policy formulation in this context. There have been two streams of my activities, one relating to water and the other relating to technology policy formulation, both in a way interacting as my activities advanced. Each will therefore be presented separately first, with the integration brought out at the end.

Involvement in the Development of Water Resources

I passed from Thomson College of Civil Engineering, Roorkee (now IIT Roorkee), in 1946 and was engaged in the Irrigation Department, Government of the United Provinces (now Uttar Pradesh), under the British regime. I was associated with India's water resources at increasingly higher levels of activity, starting with the

M.C. Chaturvedi (✉)
323, Nilgiri Apartments, Alakhnanda, New Delhi 110 019, India
e-mail: mcchaturvedi1@gmail.com

P. Ghosh and B. Raj (eds.), *The Mind of an Engineer*,
DOI 10.1007/978-981-10-0119-2_10

construction of one of the first hydro-engineering station, Khatima Power House, and next the first major dam in Ganga basin-Rihand.

I and my wife, who had also been working in the area of education, decided to advance our knowledge and go to USA for higher studies. Fortunately, both were awarded Fulbright Travel Grant and Assistantships at the university. As we completed our doctoral studies, both were offered jobs at the US University, but we preferred to return to our country and serve her. Indian Institute of Technology (IIT), Kanpur, was just being established, essentially by a consortium of US universities led by MIT. I was invited to join as head, Department of Civil Engineering, being the first professor appointed in the engineering section. I was the oldest faculty member, aged 37 years! Vipula Chaturvedi also joined the Humanities Department.

To be engaged in establishing a new IIT in a leading capacity was an exciting experience. But, it was cut short as the Government of Uttar Pradesh, to which I still belonged, recalled and appointed me as the director, Designs, Ramganga Dam, the first major water resources project on River Ganga system in the Himalayas. This also meant that I became a member, Board of Consultants of all major projects in the Himalayas, working with the leading Indian and International engineers, who constituted this forum. It was an even more challenging assignment.

As the Ramganga Project design was completed, I got an invitation from IIT Delhi, which was being established, to join it as one of their senior most faculty members. This was also exciting, but what made it even more so was that it led to collaborative research with Harvard regarding the development of India's waters, which continues to this date.[1] This came through an interesting chance international interaction, both for Harvard and me.

Harvard is hardly a leading engineering institution, but development of the art and science of water resources was being revolutionised by the emerging science of system analysis, undertaken collaboratively by engineers and economists [1]. Harvard was the leader in this emerging area. At this very time, Harvard became involved in the development of the Pakistan waters during President Kennedy's regime. He was willing to give money to General Ayub Khan, her NATO partner, but not for purchase of military equipment. It was brought to his attention that West Pakistan was suffering from a serious salinity problem of her irrigated areas. This was offered by Kennedy and was gladly accepted by the general. This task was undertaken by Roger Revelle, an important scientific and political person, serving Kennedy at the highest level [2]. After its completion, the Harvard scientists undertook the challenge of the management of the waters of East Pakistan (which ultimately became Bangladesh). Harvard scientists suggested that the development of Ganga waters should be planned in terms of this emerging science of systems analysis [3]. Dr. K.L. Rao, the Minister of Irrigation, also got the same message

[1]We have spent several years at Harvard and several US universities as Visiting Scientists or Professors. It continues by our spending the summers at Harvard in informal collaborative research now.

from Prof. Warren Hall, whom he met in California, as he had gone there to study its water development in the context of his scheme of transferring waters from Ganga to Cauvery. Nobody in the Government of India (GoI) knew what systems analysis was and what was meant. Close interaction developed between Dr. Rao and me. He was keen on the modernization of water resources activities of India, but the bureaucracy was indifferent, a great problem in the area of public service technology modernization, which continues.

Continuing the story, close collaboration developed between me and Harvard, thanks to the support of the Ford Foundation, which continues. (My wife has also been active in her field of education, thanks to Roger Revelle, who was also a leading educationist, working as a member of the Kothari Education Commission.) Roger Revelle and me got engaged closely in the development of India's waters since 1971, first independently and later collaboratively, as both of us considered that the development of India's waters can be totally transformed. We demonstrated that the monsoon waters could be stored by charging the groundwater. Thereby, we could more than double the water availability and, even more importantly, put water in the hands of the farmer, making him independent of the bureaucrats and gods! Roger called his proposed revolutionary technology Water Machine and later adopted my independent development, Chaturvedi Water Machine. We worked closely and organized workshops of Indian and US scientists to have our proposals discussed.

Ultimately, we submitted them personally to the then Planning Commission Member in this area, Prof. M.S. Swaminathan, to which the then Secretary, Ministry of Water Resources, Government of India, Patel, was also invited. He took the responsibility of getting our proposals examined, but instead, put them aside. He was hooked to his 'absurd' scheme of Interlinking of the Rivers of India, which unfortunately still dominates the official thinking and even gets wrongly reported even on behalf of the academy.[2]

Dr. K.L. Rao had proposed transferring of Ganga waters to South India, joining it with Cauvery ultimately. He had proposed pumping of the waters up to cross the Vindhyas. I had been working closely with him, considered use of hydroelectric energy for this purpose, generated in the Himalayan Rivers. Working on this idea, a novel way of developing hydro-energy, storing it and transferring the waters to south occurred to me. This has been called Chaturvedi Water Power Machine.

Another idea regarding provision of water in the Gangetic basin has been proposed by a US firm which was involved in exploring oil in it, in collaboration with the concerned Government of Indian agencies. They did not find oil, but artesian

[2]I was invited by the Parliamentary Committee, Government of India, to obtain my views on the subject. I demonstrated that it is an 'absurd' scheme. The Working Group of the National Commission on Integrated Water Resources Development of India (1998), of which Prof. Bharat Singh, me, Prof. A. Chawla were Chairman and Members, respectively, has also demonstrated in its report and categorically stated that it is totally unwarranted. Unfortunately, this is overlooked on account of certain reasons.

waters were noted. The US Company later carried out detailed investigations regarding the availability of water. On completion, they proposed its development. The World Bank organized a meeting in Washington to which the former Secretary, Ministry of Water Resources, C.C. Patel and me were invited. I considered it an excellent idea, but Patel was against it, perhaps because it ran foul with his absurd scheme of Interlinking of Rivers of India. The scheme is not talked of any more.

I had been proposing to the Government of India, Ministry of Water Resources, since the beginning of this century that the development of India's water should be undertaken in terms of these proposed advances in a coordinated integrated manner. Presentations were made to the Central Water Commission (CWC) but nothing happened. But I persisted. I brought out the subject again to the attention of the minister. He desired that a workshop on the subject be organized by the Ministry. The proposed revolution was again brought out in the workshop, held by the Ministry on 10 November 2009. We demonstrated that the development of India's waters can be revolutionized through several novel technologies that we have been developing: (1) Chaturvedi Water Machine, (2) Chaturvedi Water Power Machine, (3) Development of Artesian Waters and (4) Integrated Surface Water, Ground Water and Power [4]. This has been brought out comprehensively in three books [5–7]. They have been submitted to the Government of India. But action in the area, which is being suggested to the government at the highest level for a very long time, continues to be ignored!

However, a chance event has provided hope. The Chief Minister of Uttar Pradesh, Akhilesh Yadav, happens to have environmental engineering specialization, up to masters level, from an Australian university. The three recent books by me bring out the formidable challenges India will be facing in the management of her water resources as India undertakes her development and the revolution needed and possible as proposed by me came to his attention. He could appreciate the subject, in view of his personal enlightenment, and he invited me to guide the development of Uttar Pradesh waters. Hope some action follows.

Establishment of the Centre for Policy Studies at IIT Delhi

I got involved with the establishment of a Centre for Policy Studies at IIT Delhi concurrently with my involvement in the advancement of water resources, the two interacting a bit too.

Prof. Subramanian Swamy joined IIT Delhi from Harvard. He is very dynamic person. Coming from Harvard, he was aware of the developments in the area of systems engineering. He organized a seminar in which interested faculties from all disciplines were invited to participate, to share their ideas and experiences in the context of the challenges that they may be studying, with particular concern about the evaluation of their developments from the societal perspective, of which economic valuation is an important concern. This brought Prof. Satsangi from electrical engineering, Prof. Prem Vrat from mechanical engineering, and me, a civil engineer, besides Prof. Swamy, together, all of whom were a bit knowledgeable and interested in this area. Unfortunately, Prof. Swamy was lost to IIT Delhi for some reasons but we continued, though not with that much dynamism. A Centre for Systems Studies was established at the IIT Delhi.

It is increasingly being emphasized that a new perspective and approach to human development have to be considered. Society, economy and environment are one integrated system, and their management has to be considered integrally. Advances have been made to develop the appropriate policies, based on participative interactive management, backed by system dynamic modelling. We have extended it in Indian perspective [8].

An important extension follows, which is directing the establishment and development of an important aspect of technology study at some major international educational institutions, Technology Policy Study Programmes. I got involved in this area through my own experiences. Some of my friends at international engineering institutions were also developing similar ideas. They have much more opportunity of experimentation and established Centre of Policy Studies, MIT, with which I interacted closely, during my summer sojourn at Harvard. An Indian scholar, at Harvard, Ambuj Sagar was actively involved in it.

Fortunately, three Indian scientists at Harvard conveyed to IIT Delhi to come over and establish the programme in this area. IIT Delhi concurred, after considerable hesitation. We lost two of them but one, Prof. Ambuj Sagar, did come and join.

Another development was taking place at IIT Delhi, thanks to the vision of its Chairman, Prof. M.G.K. Menon. He proposed establishment of professorial chairs, whereby an identified professor gets additional funds equivalent to his salary, from endowments made by people to establish a chair at IIT Delhi, to undertake his activities with freedom. He can only take about 20 % of it personally, but he gets freedom to travel around, hire research scholars and so on, without any questions asked! This also entitles the donor to participate in these activities closely, thereby enabling additional support and insight in the subject to the IIT professor and the opportunity to the donor to contribute.

IIT Delhi has established Centre for Policy Studies, the first anywhere in India. I donated the required sum of Rs. 40 lakhs for the establishment of a professorial chair in this area. The identified professor, one of the three who had wanted to come back to India to establish this centre, Prof. Ambuj Sagar, has been appointed Vipula and Mahesh Professor there. He is undertaking the activity with commitment, in collaboration with MIT and some other leading international institutions, and me.

Advancing Technology in the Water Area in India

We brought out above the challenges that India faces in the development and management of her waters and the tremendous opportunities in revolutionizing the area. We also brought out how some action has got started of late, thanks to the interest of an enlightened young political leader Shri Akhilesh Yadav. His vision and support, as discussed above, are inspiring. India's water problems abound, and the academicians have to take dynamic initiative.

History also demands it. Thomson College of Civil Engineering was one of the world's first engineering college, established in 1846, in the context of the construction of the Ganga Canal [9]. Thomson was the chief engineer and later the Governor of UP, in whose name the college was therefore established.

Some examples of the problems and suggestions for their resolution may be given. India's water development depends greatly upon Nepal as most of the Himalayan Rivers originate there. The Nepal Ambassador happened to be a doctoral student of the Prime Minster, Manmohan Singh. He was tremendously knowledgeable, and he appreciated the revolutionary India–Nepal collaborative development of Ganga waters through the Chaturvedi Water Power Machine, proposed to him by me. It can be undertaken only if Nepal undertakes collaborative development with India. It will increase the hydroelectric potential several fold, with benefits equally to both countries. This will easily resolve the current stalemate as it brings tremendous advantage to Nepal (as well as to India). It will greatly contribute to the development of both the countries. Action is being taken by me to implement the idea.

To give another example, the states are quarrelling amongst themselves about their water entitlement as most of the rivers are interstate. It can be easily resolved if we develop and manage water scientifically, as proposed by us, instead of continuing to follow the age-old technology of canal irrigation. We have enough water to meet every state's needs, if it is developed scientifically.

Another interesting example is the development of Brahmaputra waters, leading to the resolution of China–India dispute [10]. Indeed, our activities have been appreciated by China, and they have shown interest in them.[3]

We proposed to IIT Roorkee that they had the world's largest faculty in the water area and they have the opportunity to contribute in the area in one of the most challenging manner. The director appreciated the suggestion and invited me to implement my ideas, appointing me Distinguished Visiting Professor. We have established a Vipula and Mahesh Chaturved Professor in water at IIT Roorkee, making a donation of Rs. 45 lakhs, and are attempting to help them to establish themselves as the leading world institute in this area.

[3]India and China have dispute about Brahmaputra waters. We have proved that it can be easily resolved [10]. I was contacted by a Chinese embassy representative in this context. Interestingly, the last Chinese President was a water resources engineer, engaged in the development of Brahmaputra waters in Tibet!

Conclusion

As we stated at the outset, engineering and technology will be the central activity in building the 'noble mansion' of India, on which India embarked after achieving independence, under the leadership of Nehru. In view of the vast challenges and some severe constraints, determined efforts with vision and discretion are needed. Indian efforts, so far, have not been very pleasing. Korea, starting at the same level as India, has forged far ahead. China is also doing better. We have to commit ourselves to attain the highest levels, far surpassing any other country, as Japan decided when she undertook her development [11].

References

1. Maas A, Hufschmidt MM, Dorfman R, Thomas HA, Marglin SA, Fair GM, editors. Design of water resources systems. Cambridge: Mass Harvard University Press; 1962.
2. Revelle R, et al. White House-Department of Interior Panel on Water logging and Salinity in West Pakistan, Report on Land and Water Development in the Indus Plain. Washington D.C: The White House; 1964.
3. Rogers P. A systems analysis of the Lower Ganges-Brahmaputra Basin. UNESCO, International Symposium on Floods and Their Composition, Leningrad; 1967.
4. Chaturvedi MC. 2009. Revolutionizing the development and management of India's waters. Ministry of Water Resources. Government of India. New Delhi. 10 Nov 2009.
5. Chaturvedi MC. India's waters-environment, economy, development. Boca Raton: Taylor and Francis; 2011a.
6. Chaturvedi MC. India's waters-perspectives in development. Boca Raton: Taylor and Francis; 2011b.
7. Chaturvedi MC. Ganga-Brahmaputra-Meghna Waters—advancing the development and management. Boca Raton: Taylor and Francis; 2012.
8. Chaturvedi, MC. Environmental systems management for sustainable development. In Process of Publication; 2013b.
9. Cautley PT. Completion report. Ganges canal. London: Smith Elder and Co; 1860.
10. Chaturvedi MC. Revolutionizing the development and management of India's waters. New Delhi: In Process of Publication; 2013a.
11. Yamamura K. Bridled capitalism and economic development of Japan 1880–1980. In: Myers RH, editor. The wealth of nations in the twentieth century. Stanford: Stanford University Press; 1996.

Further Readings

Chaturvedi MC, Chaturvedi V. India-tryst with destiny-eradicating poverty and attaining globalequity. In Process of Publication; 2013.
Michel AA. The Indus Rivers. New Haven: Yale University Press; 1967.

Glimpses from a Lifetime in Teaching and Research

S.C. Dutta Roy

The First Two Years

I joined the profession of teaching and research, by choice, in December 1961 and continued regular teaching till December 2010. That makes a total of five decades in teaching. The research activities started in early 1960 and I am still continuing, which makes five decades and a half in research, at the present time. Immediately after obtaining my M.Sc. (Tech.) degree in Radio Physics and Electronics (RPE) from the Calcutta University in 1959, I joined the River Research Institute (RRI), a West Bengal Government Organization, as the research officer (electronics). My job was to conceptualise, design, and fabricate transistorized instruments for measurements in hydraulic research. Transistors did not figure at all in the syllabus of RPE those days, and all that I knew about transistors in student days was from a few lectures delivered, at the request of students, by Dr. A.N. Daw, who was a lecturer at that time and had done his Ph.D. on transistor theory. At the RRI, the new job, the first in my life, forced me to learn transistor circuits by studying very hard, experimenting with transistors and, of course, burning a few of them. The only transistors available in the market at that time were OC71 and OC72, made by Philips, and I had to be content with whatever their capability was. I did design and fabricate a transducer for water-level recording, and the associated signal processing circuits, by using transistors and diodes; a complete electrical analogy experimental set-up; and a transistorised set-up for field measurement of soil permeability. All the three developments appeared in the journal *Irrigation and Power*, published by the Central Board of Irrigation and Power, New Delhi. A theoretical analysis of the water-level recorder and a method for linearizing the response also appeared in the *British Journal of Applied Physics*.

S.C. Dutta Roy (✉)
164, Hauz Khas Apartments (S.F.S.), New Delhi 110 016, India
e-mail: s.c.dutta.roy@gmail.com

P. Ghosh and B. Raj (eds.), *The Mind of an Engineer*,
DOI 10.1007/978-981-10-0119-2_11

At RRI, I also started research on network theory and transistor circuits. The initial studies were concerned with analysis and design of frequency selective RC networks and transistor oscillators. In these efforts, I received tremendous support and encouragement from Professor A.K. Choudhury of the Institute of Radio Physics and Electronics (INRAPHEL), fondly known as AKC by his students. Some of this work appeared in the *Indian Journal of Physics*[1] (numbers within square brackets refer to the corresponding numbered end note) and *Journal of the Institution of Telecommunication Engineers*.[2]

At the University of Kalyani

Towards the end of 1961, I realized that my interest in network theory, and my ambition of acquiring a doctorate degree, could not be pursued with full vigour in parallel with the instrumentation work at the RRI. This got coupled with my life's ambition of ultimately joining the teaching profession, and I started exploiting the possibilities. On advice from a friend, I met Professor S.K. Mukherjee, the highly reputed and respected chemistry professor, who had just left the Indian Association for the Cultivation of Science to take up the position of professor of chemistry and dean of the Faculty of Sciences at the newly established University of Kalyani. This meeting is memorable because the next day he sent me to meet the vice chancellor, and a few days later, I received an offer of lecturership in Physics at the University. I of course accepted it with pleasure and joined the job soon thereafter. I had to teach Physics as a minor subject to B.Sc. Chemistry

Honours students. This was the first year of the first undergraduate programme started by the University. Physics Honours programme started a couple of years later when Dr. D.P. Ray-Chaudhuri joined the University as a reader. While there were only three students in the B.Sc. Class, I also had to teach a Physics course to 60 odd students of the first year B.Sc. (Agriculture) programme under the Faculty of Agriculture. The Faculty of Agriculture was inherited by the university from the erstwhile Birla College of Agriculture under the Calcutta University and was well established and reputed with undergraduate and postgraduate courses in all branches of Agriculture. I was only 24 years old at that time and had a dozen or so young and freshly graduated lecturers in the whole university to keep company. While my Chemistry boys were overly quiet and eager to learn, the agriculture students were rather noisy and disinterested in the subject. Being the only Physics teacher in the University, my teaching load amounted to 16–18 h of contact per week, typically 6 h of theory and 10–12 h of laboratory; however, coping up with this load was not a problem. Soon I devised, in consultation with some colleagues

[1]Somehow, the INRAPHEL faculty those days had a great obsession for publishing in this journal.
[2]This later became the *Journal of the Institution of Electronics and Telecommunication Engineers* (JIETE).

of the same age group, several techniques of controlling a large class, the foremost being to build a personal rapport with individual students. I soon developed the reputation of being able to address each student by his first name when asking questions in the class.[3] The other thing that I practised is that since I was the one who set up the question paper and also grade the answer scripts, I was not obliged to cover the total syllabus and therefore not in a hurry. Instead, I took time to explain the concepts in as simple a manner as possible in the class, without referring to notes or books, and adding wit and humour to the rigours demanded of a Physics course. I worked out, in the class, problems which cannot be solved by routine application of formulae, but required an understanding and application of the basic concepts.[4]

At the end of the first semester, the informal feedback from the students to the dean and other senior faculty was appreciative of my teaching and I felt happy. I started the second semester without any feeling of strain, and with enjoyment of what I was doing.[5]

At the University of Kalyani, I continued my researches in network theory and transistor circuits. At that time, the problem of isolation between two components in an integrated circuit (IC) fabrication was being seriously investigated. The limitation of space gave rise to the isolation behaving as a distributed RC ($\underline{RC}$) network. This was considered as a parasitic and attempts were being made to minimize its effects. I have always been fascinated by distributed networks, transmission lines in particular, and studied the few papers which appeared in IEEE and other journals on the analysis and behaviour of $\underline{RC}$ networks in ICs, and the possibility of using this

[3]This reputation lasted throughout my teaching career. Even now, after many years, when I meet a grown-up person, who does a pronam after announcing that I had taught him or her, I can usually come up with the name quickly by analysing the smile, the manner of talking and the body language. It is not that I do not make mistakes, but they are not many.

[4]Throughout my career, I was lucky enough not to be constrained to cover a certain syllabus, and these simple steps proved to be useful in later life also. Much later, at the University of Minnesota, one of my colleagues who studied at MIT, told me that Professor E.A. Guillemin, a pioneering researcher and teacher of circuit theory used to say, "I am not here to cover a certain course, but to uncover certain parts of it as best as possible so that what you learn remains with you throughout your future career". I found this matching my ideas one-to-one and used to mention this at the first lecture of every course I taught.

[5]This feedback, even though informal, was very important for me and provided incentive for doing even better. I have emphasized this at every organization I served. At Minnesota and Iowa, student feedback was a long-standing practice, and I learnt there how to do it without any involvement of the teacher concerned. At IIT Delhi, I succeeded in getting this done first in my department and later adopted throughout the institute, but only as feedback to the teacher concerned. Many years later, the director started issuing appreciative letters to the concerned faculty, but there was no public recognition. Public recognition had to wait for many more years, and I understand that today, the institute recognises outstanding teachers in every department every semester through an award. I take this opportunity to emphasize a general principle, viz. that a proper recognition of the good and dedicated work of any individual at the appropriate time goes a long way in shaping the future work of an individual. I remember that at the felicitation ceremony of Professor S.N. Bose after his election to the Royal Society, Professor Bose said, "What use is this recognition at the fag end of my life?".

network for useful purposes. I did some work on this topic along with transistor circuits and circuit modelling of transistor action. Interestingly, I established that transistor action can be modelled as that of an active $\underline{RC}$ network. This work was published in the *International Journal of Electronics.* I also found that inductance, which could not be fabricated in ICs due to space limitations, could be realized by using the input impedance of two transistors in the Darlington connection and that along with the inductance, there also appeared a negative resistance in series with it under certain conditions. This opened the possibility of realizing an oscillator by connecting a capacitor across the input. I was excited with this discovery and tried to verify the phenomenon experimentally. Since there was no facility at the University of Kalyani, I used to go to Calcutta every Saturday afternoon (Saturday used to be a half working day those days), work in the Solid State Circuits Laboratory of Professor A.N. Daw of the INRAPHEL Saturday evening, spend the night in the Law College Hostel on College Street as a guest of one of my Ramakrishna Mission friends, work the full day of Sunday and take a late evening train back to Kalyani, to be present in the class on Monday morning at 10 AM. This routine continued for next few years.

As is well known, making a sinusoidal oscillator is rather tricky, particularly with a two terminal circuit. After a few initial problems of biasing and adjustments, one late Sunday afternoon, it was a moment of great joy for me when I observed beautiful sinusoidal oscillations in the oscilloscope connected across the external capacitance. The parameters of the active inductance were strongly dependent on the temperature. I put the whole circuit in a glass test tube and immersed it in a temperature controlled water bath and took data on the frequency of oscillation versus temperature. The results were published in *Proceedings of the IEEE* as a letter.

At this time, there appeared in the literature a few papers on the theory and applications of uniform as well as non-uniform $\underline{RC}$ networks, to be abbreviated henceforth as U$\underline{RC}$ and NU$\underline{RC}$ networks, respectively, and it was shown that using NU$\underline{RC}$ networks, one could achieve several advantages over the U$\underline{RC}$ networks. I investigated the phase-shift oscillator application of the exponentially tapered NU$\underline{RC}$ network, theoretically as well as experimentally (the latter by modelling the NU$\underline{RC}$ network by a multi-section ladder network) and published the results in *Proceedings of the IEE.* I also investigated null networks using NU$\underline{RC}$ networks and compared the results with bridged and shunted RC ladder networks, which acted as the inspiration for replacing RC ladders by $\underline{RC}$ networks. A paper on this investigation was prepared and published in *IEEE Transactions on Circuit Theory.*

I also investigated some specific exactly solvable NU$\underline{RC}$ networks and formulated a general procedure for obtaining the matrix parameters of a general NU$\underline{RC}$ network in terms of the two independent solutions of the second-order differential equation with variable coefficients governing such a network. These results were also published in *IEEE Transactions on Circuit Theory.*

I compiled all this work in a thesis entitled "Some Studies on Lumped RC Networks, Distributed RC Networks and Solid State Circuits" and submitted the same for the D.Phil. degree of the Calcutta University in 1964. The degree was granted in the middle of 1965, but before that another major event occurred in my life.

Distributed RC Networks formed a hot topic at that time. My publications in IEEE and IEE journals attracted considerable attention amongst the circuits community. When I was busy compiling and typing out my thesis, a pleasant surprise waited for me one evening when I returned home after attending college. This was an aerogramme from the USA written in long hand by a young assistant professor of Indian origin, B.A. Shenoi (BAS) of the University of Minnesota, exploring the possibility of my joining him on a research fellowship to assist him in his US National Science Foundation project on RC networks. Since I was already preparing my doctoral thesis, and I would not like to go abroad for another doctorate degree,[6] the only way I could honour this unsolicited offer was by accepting a Post Doctoral Fellowship (PDF). However, my sixth sense told me that PDF would be totally under the control of a particular faculty, in which situation, the possibility of intellectual exploitation cannot be ruled out.[7] In reply to BAS, I wrote that I was

[6]Somehow, I felt that this does not speak well for the state of education in India, specially the Calcutta University, even though I know many academics in India who take pride in their double doctorate. After all, how many American or European Academics are there with two doctorate degrees, excepting the ones whose second doctorate, usually a D.Sc. is considered a higher degree than the first doctorate, i.e. Ph.D. or D.Phil.?

[7]My decision not to go abroad on a PDF, which, as I said, was dictated by my sixth sense, was proved to be a right one by the numerous cases I observed at the University of Minnesota and also later at the Iowa State University. I made acquaintances with many Indian PDFs, mostly in Chemistry, Life Sciences and Physics, and only a few in Engineering. American PDFs were few and far between; PDFs mostly came from under privileged countries such as India, Pakistan, nationalist China and South Korea. For an American Ph.D., one or two years as a PDF under the Ph.D. Supervisor was not uncommon, but the longer the person stays as PDF, the less becomes his or her chances of a regular employment as a faculty. For the PDFs from outside USA, it is considered as a win-win situation, because they wanted to come to USA on some pretext or other, mostly pre- and postdoctoral fellowships, and ultimately settle there for a better life than is possible in their own country. On the other hand, US faculty needs assistance in research to increase their publications count and brighten the prospects of getting tenure/promotion/annual increment. In the USA, the amount of funds brought by a faculty to the university is a strong parameter in their career progression. They have to depend on talents from abroad, because the percentage of American high school graduates going for higher studies is very small, and the American Graduate Schools survive because of foreign students. Most American students who pursue higher studies quit after the bachelor's degree. This is perhaps one reason why the master's degree program there is not as rigorous as it is in India either in duration or breadth. One could obtain a master's degree in engineering in the USA in 9 months and only on the basis of course work and little or no emphasis on laboratory practice.

The second reason is that the amount of fellowship to be granted to a PDF depends solely on the principal investigator; there is hardly any university or country norm. As a result, the money required to be given to an American PDF can fetch two from India, and the more the manpower, the more will be the volume of the output in terms of research publications. I know of some science faculty in USA, who employed a large number of PDFs to carry out research for them, and as a result, their publication count was close to, or exceeded a thousand. This phenomenon is of course not rare in other countries and even in India. The other aspect of Indian PDFs in USA I observed is that scientific experiments which required overnight running of a process was mostly

about to submit my doctoral thesis and that I would be agreeable to collaborate with him in the project as a co-investigator only if I received an offer of a visiting assistant professor from the University and I underlined the word "only". BAS was a very nice human being, and foreseeing that processing such an offer by the university may not be easy and may take time, he was generous enough to suggest that I explore such a possibility with a few other US Universities and he suggested three, in each of which there was at least one faculty member who had research interests in RC networks. I of course knew about their work through their publications. I wrote to the department heads of these three universities, and surprise of surprises, all of them responded positively and enthusiastically within a span of two to three months. Two of them made concrete offers, while the third said the offer was under processing and would be mailed in a few weeks' time. It was at this juncture that the Minnesota offer came, but the salary there was the lowest of the three. However, in appreciation of BAS's generosity, I accepted the Minnesota offer and ended up at the Minneapolis airport on a cold night of September in 1965.

At the University of Minnesota

Going abroad was not an easy job for me because I had no reserve money to pay for the airfare. I had to borrow money from a friend and a relative, which of course I paid back within a span of one year. This was my first journey out of the country and that too, to the most advanced and glamorous country in the world. There were many cultural shocks waiting for me, the first one being that the door opened automatically as soon as I stood before it at the John F. Kennedy International Airport, New York! This was a kind of magic to me who had never witnessed anything of this sort earlier in India.

It is important to mention here that I never had settling in the US in mind and this is why I went on two years' extraordinary leave from the University of Kalyani. I extended my stay by another year, still on leave, for reasons of completion of the project and completion of M.S. and Ph.D. Dissertations as well as term papers of some non-thesis option M.S. students I was involved in supervising, some jointly with BAS, and a few singly. One of these term papers on inductance simulation by

(Footnote 7 continued)

carried out by Indian PDFs. I know of many Indian PDFs who run from one university to another again as PDF because the project in which they were employed was over, and there was no guarantee as to when the next project would come. I have known Indians who have spent their lifetime as PDFs and could never get a regular university employment. These are kind of professional PDFs, and I have met some of them, who, out of frustration, returned back to the country at a much lower level of appointment than what they deserved. However, some of them did very well in their later life and rose to the top echelon in their organization, including the top positions. Many of my readers may differ with regard to their experience of PDFs; what I have stated here is solely based on my experience.

a single operational amplifier, by D.F. Berndt, was published in the *IEEE Journal of Solid State Circuits;* this paper is now considered as a classic, widely applied and cited. On the basis of the research done at the University of Minnesota, I published, singly as well as jointly, papers in *Electronics Letters, Proceedings of the IEE, Proceedings of the IEEE, Journal of the Franklin Institute, IEEE Transactions on Circuit Theory and IEEE Transactions on Education.*

At Minnesota, I taught courses on Linear System Analysis and Circuit Theory at the UG level and Network Synthesis and Distributed Networks at the PG level (called graduate level in the US). While I was quite comfortable with circuit theory and distributed networks, I had to learn linear system analysis and network synthesis virtually from scratch. I was taught transform techniques at the INRAPHEL by a Mathematics teacher from outside the institute, but he could not create any interest in the subject because he virtually copied the material from the book of L.A. Pipes on the board, in the name of teaching. I was also familiar with elementary network synthesis taught by Prof. A.K. Choudhury in a span of ten lectures, but the University of Minnesota course was a three quarter sequence, going into details and depths. While I was forced to learn the two subjects, I started enjoying the concepts and innovating upon the proofs and application-oriented examples.

At Minnesota, I also attended the lectures of a few reputed teachers of the department, not only to broaden the horizons of my knowledge, but also to become familiar with how teaching is done in the USA. I learnt a few novel methods of teaching and evaluation. One of them is the practice of daily quizzes at the beginning of every lecture, lasting for not more than ten minutes. True, they consumed part of the class time, but the benefits to the teacher and the taught were disproportionately large. In every class, the first thing the professor did was to pose a problem on the board and ask the students to solve it in a given span of time, typically ten minutes or less. While the students were busy in tackling the problem, the professor would distribute the graded answer papers of the previous quiz and any handout he or she wished to give to the students. The daily quiz problem was not a routine one, but could be solved easily if the student had followed the previous class lecture seriously. This practice could be implemented in a small class of ten to thirty students, because a large class would mean loss of some more time in collecting the answer scripts (typically a single page, which every student was expected to keep ready); also grading them before the next lecture may pose a burden on the teacher. I practiced this daily quiz routine in small classes later at IIT Delhi with very satisfactory results. Initially, the students did not like it, but as the semester progressed, they came to realize the benefits of continuous preparation and serious attention to the class material made them learn the concepts behind every development and enjoy the subject. For the teacher, it is a boon because it gives instant feedback and almost 100 % attendance. Nobody can afford to miss a lecture because the daily quizzes carried 15–20 % weight in the final grading of the course.

Another new thing I learnt at Minnesota was that of open book, open notes examination. I adopted this in my teaching immediately and since that time I do not remember to have given a single closed book, closed notes examination. This was unheard of in our student days, and at IIT Delhi, I had to face much criticism and scepticism from my colleagues on this count. Some examinations at Minnesota were take home types, i.e. the student takes the question paper home, consults freely source materials and, even people, prepares the answers and submits the next day. This entails hard work on the part of the teacher and can be practiced in small classes, like an elective or a graduate course with no more than ten students. While take home examinations could not be practiced at IIT Delhi because of many reasons, including larger class sizes, open book, open notes examinations became the SCDR brand. Naturally, the questions had to be set in an innovative manner involving application of basic concepts rather than calculation from a formula or following an algorithm. I also practiced setting up new problems every time I repeated a course. Thus, solving a collection of problems set up in the previous sessions gave the students a feel of the type of questions but not any specific question which may figure in the next question.

Leaving the University of Minnesota

In the middle of my third year at Minnesota, the head of the department, Professor R.J. Collins (RJC) asked me whether I would like to be tenured as an associate professor in the department and I told him "Thank you Bob, for your consideration, but I have to return to India and struggle it out there". He appreciated my sentiment, but also told me that his door will always be open in case I wished to return. Soon after this, some quick developments took place. One day RJC brought to my notice an advertisement for faculty positions at IIT Delhi (IITD) and a letter to him from Professor P.V. Indiresan (PVI), the head of the department there. RJC asked me whether I was interested, and I said, "of course", because getting to teach at an IIT would be a great opportunity for working in the mainstream of my background as compared to going back to Kalyani and continue teaching Physics there. Things moved very quickly thereafter, ending up in an offer from IIT Delhi as an associate professor in about three weeks' time. By that time, I had already written to three other IITs and the University of Roorkee. It was my good luck that I got a telegraphic appointment from one IIT and a positive response from the other two, one short of making a formal offer and the other saying that the processing would take some time. I also got an offer from the University of Roorkee, and a surprise offer from the Calcutta University. It was a surprise because I had never written to Calcutta University, the reason being that I never wanted to go back to my alma mater where I would be amidst all my former teachers. When I was tossing with the various offers, a personal letter came from PVI urging me to accept IITD offer and join as soon as possible. This was followed by an aerogramme from him almost every week. Even though the IITD offer was attractive, I hesitated because I came to know that one of my seniors at INRAPHEL, A.K. Mahalanabis (AKM) had joined IITD as an assistant professor. A letter from him dispelling my discomfort and urging me to help him in initiating research at IITD settled the issue, and I landed at the Delhi Airport in the early morning of 23rd September 1968. I was received at the airport by PVI himself with a staff car, and after dumping the luggage at the institute Guest House, I went to the department to give the joining report.

This was the beginning of a relationship for the next 42 years. I have been associated with IITD, initially as an associate professor (September 1968–December 1969), then as professor (January 1970–July 1998), head of the department (1970–1972), dean of undergraduate studies (1983–1986), Emeritus Fellow (1998–2005), Indian National Science Academy (INSA) Senior Scientist (2005–2007), and finally as INSA Honorary Scientist (2007–2010). For the last two positions, there was no financial obligation of IITD. In between, I spent a year as visiting professor at the University of Leeds, on deputation from IITD under the Imperial College-IITD Exchange Programme and a year at the Iowa State University, USA, on sabbatical leave.

Because of the space limitation, it is not possible to give a total and detailed account of my teaching and research experience at IITD for such a long time in this presentation. I shall therefore record here only some interesting events and experiences during the initial years. I hope to supplement this narrative in a future article if such an occasion arises.

The Initial Period at IIT Delhi

The first thing that struck me odd at IITD was the B. Tech. Curriculum. Having been familiar with subject-based courses such as circuit theory, E.M. theory, and communication engineering, it was a shock to me to see that most of the courses taught at IITD were named Electrical Engineering I, II, III, etc., each course being a mixture of disjoint topics and taught by three or more faculty members. For example, the first course I was assigned was one in the third-year second semester which contained network synthesis, electronic circuits, control and communications, and was taught by four teachers. I was asked to teach the network synthesis part. This combination was a continuation of the third-year first semester course. Later, when I was in Leeds for a year, I found the same pattern of courses, and I realized the reason for the curriculum at IITD. Having been started in collaboration with Britain (Imperial College, in particular), the British pattern was followed in toto. In fact, all the founding heads of departments as well as other professors (one or two in each Department) at IITD were Britishers. Not all of them were good academics; in fact, the rumour at IITD was that this place was the dumping ground for UK Academics who could not make much dent in their own country. Many of them did not even have a Ph.D. degree, and even the ones who had a Ph.D. degree had poor research credentials. The background of one UK professor was high school teaching, and when he returned to UK after the IITD assignment, he had tough time in getting a suitable job. This explains the poor research ambience prevailing at IITD at that time.

In the first faculty meeting that I attended, I expressed my disappointment at the course curriculum and emphasized the need for a revision so that the curriculum falls in line with modern ones followed globally, even in other IITs in India. Most of the faculty agreed with me outside the meeting, but they hardly spoke at the meeting. Needless to say, I felt uncomfortable with their silence in the official forum and kept repeating the need for restructuring in subsequent meetings. An exercise in this direction started and, when completed, had to face opposition from the senate for approval. Many senators, particularly the Britishers, and the UK-trained Indians argued that there was nothing wrong with the existing curriculum. Anyway, ultimately it was implemented after about two years, in the electrical engineering department to start with, and gradually, other departments fell in line.

The research ambience in the department was also very disappointing. At the time I joined, there were only two full-time research scholars, both under the supervision of PVI, although there were twenty five or more faculty members in the department. Besides PVI,[8] there was only one other Professor, C.S. Jha (CSJ)[9]; there were a few assistant professors, but the majority were lecturers. Most of the lecturers were master's degree holders and a few only had a bachelor's degree. Some lecturers were registered for their Ph.D. with AKM in the area of control.[10]

I was promoted to full professorship in January 1970, and soon thereafter, I was appointed head of the department at the age of 32 years! I was the youngest faculty in the department, and there were quite a few faculties who were much older than me. I was then deeply involved in research with a faculty member and a regular research scholar who registered with me for Ph.D. Also, I had no administrative experience at all. I therefore requested the Director, Professor R.N. Dogra (RND) to excuse me and ask either PVI to continue or give the responsibility to CSJ who had done it earlier as the second Head of the Department after Professor John Brown, the founder Head. CSJ also served as a dean and was responsible for introducing (rather kind of forcing) the rotating headship system in the EE Department at IITD (as well as in IIT Kharagpur when he went there as the director), which was later adopted as the institute policy. He as well as PVI simply refused to step in as head for another term; they in fact insisted that I should take over. AKM was the next senior most in the department, but RND would not simply give it to an assistant

[8]PVI later founded the Centre for Applied Research in Electronics (CARE) at IITD and became the Director of IIT Madras.

[9]CSJ later became the director of IIT Kharagpur, followed by Technical Education Advisor in the Ministry of Human Resource Development (MHRD), Government of India, and finally the vice chancellor, Banaras Hindu University.

[10]AKM was promoted to associate professor during my headship, and he took over from me as the head of the department. He was later promoted to full professorship. Unfortunately for IITD, however, he left the country and settled in the Penn State University. He died an untimely death in the prime of his career due to cancer.

professor because in his opinion, that would create a bad precedent.[11] I therefore found no escape route and had to accept. CSJ and PVI fully cooperated with me and all major decisions for the department were taken by the three of us (this, incidentally, was the beginning of the concept of a Professorial Committee in the institute, which was later adopted formally by a resolution of the senate). If I needed some extra funds for the department beyond the budgeted amount (which was a petty one because the country was undergoing financial crisis for quite some time), we three used to go together to RND for sanction. After a couple of occasions like this, RND would listen to one of us and simply ask for the paper to put his signature if he considered the amount reasonable. He branded us as the "Trimurti" and would quote this harmony amongst the professors as exemplary, to other departments as well as in the senate. This harmony was not there in most of the other departments, and in some departments, a change of headship was usually accompanied by pinpricks and sufferings for the previous head! In fact, this, coupled with the charms of power and authority was responsible for many ex-heads to start new centres in the institute, in the guise of intensive research and development in a focussed area. As is well known, most centres either became departments or duplicated all the activities of a department including teaching programmes, in due course of time.

Amongst all this craze for power and authority, I appeared to be an exception and was mostly pitied upon by my colleagues, but also appreciated by some, particularly younger faculty because I had no such ambition. The headship was imposed on me due to circumstances prevailing at that time. I had to accept the deanship at the persuasion and insistence of the then Director, Professor N.M. Swani who was a dear friend and colleague, and head of the Textile Engineering Department when I was the head. I was offered deanship and deputy directorship several times in my career by successive directors, but except for this one time, I politely declined every time. The exception I made was at the advice of my wife who told me, "Why don't you do it once so that nobody can say that you shirked responsibility?" The reasons for my averting administration are twofold. Firstly, I feel that I do not have the aptitude or the tact and diplomacy to carry out an administrative responsibility successfully. Secondly, I always felt frustrated when administrative duty interfered with my passion for research and interaction with students. Successive directors of IITD as well as some other IIT directors nominated me for directorship of IITs without my knowledge several times, but every time a query came from the MHRD regarding this, I politely told that I am not interested. I have no repents on this count because I feel that I have lived a full and enjoyable life in the teaching and research profession. I also feel that even without holding an administrative office, I have been able to contribute to streamlining some aspects of administration, particularly for enhancement of the quality of education

[11]It may be mentioned in passing that IIT Kanpur had carried out this experiment of making an assistant professor the head of the department, but the results were not satisfactory, for various reasons.

not only in IITD, but also in several other national institutions, without expecting any returns in terms of power or fame.

Returning to my headship days, I must admit that I had a tough time in keeping the faculty together and carry out academic reforms. Right or wrong, I used to encourage every faculty to speak frankly, even criticizing me, in faculty meetings, as well as in personal interaction, but insisted that all differences should be sorted out in the department itself and not spread outside. By and large, I had moderate success in persuading many of the colleagues to fall in line with the culture of a modern department. Let me cite an interesting example.

Immediately after taking over as the head, I made a proposal to RND to allocate a dozen of research scholars for the department. The first question he posed to me was “Who is going to supervise them? After all, you are only three professors!” I had hard time to convince him that any Ph.D. qualified faculty should in theory be able to guide a research scholar, and I would assign research scholars to every such faculty, irrespective of whether he is a lecturer or a professor. It is worth mentioning here that RND was a chief engineer in PWD Punjab and was deputed first to Punjab Engineering College as principal, and then to Delhi College of Engineering, which gave birth to IIT Delhi, where RND became the director. He did not have deep knowledge of the academic world, and my arguments were not successful in convincing him fully. After some more discussions, he agreed for ten scholars. I recruited them in due course and assigned them to each Ph.D. qualified colleague of mine, by trying to match the interests of the faculty and the student. Would you believe that most of these colleagues were unhappy and considered this step as imposition on their academic freedom? One of them stormed into my office and shouted, “Who do you think you are that you can order me to guide a research scholar?” I tried at argue with him that IIT is not an ordinary engineering college and that research and development are essential mandates of the IIT faculty, but failed to convince him. Several other faculties followed his step, and I succeeded in getting six scholars accepted. The rest four had to get registered with me, making a total of six or seven Ph.D. students. Coupled with administrative responsibilities and regular teaching, this indeed posed a heavy load, but I somehow survived and gave no opportunity to my research students to feel that I did not give them adequate time. I could feel the good effects of the so-called imposition by the time I relinquished headship. Because of the compulsion, new supervisors had to get interested in research and started publishing in IEEE and other reputed journals. By this time, the number of regular research scholars in the department had doubled, and the initially reluctant faculty also started guiding research. Immediately after taking over as head, I had put up a notice board for posting recent journal publications of the faculty and put up the first page of some of my recent publications there. By the time I left the head office, the board was already full and we started removing some of the oldest ones to make space for the new ones.

At IITD, one of my lecturer colleagues, S.N. Hazra (SNH) who had a B.Sc. Degree of the University of Edinburgh, wished to register with me for the Ph.D. degree. I agreed and advised him to start studying active RC filters, a topic in which I and most of my research students were working. No, he did not want to work on this topic, but his choice was digital signal processing (DSP) and digital filters, in particular. Except for a pedestrian acquaintance with sampling and z-transforms, I knew very little of the subject. At that time, DSP was in its infancy, and I declined the request of SNH. He was not prepared to take a "no" for an answer and kept on insisting. He assured me that we would learn the subject together and that he would not demand much of my time. I finally gave up and agreed. In retrospect, I think this was another turning point in my research career because the subject of DSP occupied most of my attention in later life and continuing even today.

There was only one book at that time on the subject, written by C.F. Rader, which was not a comprehensive one, and we had to do a lot of library work to search for and collect research papers on the subject. The collection was a fairly difficult task because there was no photocopying facility either in the institute or outside, and the library collections were limited. We had to make requests to the authors by airmail, and fortunately, most of them responded positively. Compared to today's online availability of research papers even before they are published and almost instantaneous transmission by Internet and email, we appear to have used prehistoric methods, but this had its own positive points. We valued the knowledge earned through such hard work, and the retentivity was unbelievably large.[12] Ultimately, we could find some problems to work on, some being extensions of analog filter work I had done earlier.

I organized a Summer School on Network Theory in which SNH delivered some lectures on DSP, and this matured into a full fledged course on DSP offered by the two of us as an elective for the fourth year B. Tech. Students and beginning M. Tech. Students. We learnt more as we taught, and I have no hesitation to confess that I learnt much more from SNH than what he learnt from me. This was perhaps the first formal course on DSP offered in India. SNH got the Ph.D. Degree of IITD in three years' time; again, this was the first Ph.D. granted in India in the subject.

At the University of Leeds

Towards the end of 1972, the director proposed to send me on deputation to UK under the Indo-UK collaboration scheme. I agreed and chose the University of Leeds as the base station because of the presence of Professor J.O. Scanlan and Dr. J.D. Rhodes, both reputed scholars in circuit theory and distributed networks, there.

My experience in UK during the initial few weeks was not pleasant, but once I settled at Leeds, I could pursue vigorous research, making up for the lost time while doing headship. There was no email those days, and I continued guiding the scholars at Delhi through regular airmail correspondence. I had no teaching responsibility at Leeds, but gave some special lectures to the students and faculty there on my recent researches. Thanks to the British Council, my visit was widely

[12]Perhaps that is how my romance with DSP still continues!

publicized and I started getting invitations from various universities and other research organizations for lecture visits. Besides Imperial College London, I visited the University of Sheffield, City University London, University of Bradford, University of New Castle upon Tyne, Post Office Research Centre at Dollis Hill, London, the General Electric Research Laboratories and also the IEE Headquarters at London. I made personal acquaintance with many reputed scholars whom I knew and interacted with professionally through correspondence earlier. Also, on invitation from the Sir George Williams University (now known as the Concordia University), Montreal, Canada, I went there on a three-day visit and gave as many lectures.

Except for the social ambience at UK in general, and in Leeds in particular, where the skin complexion was a dominant factor in deciding how the ordinary Britisher would treat you, the Leeds visit was professionally very rewarding, in terms of the quality and quantity of research and exchange of ideas with leading experts in my field. Just to mention one example of racial bias, a doctor under the National Health Scheme who treated my one-year-old son for fever and cold saw my official address as the University of Leeds, and asked me whether I was a student. I said, very innocently, "No Doctor, I am a visiting professor for a year from IIT Delhi". He could not believe that a non-white young fellow like me could be recognized as a professor in a UK University. He raised his eyebrows and said, in Sherlock Holmes style, "Professor? That is quite singular!" and asked, "So who do you work for?" By that time, I got irritated and quipped back, "No, doctor, I don't work for anybody. I work for myself. I am an independent researcher". I then showed him my identity card issued by the university with the designation written there. He simply said, "I am sorry", and that too in a rough tone!

At the University of Leeds, I became aware of the bucket brigade device (BBD) and the charge-coupled device (CCD), which had just appeared on the technology scene with great potentials. Some work was going on at UK, and I held discussions with the concerned researchers. I studied the physics of these devices and their signal processing applications. In view of my growing interest in DSP, and finding that the mathematical techniques for handling these two charge transfer devices (CTDs) are the same as those for DSP, I got interested to work on the topic. On my return to IITD, I held serious discussions with a colleague, A.B. Bhattacharyya (ABB) of the CARE, and found that he was already interested in the device and had a project on analog signal processing systems (ASPs) which included fabrication and applications of the CTD. I became a Co-investigator in this project and took a Ph.D. student jointly with ABB to work on the fabrication of CCD, and two Ph.D. students singly to work on the compensation of the charge transfer inefficiency of the device and its applications. The first student ultimately made the first CCD in India at the IC Laboratory, and the other two succeeded in finding new avenues for signal processing applications.

More to Come Later

I have many more stories and events during the rest of the period I spent at IITD, viz. 1975–2010, but I shall reserve these for part II of this article which I plan to write in future. At the present time, let it suffice to say that even though I formally retired in 1998, I continued regular teaching till 2010 at IITD, the last few years being without any remuneration either from the IITD or INSA, when I was an INSA Honorary Scientist. Unlike many of my former colleagues, I did not accept any lucrative assignment in any private institution after retirement, and my decision seemed to be a correct one, in view of the drastic difference in the ambience and freedom existing in such institutions and the IITs, which I came to know directly and indirectly during the pre-retirement days. I now work from home with virtually pen and pencil and limited computer facilities, but I must say that I am fully satisfied with the way I have lived so far. It is not that I have not committed mistakes, but I have tried to learn a lesson from every mistake I made. Also, my Ramakrishna Mission background gave me the confidence to face difficulties in life bravely and to do good to others as best as possible without expecting any return. After my formal commitment to teaching ceased, I missed the interaction with students for quite some time, but nature took care to fill up the sense of void by various national honorary assignments in educational and research organisations, like visitor's nominee in IITs and NITs, visiting honorary professorships, academic committee memberships of a number of defence organisations, assessment and selection of scientists in CSIR, DRDO, etc. I have also found more time for participating in activities of some professional societies and academies in science and engineering and also to pursue my other hobbies such as classical music, reading poetry and sometimes writing one or two. My latest venture is to start a journal of the Indian National Academy of Engineering (INAE), to be named *INAE Letters*. I wish to record all these in future.

In Conclusion

I would like to conclude this part of the presentation here. Looking back from the current time to the beginning of my career, I feel that I have lived a full life, with contentment and happiness. I have had many honours, awards and recognitions throughout my career, some sought after, particularly at the beginning of my career, but several others coming as surprises. However, the recognition that I value most is the love and affection of a large number of students, whom I had the privilege of interacting with personally, and also an equally large (if not more) number of virtual students spread throughout the world, whom I acquired through the five video courses I developed, which are freely downloadable on the YouTube through the efforts of NPTEL, and interacted with through emails and chance meetings at odd places. This is what keeps me alive and active, despite my numerical age. Should God destine me to have another life after the current one, and ask me to choose a career, I would opt for teaching and research again, unhesitatingly and enthusiastically!

My Four Decades in BARC

A.K. Anand

I joined the Department of Atomic Energy in 1961 and superannuated in 2001. Our first year was a training period in the Training School of AEET (Atomic Energy Establishment, Trombay, later renamed Bhabha Atomic Research Centre after Dr. Bhabha's unfortunate and untimely demise). Sometimes I question myself, as to why I continued for so long at the same place? I received several lucrative opportunities to migrate, settle in the USA, France or Thailand. But I always got the same answer—a very high degree of job satisfaction; there were always new and exciting things (technical and non-technical) to do and in the process a lot to innovate at every step.

Joining Atomic Energy was not by design, though becoming an engineer was, because of the influence of my family. I did my schooling in Ambala and finished my matriculation in 1955 from the Punjab University. My uncle who was settled in Benaras (now Varanasi) proposed that after doing intermediate science, I should shift to Benaras; I would have an advantage in getting a seat in the engineering college of the BHU which was considered a prime institute for engineering during those days.

After my graduation in mechanical engineering, in fact, a day after the result was declared I joined my college as a teacher; it was a tradition to offer this position to the first few rank holders. At the same time, we started looking for suitable employment under guidance and advice from senior professors. Those who wished to remain in the academic field were sent by the university to the USA for pursuing M.Tech. I decided to join the Atomic Energy.

The first year was really a continuation of student life, attending classes and laboratory sessions, in addition to sports and cultural activities. My first interaction with Dr. Bhabha was on our annual day celebrations where he was the chief guest and his mother gave away the prizes; being the secretary of our batch, I was the compere.

A.K. Anand (✉)
Microtrol Sterilisation Services Pvt Ltd., Saki Vihar Road, 400 072 Powai, Mumbai, India
e-mail: anil@microtrol-india.com

P. Ghosh and B. Raj (eds.), *The Mind of an Engineer*,
DOI 10.1007/978-981-10-0119-2_12

After graduating from the training school, I joined the Reactor Engineering Division, BARC. We started studying different types of reactors and produced reports and also designed and established some experimental facilities, in the engineering halls and also around the three research reactors Apsara, Zerlina and Cirus. Though, the decision was taken to establish two BWRs (boiling water reactors) at Tarapur on turnkey basis by General Electric, USA, for the indigenous power reactor programme, no decision had yet been taken, although it was argued that the natural uranium, heavy water route would be more suitable for our country. Three countries, Canada, France and the UK, were pursuing the natural uranium, heavy water moderator route with different coolants. The coolant, in case of Canada, was heavy water, for France, it was gas, and for the UK, it was steam. Most of us who joined during those years were sent to Canada, but some did go to France and the UK; I went to France. Very soon, it became clear that if natural uranium, heavy water moderator, route had to be followed for power reactors, the coolant also had to be heavy water. This is how we started our PHWR (pressurised heavy water reactor) programme with Canadian collaboration for two units in Rajasthan.

When I was in France, Dr. Bhabha died in an air crash and we mourned the death of our founder. A couple of months after his death, all of us, the trainees from India in the Nuclear Energy Centre Saclay, were asked to assemble in a room and were informed that our 'Chief' was visiting the French Atomic Energy. We assembled but we did not know who the 'Chief' would be. We stood in a queue and introduced ourselves one by one. When the Chief approached, I said 'Sir I am A.K. Anand from Reactor Engineering' and the Chief extended his hand and said 'I am Vikram Sarabhai'. His humility, simplicity and smile left a permanent impression on my mind. It proved that we were very lucky to have Dr. Sarabhai as the second chairman of Atomic Energy.

I got my first challenging assignment in 1970; I was deputed to the USA for six months to participate in the 'reload fuel design' project, underway, at General Electric, San Jose for the Tarapur boiling water reactors. I was lucky to get this responsibility of learning and later implementing the designs for future 'reloads' and help our Nuclear Fuel Complex in setting up facilities for manufacturing the fuel. After I came back, I set up the required facilities for their design and also helped the manufacturing team, even when they were trying to acquire suitable machinery for fabrication. While these activities were going on, a lot of fuel rods from a number of assemblies loaded initially, started failing prematurely and increased the radioactivity levels. A big 'reconstitution' programme was launched to disassemble and reconstitute the old fuel assemblies till the new fuel was fabricated. This activity was done with remote tools, about six metres underwater. After my return from the USA and being stamped as 'the fuel designer', I was posted at Tarapur for this activity. BARC team of 'postirradiation examination' experts, wished to have a look at some of these failed fuel rods. We had to overcome an international administrative problem and a technical one to transport the active fuel rods from Tarapur to the BARC through public roads. The Tarapur reactors are under safeguards and IAEA (International Atomic Energy Agency) inspectors visited regularly while they could not visit BARC. As the enriched uranium content in the rods to be brought to BARC was small, IAEA approved their transfer. We used to have a lead shielded flask to transport Cirus spent fuel to the plutonium plant. We quickly modified the same flask, so that it was capable of being handled vertically and horizontally and also capable of being aligned with the hot cells. We brought the fuel rods to BARC for examination. While we were busy doing all this, in 1974, we had the so called 'PNE' (Peaceful Nuclear Explosion). The international community, and the USA in particular, were very sore and the regime of 'denial' to India started.

We started working on the plutonium-enriched fuel for Tarapur. My team also did some irradiation in a loop in Cirus, simulating power reactor conditions. Theoretically, we could get the required plutonium by reprocessing spent fuel; alternately, our Chairman Dr. Sethna challenged the USA to take back the fuel.

This, not only required a number of lead shielded heavy flasks which could float in case of a ship wreck, but also a 300-ton floating crane; thus, the whole idea was dropped.

At this time, BARC announced an ambitious project of designing and making, indigenously, one of the largest research reactors in the world, 100-Mw thermal Dhruva (named subsequently). The 'Group Leaders' of the design team were selected from various Divisions and I was designated the Group Leader for the 'fuel design'. It was a unique management structure; every Group leader had at least two Bosses to report to and in addition, had to get his design cleared from other Group leaders; something like matching with the other components, designs for which, were also being evolved. Every Group leader had to do some R&D and testing, modify designs, resulting in, at times, another cycle of change. Despite these limitations we did commission the Reactor, but we were in for a big surprise. Vibrations were felt when we stood on the top of the Reactor, this happened when the main circulating pumps were started. It was decided to operate the Reactor for some time, observe the vibrations and collect data. It was concluded that the fuel rods were vibrating and rubbing against one another, wearing off the cladding and exposing the uranium resulting in increased radioactivity. It was clearly a sign of flow induced vibrations (some resonance) which were not experienced during the out-of-pile testing of the fuel inside the coolant channel, simulating the reactor conditions. I started getting advice from all to redesign the fuel, I was hurt and had a very disturbing period; the rumour mongers had a field day. I returned from a few days of leave and heard 'Anand has resigned and left the country'. The other most important 'Group Leader' was Anil Kakodkar, looking after the design of all internal structures of the reactor, including the coolant channel assemblies. We were both equally concerned as our prestige and the Nation's future nuclear programme was at stake. Though we were irritated, at times with each other, we had (even now) respect for each other's technical competence. Along with my team, I started working on the design of different leaf springs to be located at the two support points to isolate the vibrations coming from the structure, while Kakodkar and his team started checking the natural frequencies of all the equipment, and trying to determine if the resonance was taking place. We both found answers at almost the same time.

While the above solution was being implemented on all fuel assemblies, I was given the most important assignment of my life; very few people are lucky to get this kind of an opportunity. Posting the assignment in Dhruva, I could not visualise what I would be doing in future. The post in Paris (DAE Science Counsellor in the Indian Embassy) fell vacant. I applied for the post along with a few others. A formal interview committee was constituted under the chairmanship of Dr. M.R. Srinivasan. Dr. Raja Ramanna was then the Director BARC; I was informed that I should not attend that interview as I am being designated as 'Project Manager' for the PRP (Plutonium Recycle Project), to be designed and built as a BARC Project in the DAE Complex at Kalpakkam. This code name PRP was for the 'Land Based Nuclear Propulsion Plant', for the future Indian Nuclear submarine programme. Dr. Ramanna had already selected the site, along with his friend, Vice Admiral M.K. Roy, C-in-C, East who retired and took over as Director General ATV programme; ATV (Advanced Technology Vessel/Vehicle) was the code name for the nuclear submarine programme under the overall guidance and supervision of the Scientific Advisor to Raksha Mantri. For this project, a new 'Reactor Projects Division' was carved out of the existing Reactor Engineering Division and I was appointed the Head of this Division with only about a dozen Engineers as the 'core' design team. This included Mehra, Grover, Basu and Yadav, who are all fellows of our Academy. It was decided to increase the strength, over the years, with input from the Training School and transfers from other Divisions as the work picked up.

The first task was to make a 'project report'. Three of us from BARC and four from the Navy, stayed at the DAE guest house in Kalpakkam for two weeks, with a make shift office room. We made the report, based on the data available, our estimates and gut feel. There was no authentic document available indicating the overall dimensions, speed and the horse power of the future nuclear submarine in mind. We arrived at the dimensions of the buildings and the crane capacities with hook heights, to house the equipment. The civil work started after the Chairman, Dr. Ramanna visited the site and signed a piece of paper which was hurriedly prepared by the Chief Engineer (Civil). Back in Bombay (now Mumbai), we started designing the reactor pressure vessel, steam generators and pressuriser and other reactor components. It was decided that BHEL, Trichy, would establish facilities and manufacture this equipment; BHEL would also import material, specifications for which were generated jointly. Normal Government rules do not permit a Purchase order to be placed even on a Public undertaking without calling for quotations. We used to have a wonderful IFA in BARC, Shri Borkar. I explained the problem and the sensitive nature of the project; I also told him that establishing manufacturing facilities, procuring of the material, etc. had to go in tandem with design. There would possibly be mid-course corrections and some rework. Time and cost estimates could not be given but BHEL had promised that they would deploy all required resources to finish the job in the shortest possible time. Shri Borkar prepared a draft and gave it to me to write a note to the Director BARC, Dr. Iyengar through the IFA. The matter was settled; BHEL would keep all the records, to be verified by IFA BARC, and get the expenditure reimbursed periodically.

Shri S. Basu, as the Engineer-in-charge, moved to the site with about ten engineers. There was a shortage of residential accommodation in Kalpakkam. This team commuted from Chennai for almost three years till we procured land and built a separate residential complex. The Team had expanded, and comprised of serving and retired persons from the Navy, a few DRDO Scientists and the majority from the DAE. For coordinating the activities from ATV, Nadaph was posted to the site, after the residential complex was commissioned. There is a lot of cultural difference between uniform-donning personnel and civilian scientists; we imbibed the best of both worlds and worked together; the motivating force being 'thrill of the project'. The design office in Bombay also expanded and a number of facilities were added at Bombay and Kalpakkam while ATV established their facilities in Hyderabad and Vishakhapatnam. A number of Scientists and Engineers from different divisions started working on the project; complete software and hardware for instrumentation and control were designed by the Reactor Control Division, headed by Govindarajan and the teams led by G.P. Srivastava and R.K. Patil; hardware was manufactured by ECIL (Electronics Corporation of India Ltd.). The equipment for secondary systems like propulsion turbine, turbo generators, propulsion shaft, screw and the dynamometer was the responsibility of the ATV office while the hull was designed by the Navy's 'Submarine Design Bureau' initially headed by R.S. Chaudhry and later by R.B. Verma.

The first major equipment to reach the site was the hull sections and the bulk heads to be aligned and joined in the main plant building. At that time, an Admiral in ATV got an idea that the 'Command and Control' of the PRP site at Kalpakkam should be taken over by the Navy. While we started discussing and arguing about the merits and demerits of the proposal, we did not let this controversy infiltrate down the line and work progressed as usual. I was asked to make a presentation to the then Chief of Navy, Admiral Vishnu Bhagwat; the only person who accompanied me was the DG ATV, Vice Admiral (Retd.) B. Bhushan. Subsequently, Chairman AEC, Dr. Chidambaram and SA to RM Dr. Kalam visited the site and the situation was diffused.

Mazagon Docks had been engaged in the manufacture of the conventional submarines and were invited to be yet another partner, for fitting secondary equipment, do the piping and weld hull sections. The remaining jobs were executed by BARC Engineers and at times, the designer came along with the equipment and returned after installation and testing.

By the time, I was due for superannuation, all the equipment had been installed and commissioned but the 'criticality' could not be achieved as the required number of fuel assemblies had not been delivered to the site. There was some mismatch in the timing of the expansion of our uranium enrichment plant. But soon after I retired, I got a phone call from Kalpakkam, that the reactor had become critical. I jumped with joy!

About six months before my retirement, while waiting to load the reactor with the fuel assemblies, I got another very important, additional assignment. For the past few years negotiations had been 'on going' between India and the Russian Federation to establish two VVER reactors of 1000 Mw each, at Kudankulam, but the contract could not be concluded as the cost was high and not acceptable to our Government. Our goal was to bring down the unit energy cost to the level comparable to the cost from conventional power station at that location. The Russians were insisting that the project would be built more or less on turnkey basis with hundreds of Russians stationed at the site and with very little Indian participation. Being familiar with LWR (light water reactor) systems, in the beginning, I had been a part of the KK team when the Technical Assignment Document was drafted. CMD NPCIL, Dr. Chaturvedi sought permission from the Chairman Dr. Kakodkar, to give me an adjunct position in NPCIL, as 'Executive Director KK Project' so that I could become a part of his negotiating team. We had a number of negotiations and convinced the Russians that a lot can be done by us, thus increasing Indian participation to an extent that we achieved the commercial target set by our Government. I was a witness to the contract, signed in the Kremlin during the last month of my service in November 2001, in the presence of our Prime Minister Shri Atal Behari Vajpayee and the Russian President Mr. Putin during a visit of the delegation to the Russian Federation.

While I was writing this article, 'Kudankulam' became critical and four days after, I finished writing and was still editing, criticality of the reactor on 'Arihant' was announced. What a satisfying span of life!

Think to Do Something

Anil Kumar Sinha

If you think and try to do something honestly, you will get everything, and if you think to achieve something, you will get nothing except frustration. I used to hear these words from my father and many other successful persons. Now I also have full faith in these miracle words.

I belong to a middle-class family from a small town in Bihar, inherited honesty, self-discipline, faith in goodness, and kindness from my parents, rose to a respectable position in Government's R&D sector where at present more than 600 people are working and have been honored with many awards. My primary education started at home along with the stories of various religious books including Ramayana and Mahabharata. My schooling started from Std. VI.

During my early school days, I was in the habit of reading various story books, and latter I started reading fascinating stories of various scientists and their achievements. These stories encouraged me to do something new. During those days, science, arts, and commerce classes were separated from Std. VIII. I joined science classes, because I was fascinated by the stories of various scientists and the words of my father to do something. After completion of my school education and Part-I of B.Sc. (equivalent to Intermediate), I got admission in Mechanical Engineering discipline at Muzaffarpur Institute of Technology (MIT) as per my choice and completed my engineering in December 1979 with Distinction.

After B.E (Mech), I joined BARC Training School as a Trainee Engineer (which was my dream, because work done by scientists always fascinating me). Getting selected in BARC was a challenge, because at that time only one attempt was allowed. After completion of my training course, I was posted in Central Workshops (CWS) currently known as Centre for Design and Manufacture (CDM). How and why the name was changed is also an interesting story which is mentioned latter. That time CWS was one of the largest divisions of BARC consisting of six independent sections (Design, Planning, Machining, Fabrication, Maintenance, and Inspection and Quality Control). More than 1000 persons were working in three

A.K. Sinha (✉)
Centre for Design & Manufacture (CDM), Bhabha Atomic Research Centre,
Trombay, Mumbai 400 085, India
e-mail: anilsk@barc.gov.in

P. Ghosh and B. Raj (eds.), *The Mind of an Engineer*,
DOI 10.1007/978-981-10-0119-2_13

shifts. But Design Section of CWS was not considered to be a very good section and officers working in that section were getting promotion with one or two years delay. Knowing the situation of Design Section of Central Workshops, I just remembered my parent's work "Think to do something better," and I opted for Design Section with a hope that I may bring up the Design Section.

For initial three years (1982–1985), my work was predominantly to assist my seniors in designing various types of instruments for conducting the experiments with neutron beam and also special purpose equipments required for research reactors and power reactors. During this period, we successfully completed one "Triple Axes Neutron Spectrometer" for KOREA.

After 1985, I started working independently as Design Engineer. My first job was to develop a multichannel Raman spectrometer, which was completed successfully. Design of this instrument with experimental result was published in "Indian Journal of Pure and Applied Physics."

Successful completion of the optical precision instrument and publication in journal gave me a lot of confidence and encouragement. After which I successfully developed many other types of monochromator for different purposes. This was the beginning of my career as independent Design Engineer. After successful completion of these jobs, the Head of the Division assigned me four more important and challenging jobs at a time. All the four were unique and of different types. These jobs are mentioned below:

(a) Developing a "Helium Cryostat of 1.0 m in diameter and 1.2 m in length" to house four RF resonators to accelerate the energy of the heavy ion beam coming out from the Pelletron Accelerator for Tata Institute of Fundamental Research (TIFR). That time this was one of the first largest cryostats to be developed in India.
(b) Developing a "Triple axes Neutron Spectrometer" for Bangladesh as per design parameters given by Bangladesh. As this job was for export, it needed to be completed within defined time frame.
(c) Developing an "Air Cushioned based Neutron Spectrometer" for Dhurva Reactor (BARC)
(d) Developing a "Neutron Radiography Rig" for neutron radiography of irradiated object for IGCAR Kalpakkam.

That time, because I did not have sufficient experience, I was hesitant and reluctant in accepting the entire design job alone. But my Head of Division having confidence in me encouraged me. He showed his faith in me by saying "you can do this." His encouragement worked like a miracle and all the four jobs were completed and delivered on time along with many other jobs. All the above-mentioned jobs were completed between 1988 and 1993. As these were of great importance, many prominent officers of BARC including Dr. R. Chidambaram (Ex-Chairman, DAE), Dr. B.A. Dasanacharya (Ex-Director Physics Group), and Shri S.K. Mehta (Ex-Director Reactor Group) visited CWS to see these jobs. All were surprised to see the design capability of CWS. Shri S.K. Mehta (Director Reactor Group) said that people used to think that CWS is just a workshop; as to reflect the kind of high-quality design work being carried out, the name ought to be changed. However, due to various reasons, the name Central Workshops was changed to CDM in 2000 only.

In 1993, one more difficult job was assigned to Central Workshops, i.e., development of a special purpose machine to remove one coolant channel from Narora Atomic Power Station (NAPS). Removal of coolant channel from working power reactor was one of the most difficult tasks, because maximum operations had to be done remotely with minimum exposure to human beings. To complete this task, a working Task Force was constituted consisting of two design engineers and other five supporting members. I was one of the engineers of that team. We faced a lot of problems, but were able to overcome all the obstacles and completed the task successfully.

As a recognition, BARCOA (Bhabha Atomic Research Centre Officers Association) had honored us with BARCOA Award and Department also honored us with Group Achievement Award in 1994. On the same topic, we had presented a technical paper in AIMTDR (All India Machine Tool Design and Research) Conference. Our paper was adjudged the Best Paper in the Conference.

After this, I spoke to my Head of the Division for making a team of design engineers, and he readily agreed. Now my next responsibility was to make a team of efficient engineers, but selection of engineers was not in my hand. So I started giving good grading to my Engineers to attract good intelligent engineers for my team. This idea worked and after a few years I started getting good intelligent and sincere engineers.

In 1994, I got one more challenging assignment which was to develop an automatic welding machine for welding the fuel tube and end cap. Wall thickness of the fuel tube was 0.4 mm and the end plug thickness was 5.0 mm. Earlier, many had tried it but their machines were not accepted for nuclear grade welding. I along with one more engineer from user Division accepted and completed the job successfully. Our machine qualified for nuclear grade welding.

By this time, I had established myself as a good engineer of BARC and may be best in Central Workshops. So new challenges started coming to me, like developing various types of optical and mechanical instruments for synchrotron beam line for INDUS-I and INDUS-II storage ring. I was given the task of designing and developing the instruments for three beam lines of INDUS-I and latter seven beam lines of INDUS-II. Synchrotron radiation is an electromagnetic radiation which comes out as energy, released from the fast-moving charged particles. Synchrotron radiation has a very wide spectrum (consisting of infrared (IR), ultraviolet (UV), vacuum ultraviolet (VUV), X-ray, and gamma ray also), so it is known as versatile tool for fundamental research and also for many other applications. Instruments of synchrotron beam lines were different from the instruments of neutron beam line. Each beam line consists of many (ultra-high vacuum) UHV compatible optical (mirrors, gratings, and single crystals) and mechanical instruments and also experimental station for conducting the experiments. As the technology was new, I was deputed to BESSY, Berlin, Germany, in July 1998 for four months to study the design of various types of instruments. There I got the opportunity to work in an operational storage ring and also got the exposure of various types of optical and mechanical instruments of very high accuracy. It was an excellent exposure for me.

In 2005, I got the opportunity to work with ISRO scientists for the most prestigious work of INDIA, which was designing, manufacturing, and installation of DEEP SPACE NETWORK (DSN) Antenna of 32 m diameter for CHANDRAYAAN PROJECT. I felt proud when my Group Director selected me for this prestigious work. This job was also completed on time and was successfully used during Chandrayaan mission. Now the same is being used for MARS mission. For this work and other challenging jobs, I was awarded DAE's most prestigious award i.e., HOMI BHABHA SCIENCE and TECHNOLOGY AWARD of 2008.

By this time, I was having a team of good Design Engineers. After completion of Chandrayaan Project, another challenging job that I took was the development of Six-Axis Parallel Kinematic Positioning System (also known as Hexapod, and Stewart platform). There was great demand in India for the development of this instrument. Many academic institutions and robotic industries in India were working on this developmental work. Many users of DAE had requested me to take up this job because the cost of importing the hexapod was very high and also it was not easily available for BARC. We accepted the challenge and completed the task. Dr. R.K. Sinha (current Chairman of DAE) had visited CDM to see this job. Till now six numbers of Hexapods have been completed and six are under process. For this also DAE had honored us with Group Achievement Award.

At the time of writing this article, I am involved in design, manufacturing, and installation of 21-m-diameter Major Atmosphere Cherenkov Experiment (MACE) Gamma ray Telescope (Second largest in world) at Hanle near Ladakh. This is being developed for capturing the image of a few nanosecond duration of Cerenkov radiation produced by the interaction of very high energy particles or photons with the earth's atmosphere.

Throughout my life, I have always thought for doing the best and enjoyed the work, never thought for achieving something or worried for promotions.

In my opinion, if you are working as a team, the members of the team should be like-minded and should have in-depth knowledge on the activities of the task as well as inclination to perform the task. The task members need not possess knowledge in all the spheres of the project activities but may be specialized in a particular activity. The decision of the project leader should be binding to all in the organization.

Success cannot be achieved by an individual alone. Credit of success should always be given to person even if he or she is at the bottom of the hierarchical line.

Always maintain a strict discipline and devotion to work. Our country needs dedicated people to do hard work to bring our country at par with other developed country.

During my service life of 35 years, I have worked with full devotion, sincerity, and integrity and department has also honored me with many Awards. For this, I am thankful to our department.

I have experienced different types of feelings, mostly sweet but sometimes sour too because of change in management. In last I would like to share a few thoughts which are mentioned below with my young friends. At the same time, I request them not to bother about these small things and try to do your best; success and achievements will surely come to you.

1. When you take up a challenging task or are selected for a challenging task, colleagues will be watching you; if you fail they will laugh at you and if you succeed they will be jealous of you and only a few will appreciate you.
2. At the beginning of any challenging work, there will be a few persons with you and when you complete the work there will be a big crowd to take the credit and you may be lost in the whole crowd.
3. When you take up a new challenging and difficult task, in the beginning the people may react asking "can these people do it?" Once you complete the task, the same people may say—"what is so great about this?"
4. When you confront a serious problem, it is only you who has to solve it. Do not expect others to solve your problem.

I am thankful to all my seniors who have always encouraged, supported, and given me the opportunity to work on various important and challenging projects.

The Milestones of My Professional Life

Arun Kumar De

> *If you want to succeed in life, you should strike new paths rather than travel paths of accepted success.*
>
> John D. Rockfeller

The above statement has guided me through my four decades of professional life to a great extent. I graduated in 1947, the year of our independence from British rule, from College of Engineering and Technology, Bengal, Jadavpur, in Mechanical Engineering. My first job was in one of the finest light engineering industries of the time, Jay Engineering Works Calcutta (now Kolkata), which used to manufacture sewing machines. The company, liquidated many years ago, was owned by late Lala Shri Ram.

Jay Engineering was reputed to have very progressive and forward looking policies of that time. During the mid-1950s and early 1960s, several West German experts were engaged as consultants to improve product quality and manufacturing productivity. Keurtz Hotz, a brilliant mechanical engineer and the head of development department then, offered me a privileged opportunity to work with him. This was a turning point in my professional life. Under his mentorship, I developed a very high degree of technical competence and self-confidence to develop new products, product improvements, and new manufacturing methods for increased productivity. I owe this mentorship to my ability to shoulder challenging problems in my future professional life.

In 1958, the second IIT of the country was established in Bombay (now Mumbai) through the UNESCO Assistance Program, largely with the help of the USSR (now defunct). After eleven years in the industry, I decided to switch from industry to academics and pure research. Though I was specifically recruited for developing the postgraduate program of machine tool design, as an assistant professor I happened to be the senior most faculty member of the mechanical engineering department of the institute at that point. This gave me the unique

A.K. De (✉)
Flat No. 301, Bldg. No. 16, Regency Estate, Off Kalyan Shil Road, Dombivili (East), Thane 421 201, India
e-mail: arun25de@gmail.com

P. Ghosh and B. Raj (eds.), *The Mind of an Engineer*,
DOI 10.1007/978-981-10-0119-2_14

opportunity to work with the experts from USSR and other developed countries who visited the institute to guide us in initiating postgraduate programs and research projects.

Besides developing the curriculum for the first two years, I also had the responsibility of planning and constructing the mechanical workshop and drawing room facilities. My industrial experience was of immense help in the challenges during the first few years of the institute. In 1962, one of the country's first postgraduate program on Machine Tool Design was started in IIT Bombay with the close cooperation of the USSR experts. The program received good response from the industry since the country's focus in the Second Five-Year Plan was in machine building.

I received a UNESCO Fellowship to complete my Ph.D. from Moscow Machine Tools and Small Tools Institute. On my return in 1965, I was appointed full professor and served as the head of the department from 1967 to 1969. The next turning point in my professional life came in the form of a summon from CSIR in early 1969 to appear for an interview for the post of director of the Central Mechanical Engineering Research Institute (CMERI), Durgapur in West Bengal. The chairman of the interview panel was Prof. V.K.R.V. Rao, the renowned economist and the then Minister of Industry and Commerce and also the vice president of the CSIR. I had no hopes of getting the job since the other candidate was my old boss at Jay Engineering and I had said "I don't know" to one of the questions the minister had asked. However, I was pleasantly surprised to receive the appointment letter. Much later, I came to know that the minister favoured me because I had not answered that question by inventing some story. Honesty triumphed.

The assignment at CMERI was daunting from three major points. One, I was expected to manage an institution with over one thousand engineering, scientific, technical people, and a large number of industrial staff. Second, at that time Durgapur was the hub of ultra-leftist political movement and radical industrial disputes. Third, my predecessor had left the institute six months earlier due to policy issues with CSIR HQ. People were waiting for the new incumbent to join. It was test for a professor of a department of an educational organization in a politically peaceful state.

Within the first few days, I had to deal with the "charter of demands" and permission to start "Employees' Association." Within the first few months, I had to deal with the transfer of technology of the 20 HP Tractor Project which was a prestigious project of the institute with a major agricultural impact. None of them were areas of my previous expertise.

Dealing with a militant but un-recognized Employees Association was valuable time wasted, including being gheraoed several times. However, I was successful in introducing a new technique of "R&D Management" and attempted to change the moribund research culture. I also concluded the transfer of technology of the 20 HP Tractor to Punjab Tractors Ltd. a PSU of the Punjab State, with a path-breaking decision to depute the project leader Mr. Chandra Mohan with his team along with the technology for commercial production. This turned out to be a successful introduction of an indigenously developed product to the market. The product enjoyed a modest market share, and CMERI earned a royalty of over rupees four crores within a period of five years. I believe my moral courage, integrity, and transparency in dealing with complex problems stood me in good stead.

In 1974, just as I was getting settled in supervising many projects under my personal supervision at CMERI, I was persuaded by the then DGCSIR, Dr. Y. Nayudamma, and the then Union Minister of Education Prof. Nurul Hassan to accept the responsibility of Director IIT Bombay. I joined on July 1, 1974, and over the next six months handed over the projects to new leaders at CMERI. Though I had approach to the minister himself, I committed a Himalayan blunder in not negotiating the transfer of my services from CMERI to IIT. This resulted in my forfeiting my rightful pension benefits. What hurt me more was the administrative support system failed me where I needed it most. I have since advised all that one should be alert to one's rightful privileges and not be swayed by entirely by emotional appeals.

In 1974, IIT Bombay did not have a state-of-the-art mainframe computer apart from burdened with obsolete scientific research equipments. Finance was scarce as the country was going through financial difficulties. Luckily, the Department of Science and Technology was created around this time. I was able to utilize the resources so available, and for the first time after the UNESCO program ended, the institute could procure badly needed equipment without depending entirely on the Ministry of Education. The computer came much later, as there was embargo on supply of such computers to the country. For future purchases, attempts were made to avoid small purchases and consolidate the institute's plan funds for major research equipments.

On the academic front, I put in special efforts to strengthen the humanities and social sciences in addition to the core engineering and science faculty with the inception of several centers of studies. To boost consultancy, the Industrial Research and Consultancy Centre was created. A new integrated M.Tech. program for JEE entrants was offered. The conscious effort was to change the perception of IIT from a teaching to a research organization. Globally, this helped the institute to rank better in prestige.

One incident in 1980 however had the potential of becoming an embarrassment. Four students of Indian origin but hailing from Malaysia and Sri Lanka were asked to leave the institute due to their inadequate academic performance as by the rules of the institute. These four students went on a hunger strike in protest with the support of a section of the local students. I along with some of the senior faculty members were confined in my office to force us to revoke the decision on the four students. Against advice to the contrary, I had the academic activities suspended and got the hostels vacated to restore normalcy. This decision was supported not only by the intelligentsia but also media and the political system. Regular activities resumed in three weeks upholding the standard of the IIT.

Atomic Energy Regulatory Board (AERB) was created in 1984 through a Government of India Gazette Notification to independently supervise all safety-related activities of nuclear power generation and help protect public from the effect of undesirable nuclear radiation. I was offered the post of Chairman of

AERB, a post equivalent to Secretary to the Government of India, sanctioned by the Union Cabinet. After an initial hesitation, I happily accepted the new challenge.

I soon discovered that neither the Department of Atomic Energy (DAE) nor my boss, the Chairman of Atomic Energy Commission (AEC) actually wanted the AERB to function as an independent body. The intended role of AERB continued to be played by the DAE Safety Review Committee, and peripheral activities of evaluating safety from radiation of medical equipments were assigned instead. The grip of DAE officials on the Government of India was very strong and efforts to apprise the PMO of the situation, since DAE reported to the prime minister, did not appear to bear any result. In December 1986, the 40 MW Prototype Fast Breeder Reactor at Kalpakkam was to be inaugurated by the then P.M. Shri Rajeev Gandhi. Mandatory clearance from AERB was needed as per the law. The DAE realized this too late and the office of the Chairman AEC sought an ad hoc approval from me without furnishing any information relating to safety issues. I refused, but was successful in getting the approval well in time by obtaining and placing the relevant safety-related matters to the AERB for due diligence. DAE officials must have been very upset at this show of independence. So much so, that the Chairman AEC personally disparaged me in the presence an officer in DAE, when I went to hand over my charge on demitting office as Chairman AERB in 1987.

However, as events unfolded, within a few months, I was asked to again take over as Chairman AERB by the new Chairman AEC, with the intervention of the PMO. I accepted the offer, only on the condition that the DAE Safety Committee is disbanded and AERB plays its mandated role as the sole safety regulator for nuclear engineering operations, independently. I am proud I helped establish the fledgling AERB as a credible and strong organization for safeguarding the public from harmful nuclear radiation.

My success was all the more satisfying that this battle happened without the media coming to know of it and without adversely affecting the prestige of any organization or person. My objective has always been to acquire rights for an organization aggressively but responsibly. One should not avoid a conflict, however strong the adversary. If the goal is right, you will prevail. In conclusion, the above facets in the shape of various “Avatars” that I played in professional career with their own pit falls, successes and failures.

Reminiscing My Days at the Vikram Sarabhai Space Centre (VSSC), Trivandrum

Dhruba Basu

My Early Life and Education

I was born and raised in a middle-class family in Calcutta and had my schooling at the St. Xavier's Collegiate School. I passed my School Final Examination from the West Bengal State Board with a rank and a scholarship in 1959 which entitled me to free education during my two years of I.Sc. studies at the Presidency College, Calcutta. I went on to complete my B.Sc. with Honors in physics from the same college and in 1963 joined the Institute of Radiophysics and Electronics at the Calcutta University Science College Campus to complete my B.Tech. and M.Tech. degrees. Thereafter, I joined a research group at the same institute working on switching theory. This was my first introduction to the study of digital computers which in those days was looked upon with both awe and fear. It may be noted that I did not have any formal course on the subject during my studies and whatever little I have achieved in life has been due to a learning process in my work place inspired by some great personalities who came into my life at the VSSC in Trivandrum. Our university education in those days taught us how to learn on our own rather than impart some knowledge by rote.

Journey to Kerala

When I moved to Trivandrum in 1968 from Calcutta as an engineer at the Thumba Equatorial Rocket Launching Station (TERLS), I must admit that I felt immensely home sick. Born and raised in an urban milieu with a strong Bengali flavor, I found it very difficult to come to terms with the way of life in Trivandrum of those days.

D. Basu (✉)
1/1B, Prannath Pandit Street, Kolkata 700 025, India
e-mail: dhruba_basu@hotmail.com

P. Ghosh and B. Raj (eds.), *The Mind of an Engineer*,
DOI 10.1007/978-981-10-0119-2_15

An intensely rural town had little space for a first-generation Calcuttan. Little did I know that TERLS would morph into the gigantic and wonderful Vikram Sarabhai Space Centre (VSSC) in its formative years. Trivandrum became my home for the next 34 years till 2002, my year of superannuation. VSSC became my work place where I found the best mentors, the most precious friends and a set of devoted and committed colleagues all of whom combined to present to me the greatest gift of life, i.e., a beautiful existence. But all of this would not have been possible without a creeping feeling of commonality with the ethos of Kerala during my first two years of loneliness. I could realize that there after all was a strong bond between my home state of West Bengal and my adopted state of Kerala. My analysis revealed that both the peoples loved three very disjoint things in life. Our food habits made fish a must in our eating patterns. The romanticism associated with a class of political belief was another strong common factor. And in sports, the love for football which I found among the Keralites of the late 1960s gave me reason to believe that if there could be a second home for me then it must be Kerala. Be that as it may, some of the luminaries of rocket science who till today have been my greatest source of influence have been Dr. Vikram Sarabhai, Prof. Satish Dhawan, Dr. A.P.J. Abdul Kalam, Dr. S.C. Gupta, Dr. U.R. Rao, Dr. K. Kasturirangan, Dr. S. Sreenivasan, and Dr. G. Madhavan Nair. I cannot complete this essay if I have to write about how each one of them inspired me in my workplace. Suffice to say that generally speaking it was their way of life more than anything else that must have encouraged me to learn new technologies and methods and thereby build the systems that I did. However, I must mention about Dr. Sarabhai after whose name the center was christened after his untimely death on the December, 30, 1971. I had seen him from a distance during his frequent visits to TERLS during the period. Clad in Kurtha and Pyjamas and with an ever smiling face, he would move about the place alone going from office to office and enquiring about the staff as if they were known to him for a long time. He would chair review meetings and try to understand what we proposed to do and how the projects fitted into his vision of making India the space power it is today. I still remember that he would drink tender coconut water during the meetings and not the usual beverages of tea and coffee. His last visit to Trivandrum was in connection with the inauguration of the Thumba railway station. On the morning of the fateful day when we went to office we were told that Dr. Vikram Sarabhai was no more and that he had passed away in his sleep the previous night after attending meetings with senior members of our staff till late at night. He died only to gain immortality through the existence of the Vikram Sarabhai Space Centre, one of the greatest and successful research institutes in the country which till today has been the bedrock of development of intricate rocket systems that have given credence to the belief that India is mature in the design and development of launch vehicles.

My First Assignment at TERLS

My first assignment in 1968 at TERLS was to design logic circuits and thereby acquire a capability to build the on-board processor (OBP) systems that would be used in the navigation, guidance, and control systems of the launch vehicles that the center would have to build in the days to come. My boss was Dr. S.C. Gupta who was the head of the Control and Guidance Division in those days. Dr. Gupta has been my chief mentor during all my years at VSSC. His analytical mind, unassuming behavior, and above all the air of being a perfect gentleman have all left a lasting impression in my mind.

It was an era when integrated circuits (ICs) in their most primitive forms (like a Quad 2-input NAND gate or a Dual J-K Flip Flop) were making waves and I can still recollect the excitement that overwhelmed us when the first consignment of ICs arrived in our laboratory. That excitement is understandable because I had by that time gone through the mill of building a transistorized version of a full adder which occupied a printed circuit board space of about 10 square inches! Digital voltmeters were just coming into the market and were replacing the more ubiquitous vacuum tube voltmeters known as VTVMs. The present-day acronyms such as RAM, PROM, ASIC, SCSI, and PCMCIA made little or no sense to the most knowledgeable and the letter C had no more importance than its other 25 peers that constitute the building blocks of the English language. PNP was a jargon used to classify a transistor by the polarity of its majority carrier. The yuppies of today's electronics industry feel that PnP is an acronym best used to describe "plug and play." Mercifully, for sentimentalists of my generation, the letter "*n*" used by them is in italics. Or, if I am permitted to borrow a phrase from my younger colleagues, the acronyms have been chosen to be made case-sensitive.

Most of my colleagues including myself were products of an indigenous education system with no formal training in an overseas university. But we had our ears and eyes firmly rooted to our environment soaking in the knowledge that was needed to do our jobs meticulously. We did not shut ourselves from the world at large as we had arguably one of the best technical libraries in the country—again a result of the vision of Dr. Sarabhai. There was neither any Internet nor was Google known at that time. But we used to get the hard copies of the IEEE Transactions and Proceedings of the IEEE regularly to keep us connected to the developed world. It was with this sort of a background we embarked on the design and development of the OBP systems which I am going to describe now.

The Story of OBP Development at VSSC

The story of the development of on-board processors (OBP) at VSSC has always been related to the decision to deploy a closed-loop inertial navigation and guidance system for a particular launch vehicle. As must have been explained elsewhere, such a system requires that the velocity and position of the vehicle are computed on board in an OBP by sensing the linear and angular movements of the rocket and determining in real time any correction that may be required to the sensed trajectory in case of any deviation. The corrections were sought to be issued in the form of commands to the vehicle autopilot system. Thus with these broad goals in mind, the development of the OBP system began in the early 1970s. Realizing that the development of such a system may evolve with the development of computer technology at the global level, the first such attempt was christened OBP MKI.

OBP MKI

This was built using medium power DTL and TTL technology for the central processing unit (CPU) and the processor bus and magnetic core for the memory having a 5-μs cycle time. The word length of the machine was 20 bits and it had about 30 odd instructions. It may seem odd that a word length of 20 bits was ever thought of. But it must be remembered that in the early 1970s, the concept of byte being a unit of data was not as well established as it is today. However, this choice of 20 bits for the word length was made after a lot of deliberations with the designers of the navigation and guidance systems who felt that a precision of 20 bits would suffice for the SLV3 launch vehicle. It has to be remembered that building floating-point hardware was not even contemplated and the legacy of using fixed point arithmetic continued even in the ongoing missions. Described in this article, The CPU was built around a microprogrammed control unit and the entire microprogram was hand coded by VSSC engineers.

The machine was built as a laboratory model in 1974, and it functioned well in the comfortable environment of the laboratory. Further development of OBP MKI was abandoned soon thereafter with the decision to have an open-loop guidance system and analog autopilot for the SLV3.

While this was a minor setback for the engineers who developed the system, it nevertheless gave them a lot of insight into the building of complex hardware. It is instructive to point out that the software for the machine was written entirely in machine language as tools such as compilers, assemblers, and linkers were yet to be developed.

However, with the decision to adopt an open-loop guidance system, the engineers involved in the design of OBP MKI had to rise to the occasion and develop an embedded system known as the vehicle attitude programmer (VAP).

VAP

This was built as a 2-unit system known as

1. a rate multiplier unit and
2. a rate storage unit.

The rate multiplier unit was used to digitally integrate its inputs obtained from the rate storage unit and to output the result to the vehicle autopilot system. The rate storage unit which was a plug in unit to the rate multiplier unit consisted of a bipolar programmable read only memory (PROM) of capacity 64 × 8 in which was stored the pitch rate segments for the particular mission. Essentially what this means is that the output of the VAP was a predefined pitching profile of the launch vehicle. Thus, by just altering the contents of the PROM, a new open-loop mission for the SLV3 could be defined. While all this may seem to be trivial from the perspective of today's hardware technology, it has to be remembered that the design, development, and deployment of the VAP took place during 1974–1983 when the technology of microprocessors even in its most primitive 8-bit form was not known.

OBP MKII

The Background

With the successful culmination of the SLV3 series of flights in 1983, the sights were set on a more advanced ASLV launch vehicle with a clear mandate that adoption of a closed-loop inertial navigation and guidance system and use of digital autopilot would be one of its most important goals. Thus, the designers of OBP MKI were asked to develop a reliable OBP system which would form the basic hardware platform on which such a system would be built. It was back to the design review committee to select one among the many options that were there. However, with the availability of 8-bit microprocessors and associated development tools such as in circuit emulators and logic analyzers, the committee did not feel it was necessary to repeat the strategy of OBP MKI. Time was short and the system had to be developed fast. However, since this OBP system would be a mission critical one, it was mandated that the OBP MKII would have redundancy in the form of a hot standby computer so that all single faults in the system would not lead to mission failure. In other words, the system was designed to be a single-fault tolerant one. It was also necessary to augment the power of the microprocessor to cater to the demands of high-speed integer multiplication and division. While selecting the technology, it was decided that reliability would be the single most important factor which would dictate such a choice. Thus, components to be used for such a mission critical hardware must conform to stringent military standards, must have had a good record of use in similar applications elsewhere, and must pass through the rigorous screening tests that they would be subjected to in the components screening laboratory at VSSC. One of the consequences of this consideration was the choice of the microprocessor and memory devices. Motorola's 6800 microprocessor may seem to be outdated today, but at the time of finalizing the OBP MKII configuration for ASLV in the early 1980s, this was the only device that could satisfy all the criteria. Similarly, the 2114 a 1K*4 static RAM was chosen as the memory device because of its relatively long and successful record of use.

The Architecture

Having decided on the redundancy mechanism and the choice of major components, it was necessary to examine the other architectural attributes for the given application. Since the dominant natural frequencies of the stabilization and control system of a launch vehicle are considerably higher than those of a navigation and guidance system, a higher computing speed was required for the auto pilot algorithms. Typically, two to four computing cycles per second were considered adequate for the ascent navigation and guidance of the launch vehicle, while fifty computing cycles per second were sufficient for the flight control. With the above considerations in mind, the main computing cycles for the OBP were split into two and were known as

- *Major Cycle*—every 500 ms. The algorithms of navigation and guidance were carried out once every major cycle.
- *Minor Cycle*—every 20 ms. The functions of vehicle control, digital filtering, and self-check were implemented every minor cycle.

The architecture which supports this type of cyclic, iterative computations was based on a distributed and multiprocessing system. It had the following major features.

- A few basic building blocks or modules,
- Parallel bus for intraprocessor data flow,
- Serial bus for inter processor data flow, and
- A set of standard software modules.

An estimate was made of the computational load both in terms of execution time and memory capacity. From these estimates, it was realized that the major number crunching operations should be carried out in a central processing system known as the Navigation, Guidance and Control Processor (NGCP). To protect the NGCP from electrical noise, it was isolated from its environment by using optical isolators at its inputs and outputs. Therefore, the basic tasks of input, i.e., the data acquisition from sensors, and output, i.e., the data distribution to the actuators, were sought to be carried out by separate stage processor modules (SPM) located close to the sensors and actuators of the rocket. The transmission of data between NGCP and SPM would be via optically isolated serial links. It is worthwhile noting that the SPM which was located close to the navigation sensors was christened the navigation electronics module (NEM) to give it a separate identity. In addition, there were two more serial links connected to each NGCP. One of them was connected to the check out (C/O) computer on ground and meant to support all prelaunch operations such as automatic check out and flight parameter initializations. The other serial link was for telemetering flight data, thus, evolved the architecture of OBP MKII. It consisted of two identical computing chains in which each computing chain consisted of an NEM, an NGCP and an SPM. A hardcore circuit variously known as a watchdog timer in the NGCP of the primary chain in conjunction with the self-check software monitored the health of the entire chain including the health of the serial links, the NEM and the SPM of that chain. In the event of a detectable malfunction anywhere in the chain, the identically designed hot standby chain was to be switched in, without causing any disruption to the flight.

The System Software

The description of the OBP MKII system would remain incomplete without a brief mention of its system software. Since the primary aim of the system software was to manage and coordinate the hardware resources in real time, a real-time executive (REX) was designed to incorporate the following features.

- Schedule execution of different tasks in real time,
- Maintain a real-time clock for the above,
- Handle all I/O operations,
- Keep a watch on the system health by means of a self-check routine, and
- Receive and execute commands from the ground check out computer during prelaunch operations.

It had three modes of operation, viz. monitor mode, preflight mode, and flight mode. The monitor mode is a non-real-time mode and is used for loading data from ground and also for testing different hardware modules under command from ground computer. The preflight mode and flight mode denote that the operating environment of the OBP is real time, and they were executing tasks in the preflight and flight phases of the mission, respectively.

The deployment of OBP MKII took place in the first two unsuccessful missions of ASLV in 1987 and 1988. Massive corrections were introduced on the various aspects of the launch vehicle control system design. One of the major changes introduced was in the introduction of real-time decision (RTD) making on board. This called for the design of RTD software as a part of REX support to other various application software tasks. There has been no looking back since. The first successful flight of ASLV D3 in 1993 was a day which all those associated with the development of OBP in VSSC would always look back upon as one of the most joyful moments in their lives. The last deployment of OBP MKII was in 1994 in the ASLV D4 mission. By this time, the overwhelming computational requirements of the more advanced Polar Satellite Launch Vehicle (PSLV) had emerged and it was time to move on from OBP MKII to OBP MKIII.

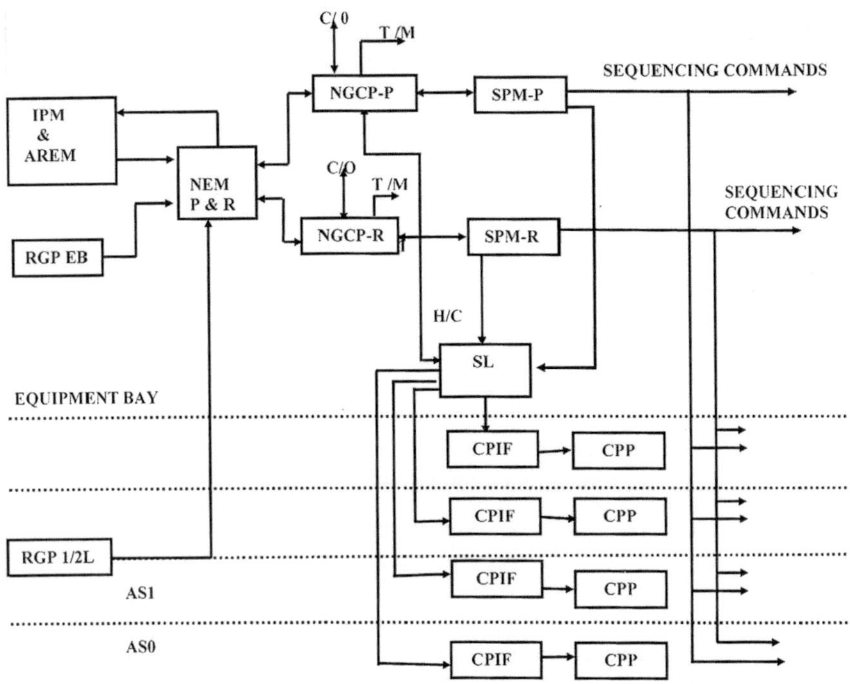

OBP MKII CONFIGURATION

OBP MKIII

Toward the end of the 1980s, it was clear that to support the more ambitious PSLV which was to have a strapped down inertial navigation system as compared to a stabilized platform based inertial navigation system in the ASLV, a more powerful OBP had to be developed. By that time, the more powerful 16-bit microprocessor M68000 from Motorola had been introduced in the market. The RAM chips also had to be upgraded to the higher capacity 2K*8, 6516. In fact, some preliminary hardware design had already been done as a part of the Automatic Payload Control Rocket Experiment (APCREX) project. A point debated over some period of time was whether the OBP MKIII like its predecessor would have a single NGCP to cater to all the major computational needs of the program or not. A school of thought, advocating a separate processor for the navigation functions alone, had emerged. This was mostly to have an autonomous inertial navigation system complete with its own computer. This would enable the navigation system to be tested on its own without being coupled with the guidance and control system of the PSLV. The OBP MKIII adopted this approach and in hindsight it seems to have been a very wise decision because over the years the computational requirements exploded and a single CPU would not have been able to do the jobs, which are currently shared between the two processors, the navigation processor (NGP) and the guidance and control processor (GCP).

The Architecture and Software

The basic attributes of the architecture such as

- use of optically isolated serial links to carry out inter processor communication;
- use of an SPM for data distribution to stages;
- two identical computing chains, one primary and the other redundant acting as a hot standby;
- two periodicities of computation, viz. a major cycle every 500 ms and a minor cycle every 20 ms;
- C/O and T/M links from the main computers; and
- REX which schedules the execution of real-time tasks among other things

were all derived from the experiences of OBP MKII.

Thus, the OBP MKIII in each chain consists of an NGP, which is interfaced with the inertial sensors at its input and connected by a serial link to a GCP at its output. These are located in the equipment bay (EB) of the PSLV. The SPM which distributes output data to the lower-stage actuators is located in the base shroud and receives its inputs from the GCP via an optically isolated serial link. The per chain configuration of OBP MKIII is therefore an NGP, a GCP, and an SPM. A hardcore circuit in the GCP along with REX self-check software determines which unit in the primary chain is faulty.

However, a significant difference between OBP MKII and OBP MKIII is with respect to fault tolerance. It can be explained as follows.

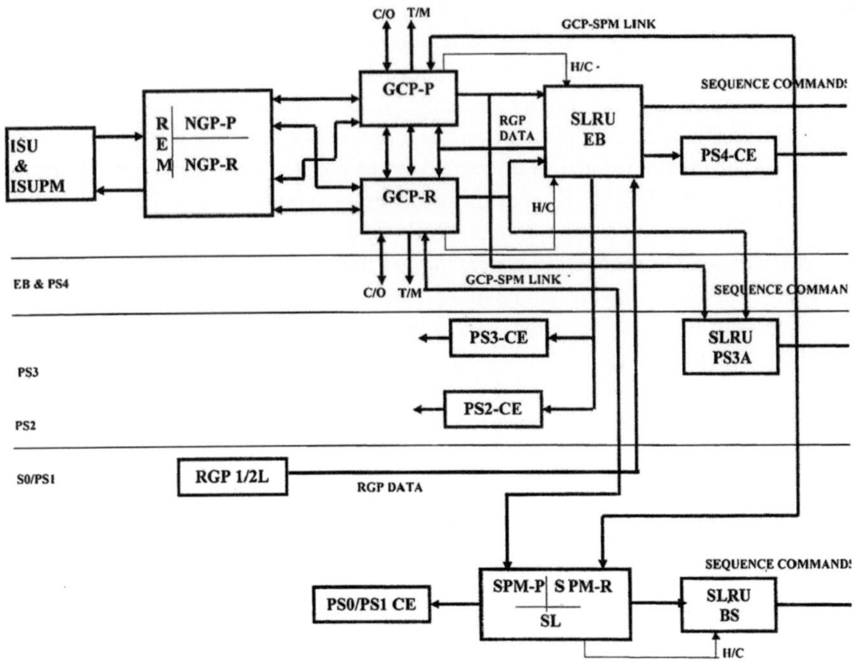

OBP MKIII CONFIGURARION

In OBP MKII, a fault in any unit in the primary chain caused the entire chain to be discarded and replaced automatically by the redundant chain. In OBP MKIII, this replacement is limited to the individual unit which has become faulty. Thus, if for example the SPM (P), i.e., the SPM in the primary chain develops a fault, then SPM (P) would be replaced by SPM (R), but the other elements of the primary chain, namely the NGP (P) and GCP (P), would be still functional. In other words, SPM (R) would receive its inputs from GCP (P) rather than GCP (R). This type of a system is known as dual redundant scheme with cross-strapped elements between the primary and the redundant chains. This increases the hardware utilization manifold in the sense that in the event of a failure of one of the elements in the primary chain only the corresponding element in the redundant chain is brought into the computational loop without losing the other elements in the primary chain. This is as if the two chains are cross-connected or cross-strapped. It is to be appreciated that this concept of cross-strapped dual redundant scheme brought along with it an order of magnitude increase in the complexity of software and test systems. But then these challenges were met and till my superannuation in 2002 barring the first unsuccessful flight of PSLVD1 in 1991, the OBP MKIII continued to serve ISRO's launch vehicle programs of PSLV and GSLV.

Epilogue

I have tried to describe the work carried out at VSSC in the area of avionics computer systems during a period when the technological advances of today were unheard of. But we knew that the USA had in 1969 used the Saturn V rocket to put astronauts on the moon. They had done so using computer technology which must have been yet more primitive. What we possessed then was the basic will to succeed. We had neither any formal training in computer science and more profoundly we were all products of an education system that was fully indigenous, a system that had taught us to learn on our own rather than committing to heart a few basic theories. We were not afraid to question our own beliefs if we were not convinced. In all this, we had a set of brilliant men who did not impose their leadership on us but showed us the way by practicing what they preached. I have no hesitation in admitting that had it not been for these superb gentlemen we would not have succeeded in our mission objectives. It is the quality ambiance created by them that has made the ISRO what it is today.

In OBP MKII, a fault in any unit in the primary chain caused the entire chain to be discarded and replaced automatically by the redundant chain. In OBP MKIII, this replacement is limited to the individual unit which has become faulty. Thus, if for example, the SPM (P), i.e., the SPM, in the primary chain develops a fault, then SPM (P) would be replaced by SPM (R), but the other elements of the primary chain, namely the NGP (P) and GCP (P), would be still functional. In other words, SPM (R) would receive its inputs from GCP (P) rather than GCP (R). This type of a system is known as dual redundant scheme with cross-strapped elements between the primary and the redundant chains. This increases the hardware utilization manifold in the sense that in the event of a failure of one of the elements in the primary chain, only the corresponding element in the redundant chain is brought into the computational loop without losing the other elements in the primary chain. This is as if the two chains are cross-connected or cross-strapped. It is to be appreciated that this concept of cross-strapped dual redundant scheme brought along with it an order of magnitude increase in the complexity of software and test systems. But then these challenges were met and till today barring the first unsuccessful flight of PSLVD1 in 1991, the OBP MKIII continues to serve ISRO's launch vehicle programs of PSLV and GSLV.

Chasing a Dream

Dipak Mazumdar

When I started writing this article, I was not sure of what to write in the beginning. Given a fair bit of liberty on the subject matter, I was almost certain that I should not take the opportunity to write about my research or my achievements. With such self-imposed constraints, trouble started almost immediately as I was lost in the ocean of thoughts about the theme of my write-up. I soon realised that writing a research article is far easier than to develop an easy flowing, convincing story that is intended to leave a mark or two in the mind of the readers. To identify an appropriate theme, frame an inspiring story and narrate the same in a creative style, thus became a full time business for me. I started to look back in time, frantically searching for a series of incidents that could be interwoven, made engrossing and narrated in a readable style. It was a remarkable and gratifying experience, as I travelled decades back to my student days to dig out moments and a series of incidents to develop the present narration. Having prepared a sketchy outline, I discussed the same with several of my friends at IIT and outside, who enthusiastically supported my idea.

I have not become a metallurgical engineering professional by accident, neither have I become a professor of steelmaking by chance. In grade X, I was a reasonably bright student and my father wanted me to take up physics and become a teacher. My father, patriarchs like most traditional Bengali head of the family in those days, wanted to see his son study physics to become a professor and daughter, well versed in music, in particular, Rabindrasangeet! I, however, had a different take on the matter and was not much interested to pursue pure science and instead, wanted to become an engineer. My father was a mechanical engineer and having spent most of his lifetime on the shop floor certainly did not want me to follow his footsteps. Naturally, therefore, he kept on seriously discouraging me from considering engineering as a possible career option beyond grade XII. After all, in mid-1970s, engineers in the country were not doing too well. Nonetheless, I was determined, though not sure of the engineering discipline that I would like to pursue.

D. Mazumdar (✉)
Department of Materials & Metallurgical Engineering, Indian Institute of Technology, Kanpur 208 016, India
e-mail: dipak@iitk.ac.in

P. Ghosh and B. Raj (eds.), *The Mind of an Engineer*,
DOI 10.1007/978-981-10-0119-2_16

Interestingly, a train ride from Howrah to Mumbai, during a trip to my maternal uncle's house in Navy Nagar, Colaba, helped me move in a direction and decide on my future career plans.

Immediately after the school leaving certificate examination, in March 1973, for about three months, there was no school and no studies! I was to spend time with my uncle and his family in Mumbai and one fine morning started my journey from a small town in Assam to proceed to Mumbai via Calcutta (now Kolkata). After a brief stopover for a few days at Kolkata, I boarded the Howrah Mumbai express, which used to leave Howrah rail station at about 12 noon to reach Mumbai after about 42 h! I had no clue about the train route or stoppages en route, but remained reasonably calm during the journey since my uncle would be there at Bombay VT station to receive me. At about 5.30 in the evening, the train stopped at Tatanagar rail station for about 20 min or so. I took the opportunity to stretch my legs and walked towards the nearby door of the sleeper coach I was travelling into catch a glimpse of the outside. It was nearly sunset time though the sky surprisingly appeared very bright and fiery as if, a huge fire had set-in the vicinity. It was a remarkable interplay of light and colour, and I got thoroughly engrossed wondering what might be happening. I had no clue of steelmaking processes then and barely knew that there was a steel plant of the Tata's, somewhere in Jamshedpur. It was years later that I came to know of the Bessemer steelmaking process and could rationalise my observation, a manifestation of blowing of air through molten pig iron. The 20-min experience at Tatanagar rail station kept me mesmerized and left an everlasting impression on my young inquisitive mind. I returned home from Mumbai after about a month and started to gather more information on the subject of iron and steelmaking from my elder brother's Inorganic Chemistry book. Around the same time (i.e. 1974), Professor H.S. Ray, who was then teaching at IIT Kanpur, wrote a feature article on blast furnace iron making in Science Reporter, a popular science magazine for college students. I read through Prof. Ray's article and found the subject fascinating. These experiences helped me understand and appreciate the science and technology of processing of metals at elevated temperatures. By the time I was in Grade XII, I made up my mind and expressed my interest of pursuing metallurgical engineering to my father. I thank him for not trying to impose his ideas on me or influence my thinking at that stage!

Living in a remote Assam town, I had difficulties in gathering information. Yet, I had reasonable idea of various engineering institutions in the country imparting education in metallurgical engineering. Things however did not work out favourably for me as all Assam College Teachers' Association (ACTA) launched a massive state-wide strike immediately before the +2 final examinations in 1975. In the process, final examination was delayed and by the time the results were announced, and it was middle of August or so. July admission session in the country was practically over; my father did not want me to waste a complete year, and thus, I was left with no option but to wait for possible December admission to BITS Pilani! I was not keen as there was no metallurgical engineering programme in BITS but eventually succumbed to the fear of losing one full year! After about a fortnight of my stay at Pilani, I received a letter from my father informing me of my selection to the four-year engineering programme at RIT Jamshedpur (now an NIT), where academic session was unduly delayed due to the JP movement in Bihar. I was delighted and saw this as a golden opportunity to fulfil my dream of becoming a metallurgical engineer. Many of my friends dissuaded me, since RIT was a lesser known college, not a preferred destination, and metallurgical engineering was not a priority choice for good students. But I was determined and took a final call myself without even consulting my father, and one early morning packed up my meagre belongings to proceed to Jamshedpur so as to be there on the day of counselling. It was an amazing experience and journey as I had little left over cash in my pocket and no clue how to reach Jamshedpur from Pilani. But my spirit was high and I made it to Jamshedpur on time and selected metallurgical engineering absolutely on the basis of my own conviction and desire to study steelmaking and nothing else. Incidentally, I was short of fund for paying my admission fees and had to enact a drama to win special favour from the principal for a deferred payment of the college fee. Thus, finally began, in early 1976, my journey to become a metallurgical engineer!

I continued to perform exceptionally well in my discipline throughout my undergraduate years. By the end of third year, I used to fancy many degrees after my name and started to even daydream, becoming a professor, teaching steelmaking someday. I was not quite happy with the academic seriousness and the quality of education at RIT and went on to express my willingness to the head of the department, Professor Mahabir Ram, to pursue higher studies. Professor Ram was a dynamic person and a teacher par excellence. The style with which he used to communicate complex issues at ease used to fascinate me. I developed a great sense of appreciation and admiration for his ingenuity in teaching. Professor Ram, who was aware of my inclination towards higher studies, only suggested that I move to IIT Kanpur to pursue graduate studies with Professor Ahindra Ghosh, a well renowned professor in the subject of iron and steelmaking. From third year onwards at RIT, I gradually geared up myself to do a master degree at IIT Kanpur and started to work seriously in that direction. I was so passionate about the idea that I did not even sit for a single job interview in the final year of my studies. Instead, I motivated many of my friends, who were good in studies, to move to IIT Kanpur for higher studies.

Graduate study at IIT Kanpur was far too intense, and we had to put in enormous efforts to come up to the expectations of the professors, who were leaders in their respective fields. Rigorous course work mixed with research helped strengthen our foundation and removed deficiencies in our training received at RIT. Working with Professor Ghosh, as a young graduate student, was extremely gratifying, as I learnt beyond steelmaking, several basic values of life. Discipline, punctuality, honesty and integrity used to be the hallmark of training under Professor Ghosh. It was working with him that my conscious journey to be a good human being truly began, and I started to learn different values of life seriously. Simple lifestyles of professors, their holistic living and the beautiful campus of IIT Kanpur influenced me profoundly. Staying at IIT Kanpur largely motivated me to pursue education farther, becoming eventually a professor and I started to consciously work in that direction. However, I was well aware that to accomplish my objectives, finishing a good Ph.D. from a well-reputed school abroad as well as postdoctoral experiences were a must. With his continuous mentoring, strong recommendation and my excellent grades at the undergraduate and graduate levels, by the end of the third semester, I was successful in getting admission with full research assistantship in two different universities in USA and Canada. I decided to move to Canada to pursue Ph.D. with Professor Rod Guthrie at McGill, who was working on the frontiers of process metallurgy.

Getting a Visa to Canada was initially difficult due to a childhood surgical intervention. The medical board created a problem, and the Canadian Embassy was not convinced with my explanation. Fortunately, my father could dig out the fifteen-year-old hospital discharge certificate, which finally did the job, and I was granted a VISA to study in Canada. I landed at McGill around early March of 1983 to pursue my dream of completing a doctoral degree and eventually becoming a professor of steelmaking. I was received at the Mirabel airport by my would be supervisor, Professor Guthrie. Remarkably, arrangements for my first night stay were made by him in the downtown Holiday Inn! Thus began my association with a very eminent professor and one of the finest gentlemen in a prestigious North American University, in the wonderful city of Montreal.

With Professor Guthrie, I enhanced my skill considerably and turned out to be an excellent researcher. The academic liberty that he extended to all his students helped them grow and mature considerably. The teacher student relationship in North American University is very different from the one in this country. We discussed everything under the Sun, exchanged views on diverse issues and even shared a drink or two occasionally during happy hours on select Fridays. Professor Guthrie and I had similar views on many fronts and thus developed liking for each other. The teacher–student relationship thus blossomed over a short period of time into a warm friendship that both of us even cherish today! It is with him I learnt to execute research work meticulously and in a time-bound fashion. He highlighted the importance of industrial research in steelmaking which I have tried to emulate throughout my professional career. It is the training that I received from Professor Ghosh and Professor Guthrie helped me carry out research with passion and remain highly focussed throughout. Within two years and four months of my stay, Professor Guthrie suggested that I should wind up and submit my Ph.D. thesis. I simply could not believe my ears. I defended my thesis in August 1985 and returned back to India for a brief holiday with my parents. I was happy to have made them proud! I continued to work with Professor Guthrie as a postdoctoral fellow for some more time before I finally decided to return back to India in early 1987. I was offered positions at IIT Kanpur, IISc, IIT Mumbai and IIT Kharagpur. I, however, decided in favour of my Alma Mater to fulfil my long cherished dream!

Although a very strong moral, spiritual and intellectual foundation of my life was laid early by my father, my teachers too, at high school (particularly, Ashish-da), at RIT, at IIT Kanpur as well as at McGill, played pivotal roles in shaping my career and personal life. They all taught, inspired and motivated me to be what I am today. It is gratifying that I continued on their footsteps to become an educator. Their high values have always guided me and helped steer my professional career and personal life in the right direction. One needs to be truly fortunate to be associated with and guided by enlightened teachers during the formative years of one's career. I feel utterly blessed and thank my stars!!

It has been and is less fashionable to study metallurgy and to make a career in metallurgical engineering. Many of us tend to believe that metallurgy is an ancient discipline, much is known and very little or no challenges exist. This is a myth and is far from the reality. In fact, this mindset has to change to perceive the reality. Once, a friend of mine, in-charge of Sony Erickson operation in the States, asked me the question sitting in his Fremont house: "Dipak, do you regret that you studied metallurgy when you had the option to go in for more popular engineering disciplines?" I spontaneously replied "Not a bit! I enjoyed and continue to enjoy metallurgy with the same enthusiasm even to this date". At the end, it does not really matter what one does but more so, how it is done. Excellence and creativity result only when there is passion and there is conviction.

My Eureka Moment

N.S. Mohan Ram

As engineers, we do a lot of solid and useful work during our working lives. The opportunity to create truly path-breaking and innovative products or services comes rarely in one's life. I have been fortunate in being blessed with several opportunities during my long career, spanning nearly six decades. Let me share one such moment.

I joined the Indian Navy after completing my degree in Naval Architecture from IIT Kharagpur. The Navy deputed me to UK to attend a prestigious four-year course on warship design at the Royal Naval College, Greenwich, UK. Five years later, I got the rare opportunity of working in the Royal Navy's Design offices Bath for a year in 1967–1968, training in design for the Leander Frigate Project, for warships to be built under license from the UK.

Six frigates of the class were built in Mazagon Docks Limited over the next twelve years. The Nilgiri classes—INS Niligiri, Himgiri, Udaygiri, Dunagiri, Taragiri, and Vindhyagiri—were the first full-fledged warships built in India. I was closely associated with the project as a designer, introducing major changes, including localization of air-conditioning in the ships and accommodation of a larger HAL-built helicopter "Alouette" on the ships. The project gave Indian designers the basic skills for warship construction and the ability to introduce major design changes in a bought out design. The next major step for us was to develop a totally Indian design warship to meet specific requirements of Indian Navy.

Work started in 1974 on the follow-on frigates (later named Godavari class), which were much larger than the Giris. The ships carried surface-to-air and surface-to-surface missiles for the first time in India. We were required to develop ab initio a larger frigate, which combined Western underwater weapons and Soviet surface weapons in one hull, for the first time in the world.

The Indian naval staff realized that the requirements for surface–to-surface and surface-to-air missiles, improved helicopter availability and greater fire power, could not just be accommodated in the Leander hull. This was apparent by the time the construction of the sixth frigate of the Leander class was taken up. The naval staff also wanted the ship to travel over knot faster, over 29 knots for strategic

N.S. Mohan Ram (✉)
TVS Motor Company Ltd., PB No. 4, Harita, Hosur 635 109, Tamil Nadu, India
e-mail: capt.mohanram@tvsmotor.co.in

P. Ghosh and B. Raj (eds.), *The Mind of an Engineer*,
DOI 10.1007/978-981-10-0119-2_17

reasons. Quick studies revealed that the ship had to be a good 12 m longer and would weigh 600 tons more than the largest Indian Giris, Vindhyagiri, and Taragiri. The need was for a bigger and faster ship.

This called for the complete design of a new major warship ab initio. I was assigned the task for which I had equipped myself over ten years. In 1974, I was appointed as the project officer in charge of INS Godavari class of ships, then code-named project 16.

The Navy then took the daring decision of importing missile systems and guns from the Soviet Union and incorporating Western underwater weapons in the ship. Such an exercise of marrying Soviet origin weapons and Western origin weapons in the same ship had never been tried before and has also never been done successfully in any other Navy except ours. Many doubting Thomases felt that the naval designers had bitten off more than they could chew and were not capable of designing such a ship. They wanted to import the ships from abroad.

As the project in charge, I was assigned the task of coordinating and producing the basic ship design. Also as a hull specialist, I was responsible for the lines of the vessel, powering, layout, structural design, propellers, and major systems such as air-conditioning.

The Giri class ships displaced about 3000 tons and had a top speed of about 28 knots. The Naval staff demanded a minimum one knot extra from the ships which were estimated to weigh about 3600 tons and be about 123 m (40 ft) longer. The Leander was propelled by two steam turbines of 15,000 shaft horsepower each. Naval marine engineers of the Navy wanted to fit gas turbines, which we had never used in a new design before.

A bigger and longer ship designed to go faster would obviously need more power, as per conventional wisdom. So the search was on for a larger power plant.

The Navy was debating introduction of different propulsion options such as gas turbines, combined gas turbine, and diesel propulsion.

On a wet Saturday afternoon, I was doodling on a piece of paper at home and idly wondered "What would happen, if I powered the new ship with the same power plant—two turbines of fifteen thousand horsepower each—without any change. How much would the speed drop? Was there any chance of convincing the naval staff that a small sacrifice in top speed would make the ship more economical and easier to construct?"

I performed a back-of-the envelope calculation to estimate the speed loss. To my utter surprise, the answer came out that the ship did not lose speed at all. On the contrary, it would go a full knot faster, more than 29 knots which the naval staff wanted! I checked the numbers again and again and could not find any mistakes in the calculation. The power required and found that the answer came out the same.

I was elated. Perhaps, this was a brilliant solution for meeting the Navy's requirement without any additional investment, using equipment being manufactured in India. I was so excited about my discovery that I could hardly sleep the whole weekend.

I rushed to the office on Monday and announced my discovery. No one believed me at first. I was greeted with a stony silence. Most of my colleagues thought that I had gone out of my head. I could not blame them, as my findings were totally counterintuitive. My boss insisted that another officer should check the numbers. Lo and behold, the answer was the same. We realized that we could preserve the steam turbine power plant of Leander class frigates and meet the naval staff's requirement for higher speed.

We had to find the reason for this windfall benefit. A detailed analysis showed that below 22 knots speed, the larger ship required more power for the same speed as the Leander due to increased friction. At around 22 knots, both ships required the same power but above 22 knots, the bigger ship required less power. Again above 31 knots, the bigger vessel again as at a disadvantage compared to the Leander class ships. But happily, at 29 knots plus, the larger ship needed only the same power as the Giri's power of 30,000 HP. Fortunately, a happy combination of the laws of hydrodynamics was working to our advantage.

Above 22 knots, the resistance to motion from wave-making due to the ship cleaving through the sea became much more prominent than friction. At higher speeds, if the interference between the waves created by the bow (front) of the ship and the stern (rear) of the ship were positive, resulting in a crest at the rear end, resistance due to wave-making would be lower. If the interference between the bow and stern wave systems resulted in a trough at the stern, the resistance due to wave-making would be higher. The interference is a function of Froude number, which related the square of the speed of the ship to the length of the ship. In the case of the Leander at 28 knots, the interference caused a trough at the stern increasing the wave-making resistance. But in the new longer ship, the interference resulted in a crest. This resulted in a lower wave resistance in the bigger ship, which more than compensated the increased drag due to greater area.

Once we had done this analysis, the picture became clear. In nature, as in life, however, there is no free lunch. While we could reach the top speed comfortably with the same engines, at the normal cruising speeds the ship consumed twenty percent more fuel than the Leander. We had to increase the fuel tank capacity of the ship.

By discovery enabled us to use the same power plant, gearing, transmission, and even propeller as the Leander in the new ships, saving immense design effort and costs. Moreover, Mazagon Docks had experience of construction of six Leander class ships with the same power plant and the learning.

When we announced our findings to the naval staff, the marine engineering fraternity of the Navy was up in arms questioning the validity of my calculations. I stuck to my guns staking my professional reputation on my calculations. In a meeting chaired by the chief of naval staff, Admiral Kohli, the top naval brass, supported me, and the Godavari class frigates were designed round existing steam turbines only.

By physically locating all machinery spaces, diesel alternator rooms, boiler room, engine room, stern compartments, and propeller shaft supports and the propeller supporting bracket at the same relative locations and distances as in the Giri class, substantial amount of engineering redesign was averted. These were major decisions, which I was personally responsible for. To some extent, my lack of experience and data turned out to be a major asset, as I was not constrained by past practice, which often works against innovations.

The Godavari design group had a great time designing the ship. Each day threw up fresh challenges and problems and revealed exciting solutions. The designers were young and inexperienced but incredibly committed. The whole group was working in a fever pitch of self-actualization. The mood was similar to a *jugal bandi* of Ravi Shankar and Ali Akbar khan or a tabla duo performance of Allah Rakha and Zakir Hussain.

To confirm our findings, we carried out scale model tests for propulsion and sea keeping of the hull form at the renowned National Physical Laboratory in the UK. The tests were successful. They vindicated my original assertion that we could achieve over 29 knots speed in the larger ship with the same power plant as the Giris. The sea-keeping qualities of the ship were also excellent, providing a very stable platform even under rough sea conditions. I completed the structural drawings and major layouts of the ship during my tenure.

For the work I had done on the project, I was recommended for the Vishist Seva Medal (Distinguished Service Medal) and was awarded the same in the republic day Jal Cursetjee, in an impressive ceremony held in Bombay.

Six years later, I was in the navigating bridge of INS Godavari achieved well over 29 knots during sea trials. The ship participated in the US celebrations of two hundred years of independence and was feted by the world's press as a remarkable achievement in combing Western and Soviet weapons in one warship hull, a feat never before attempted.

Six ships of the class have been delivered to the Navy. The savings in machinery costs alone were over two thousand crores of rupees. Precious Forex was saved at a time of low reserves.

Using known aggregates speeded up design, and I had proved that a heavier faster ship could travel faster with the same power plant. My innovation was vindicated. The ship also built the confidence of Indian Naval designers who went on to design more complicated warships, culminating in a nuclear submarine and an aircraft carrier.

Godavari commissioned in 1983 December has served the nation for over thirty glorious years. Among its achievements were the pivotal role in preventing an armed rebel take over Maldives and rescue of personnel from pirates near Somalia. The ship participated in the US celebrations of two hundred years of independence and was feted by the world's press as a remarkable achievement in combing Western and Soviet weapons in one warship hull, a feat never before attempted

My Eureka moment has served the Navy and the nation well.

My Reminiscences

Ahindra Ghosh

The Early Life

I was the youngest child of my parents and was born in the city of Howrah, West Bengal, in 1937. My father retired when I was only few years old. From 1942 to 1948, we witnessed large-scale turmoil. Japanese invasion and air raid over Calcutta, food shortage and inflation causing the Bengal famine and millions of starvation deaths. Within a few years, large-scale communal riots, massive refugee influx from East Bengal before and after independence, took place. The social and economic fabric of Bengal received serious setback as a consequence.

It is during that period, my father lost all his savings kept in a private bank due to its liquidation. Since my elder brothers and sisters had not yet completed their studies and were not earning members, somehow the family was sustained through sale of mother's ornaments and some help from relatives.

I was not a good student at primary school level. It was at the secondary school that my performance started gradual improvements. I developed special interest in mathematics. I also started to take interest in extracurricular books in a variety of subjects. My father was not directly teaching me, but used to tell lots of stories, especially on history. Thus, he was instrumental in creating academic interest in me. I was a rank holder at the district level in School Final Examination, making me entitled for scholarship for 2 years.

I was a student of the 2-year program of Intermediate Science (I.Sc.) at the college. Then, I became very much interested in physics and chemistry, besides mathematics. Even before the classes started, I had covered the textbooks substantially through self-study without any help. I even studied some advanced topics in physics which were not part of curriculum. My performance was so good that I

A. Ghosh (✉)
Flat 3C, 'Shatabdi', 54/2/1/1/1, Andul Road, Howrah 711 109, India
e-mail: ahin1937@yahoo.in

P. Ghosh and B. Raj (eds.), *The Mind of an Engineer*,
DOI 10.1007/978-981-10-0119-2_18

was a rank holder in the I.Sc. examination at the Calcutta University, which entitled me for freeship and scholarship for further studies.

I was also a rank holder in the entrance examination of the then Bengal Engineering College (now Bengal Engineering and Science University). This entitled me to receive the merit-cum-means stipend, which was sufficient to cover all my expenses including boarding and lodging in hostel. As it is now, then also metallurgy was almost the last choice of students. I could have joined any branch, but voluntarily opted for metallurgy. I received B.E. degree at Calcutta University in 1958 with first rank among the Metallurgical Engineering graduates. After a few months, I joined the then Hindustan Steels Ltd., which is now known as Steel Authority of India, as a trainee.

At that stage in ever thought of higher studies abroad, but had decided to be an earning member of the family. But my B.E. college professors suggested that I should go abroad for higher studies. Since my preference was to specialize in extractive metallurgy and the MIT had the strongest group, I decided to join MIT and landed in Cambridge in 1959 with an offer of admission and financial assistance. I received the Sc.D. degree in 1963 and was at the Ohio State University as postdoctoral fellow for a year.

During that period, the Indian Institute of Technology Kanpur was being set up with collaboration of 9 US universities. For faculty position in an Indian University, according to the then prevailing norm, a candidate had to apply against advertisement and appear in interview in India. This was a deterrent for qualified professionals abroad to return to India for a career.

IIT Kanpur was the first institute in India, which decided to advertise abroad inviting application in plain paper and selection directly. So many professionals, including myself, joined IIT Kanpur. Also the institute opted for science-based engineering education like the US Universities. We also were given freedom in research efforts and liberal dollar grants under the Kanpur Indo-US program for import of sophisticated equipment and materials, not available in India.

My Professional Mission

From technological point of view, my broad area of specialization is extractive metallurgy, which is concerned with extraction and refining of metals. Extractive metallurgy, again, has 2 subdivisions—ferrous and nonferrous metallurgy. I have specialized in ferrous metallurgy, and I am concerned with iron making, steel making and casting of liquid steel. These processes are carried out at high temperatures in liquid or solid state.

The relevant scientific fundamentals are: (i) chemical metallurgy, dealing with thermodynamics, chemical kinetics, physicochemical properties of high temperature materials including non-aqueous melts, (ii) process engineering, dealing with momentum, heat and mass transfer; mathematical and computer techniques.

I joined IIT Kanpur at the end of 1964. From then up to about 1990–1995 was a dark period for the country's economy as well as for steel and metals industries. Growth rates as well as infrastructure developments were marginal. Sickness in manufacturing sector was rampant. It is an attribute to the Indian steel makers that, even then, they made reasonably good progress in modernization of plant practices.

Undergraduate teaching in extractive metallurgy was the primarily descriptions of the processes. Emphasis on quantitative metallurgical and engineering sciences was almost absent. Postgraduate programs almost did not exist.

So far as research is concerned, steel plants were doing some tinkering. You cannot call them as serious research. NML was fairly active in process research. But it was mostly on pilot plant process development. So far as academic institutes are concerned, BHU was active, but mostly on pilot plant studies. Prof. K.P. Abraham of IISc Bangalore was doing some fundamental studies on thermodynamics and kinetics. Prof. V.A. Altekar of chemical technology of Bombay University and Mr. C.V. Sundaram of BARC were also doing something.

Surveying the situation, I recognized that my principal efforts should be to promote nurturing of metallurgical sciences as well as application of the same in teaching, research, and industrial practices. Also, perhaps it has to be a lifelong mission. I am still pursuing this mission. However, the natures of activities have been different in different phases. For example, one of my major activities for the last 20–25 years has been the publication of text books either exclusively on quantitative metallurgical sciences, or on technological topics with considerable coverage of theory. Five books have been published. The last one on kinetics is expected to be published in the next year. The yare low-cost paperback books can be acquired both by teachers and students. There have been almost no such equivalent books in the Indian market. An example is the book on "Iron making and Steel making-Theory and Practice" by Ghosh and Chatterjee [1].

So far as research is concerned, a major activity was the laboratory research. Process research, by and large, is not possible with standard equipment. It required design and fabrication of apparatus for object-oriented investigations. Several such apparatus were setup from 1965 to about 1990 for high temperature thermodynamic, kinetic and some other miscellaneous studies, as well as experiments in room temperature transparent water models for fluid flow, mixing and kinetic studies.

As example, Fig. 1 shows the photograph of an apparatus for continuous measurement of weight loss/gain of sample kept in a vertical resistance furnace under flow of chosen gases. Only the Cahn Electro-Balance unit was imported. Rest of the setup consisting of furnace, the framework assembly, gas train was indigenously designed and fabricated.

Figure 2 shows another apparatus for the study of solidification of alloys in metal mold [2]. Again, it was indigenously designed and fabricated. Solidification was vertically upward from the bottom and the rates were controlled. The solidified ingots were vertically sectioned into 2 halves, which were subjected to a variety of physical, chemical, and metallographic examinations.

Fig. 1 An apparatus for continuous measurement of weight loss/gain

Figure 3 schematically shows the main steps for the individual experiments. Elaborate mathematical modeling was carried out for quantitative interpretations of data [3].

Interaction with steel plants and R&D organizations was an important parallel activity, on and off, from 1970 to 2007. Programs consisted of visits, delivery of lectures, and collaborative R&D. For example, significant part of collaborative R&D was the fundamental studies on fluid flow, mixing, chemistry, and heat transfer in ladle furnace refining of molten steel. Data collection and experiments were carried out in the LD-2 shop of Tata Steel, Jamshedpur. Scientists of the Tata Research, Development, and Design Center, Pune, actively participated in this program and also carried out the mathematical modeling exercises. I acted as the adviser for the entire program.

Finally, based on these studies and further work, an online reckoner for prediction and control of steel temperature and composition was developed and installed in the LD-shop of Tata Steel [5].

Figure 4 shows the sketch of a ladle furnace.

Figure 5 schematically presents the control strategy.

Mathematical modeling is a powerful tool for R&D. In several research projects, we carried out mathematical modeling of process in addition to experimental work. This combination gave us significant results. I have already cited two examples [4, 5]. Ghosh [6] has reviewed some of these, which has the list of original publications. It was possible to do this only by collaboration with some others, some faculty colleagues or scientists from other organizations who had expertise in modeling.

Fig. 2 Apparatus for study of solidfication of alloys in metal mould

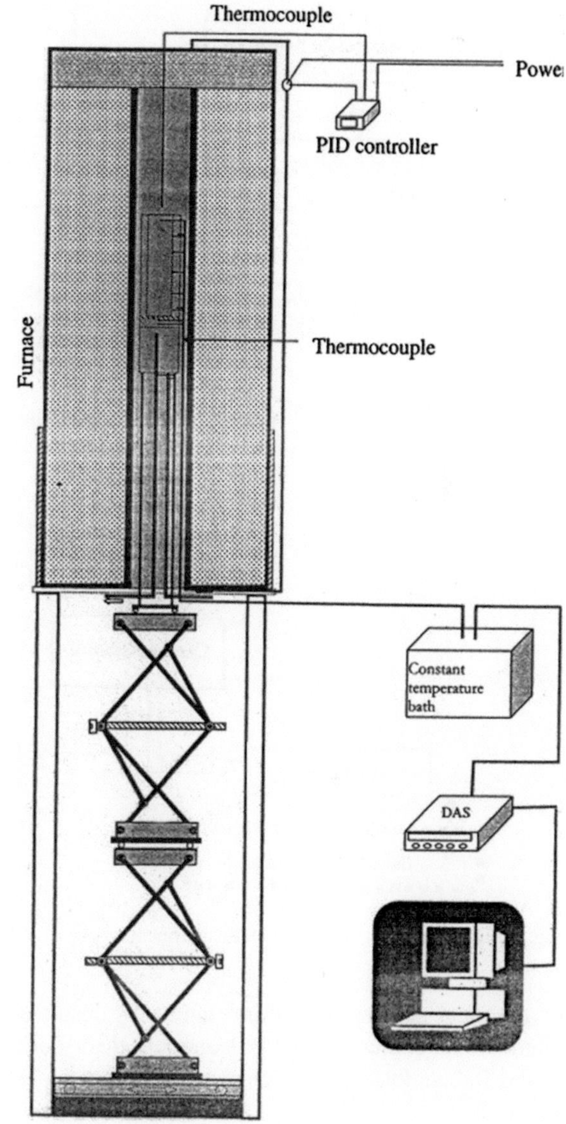

At this juncture, I wish to state that during my interactions with industry, I had opportunities to meet and have discussions with many professionals who were very knowledgeable with plant practices. It was a great education tome.

However, importance of metallurgical and engineering sciences in design, development, innovation and process control has not yet been recognized by the steel-making community. Some of us in the academic institutes became concerned with this and thought that something had to be done about it for the future viability

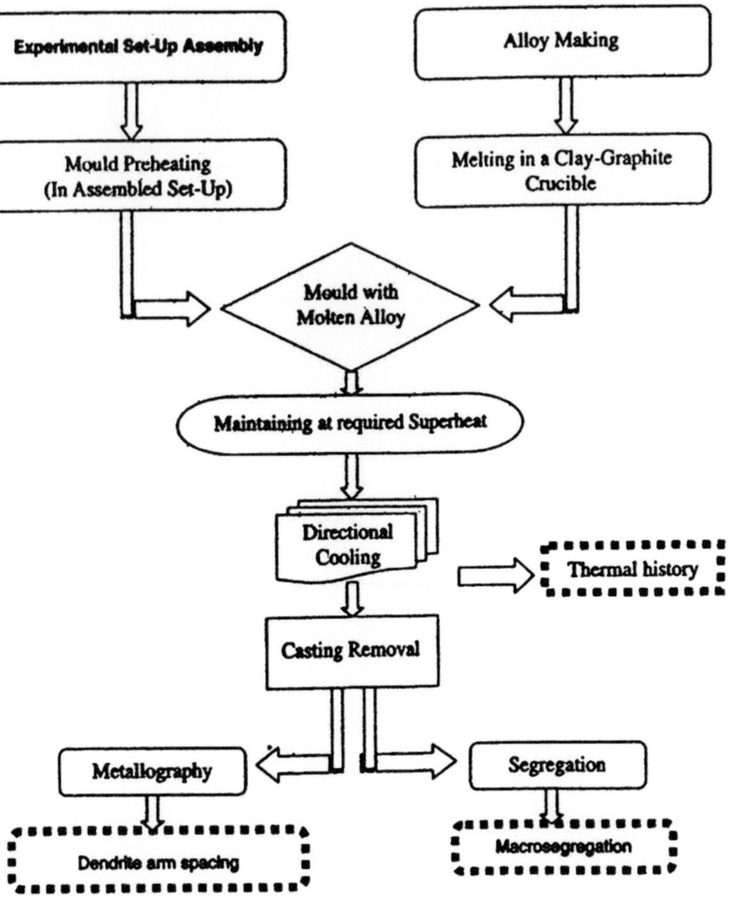

Fig. 3 Main steps for the individual experiments

of the steel sector. Our efforts ultimately resulted into organizing a successful international conference on "Advances in theory of iron making and steel making" at the Indian Institute of Science, Bangalore in December 2009. I acted as the Chairman.

On Science, Religion, and Society

For World peace and healthy environment on the Earth, we need awareness of current ethical challenges. The essential role of ethics is to promote social consciousness for our harmonious existence with the society. The term SOCIETY is

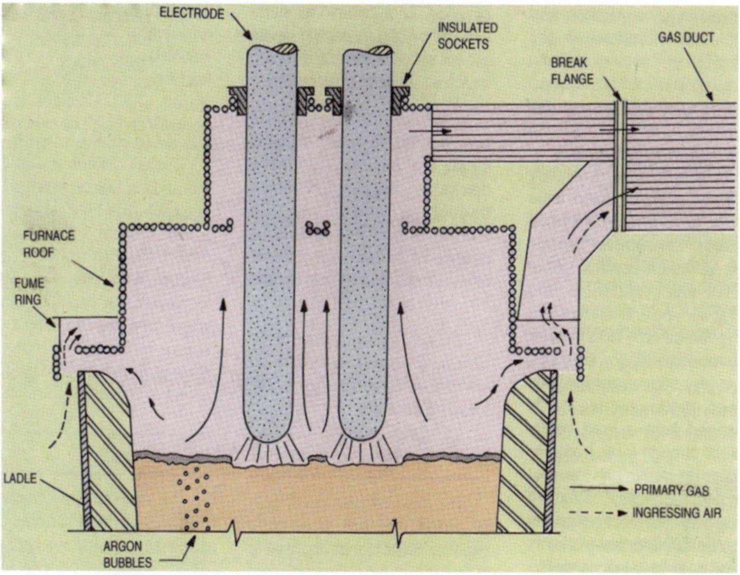

Fig. 4 Sketch of a ladle furnace

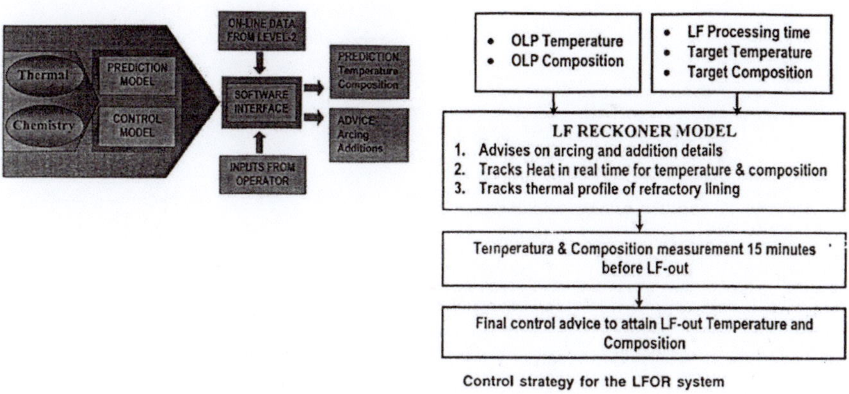

Fig. 5 Schematics of the control strategy

being increasingly replaced by a more broad-based term ENVIRONMENT, which includes not only human beings, but also our surroundings on the Earth as well. An important cultural issue, which is very relevant in general and in this connection, is harmony between science and religion.

I have been concerned with this and have done some reading and writing, such as my paper on "Peace and Universal Religion" [6].

References

1. Ghosh A, Chatterjee A. Iron making and steel making—theory and practice. New Delhi: PHI Learning (P) Ltd.; 2008.
2. Singh AK, Basu B, Ghosh A. Experimental investigation of macro segregation during vertical solidification of lead-tin alloys. Trans Ind Inst Metals. 2012a;65:85–96.
3. Singh AK, Basu B, Ghosh A. Numerical simulation of macro segregation during vertical solidification. Trans Ind Inst Metals. 2012b;65:179–90.
4. Nath NK, Mondal K, Singh AK, Basu B, Bhanu C, Kumar S, Ghosh A. Ladle furnace on-line reckoner for prediction of temperature and composition. Ironmaking Steelmaking. 2006;33:140–50.
5. Ghosh A. Basic research on high temperature metallurgical processes. Trans Ind Inst Metals. 2002;55:579–89.
6. Ghosh A. Peace and Universal Religion. Bulletin of the Ramakrishna Mission Institute of Culture, Kolkata, Sept 2013, pp. 413–17.

From 'Nowhere to Somewhere'

D.N. Singh

Whenever there is a cyclone, the tallest of the trees get uprooted first; never do the small grass, one should always be like the small grass… this philosophy was implanted in me by my parents, (Late) Sri Dharmendra N. Singh ji and (Late) Smt. Lakshmi R. Singh ji. I try following this philosophy in my day-to-day life. I would like to share a bit of what I have learnt from my life till date, and you may call it a journey from 'nowhere to somewhere.'

In my early life, I was very well groomed by my mother, an educationist who always motivated me 'to do something new' and kept on doing so until she breathed her last in 2010. Also, I used to imitate my maternal uncle, Professor Ram Dayal ji, Banaras Hindu University, B.H.U., a renowned plant pathologist. His laboratory at B.H.U. was my destined place whenever I would visit Varanasi during summer breaks. I remember using the ovens in his laboratory to prepare 'pancakes' and distributing it amongst his students and co-workers. This might have got imbibed in me to have later motivated me to develop my own laboratory and a research group at the Indian Institute of Technology Bombay (IIT Bombay).

Mentored by Professor Madhav, I joined Civil Engineering B.Tech. program at IIT Kanpur in 1982. I took up several jobs as a B.Tech. graduate, but soon I realized the redundancy of the work. The drudgery eventually led me to the doors of higher studies and Professor Yudhbir and Professor Basudhar introduced me to this amazing world of research, where my thinking knew no bounds, and I could develop an urge to explore the 'remaining 99 % of the Nature,' which is unknown. Interacting with such great gurus made me determined to reach somewhere, by choosing a path that appealed to my sense of productivity the most. As a child, I dreamt to be sailing in the air as a pilot; the vision was always to scale the infinite, which I am still doing. I experience similar thrill each morning when I am served 'breakfast of ideas' by my research group. I believe I am riding all their ideas right up to the ninth cloud. These informal interactions help me and my research scholars to take up several challenging and trivial projects, and help us in 'growing

D.N. Singh (✉)
Department of Civil Engineering, Indian Institute of Technology, Powai, Mumbai 400 076, India
e-mail: dns@civil.iitb.ac.in

P. Ghosh and B. Raj (eds.), *The Mind of an Engineer*,
DOI 10.1007/978-981-10-0119-2_19

together.' This also reminds me of the education that I received at Colvin Taluqdar's College, Lucknow. Incidentally, Mrs. Kaur, Principal, City Montessori School, Rajendra Nagar, Lucknow, was instrumental in inculcating in me an unimaginable capacity to 'concentrate on our work' for several hours, without thinking of 'what is happening elsewhere.'

I have believed that the 'new-generation researchers' are mature and hold a staunch individuality, which should be respected. It is an entirely different era, without the conventional inhibitions and tameness. We should only observc them, interact with them regularly and intensely, and create role models; alternative to the hyped media-clad glamour achievers, reeling in the young minds due to the social media. Internet has brought everything at their fingertips, which has its own pros and cons. There appears to be an exponential decrease in the tendency to figure out things on their own in the generation of 'smart phones' with the 'less smart researchers.' Any scientific technological advancement could be a boon or a bane subjectively, depending on the amount of judiciousness shown by the user. I frequently question my students, 'Don't you always choose to email your professor instead of directly meeting him/her?' The fading bond between a teacher and students is one of my major concerns, for which I consider both the sides as equally responsible. It reminds me of my Ph.D. days, when I would generally start discussion with my supervisor close to the dinner/lunch time, at his apartment, fully aware of the fact that 'food will be served very soon!' by his mathematician wife. Having played both the student and teacher, I hold a strong opinion that such a bond fetches remarkable benefits to both the parties. As a student, I enjoyed the companionship of my teachers, and the chemistry, I feel, has social benefit only as a by-product. The major product of the irreversible reaction is the magnitude of achievements.

Professor Chandra S. Desai, USA, motivated me to hold the international conference of the IACMAG, the International Association for Computer Methods and Advances in Geomechanics, at Goa, during 2008. This turned out to be a huge success. I came to the international front for the first time, due to this event; I am so glad that I did it. Delegates from about 70 countries participated, as a big and the most memorable event in their lives. Professor Shamsher Prakash is another name in the list of the influential people in my life. I won 2003 SP Research award instituted by Shamsher Prakash Foundation, MO, USA. A few years later when he visited my laboratory, he was impressed and satisfied that the award fell in the right hands. Professor Sukhatme, former director, IIT Bombay, also has a very special place in my mind and heart. Without his help and support, I could have never been able to venture in the geotechnics of atomic waste disposal, a multi-disciplinary topic dealing with a substantial amount of nuclear chemistry. This interaction brought me close to the officials at Atomic Energy Regulatory Board (AERB), Bhabha Atomic Research Centre (BARC), Indira Gandhi Centre for Atomic Research (IGCAR), and Board of Research in Nuclear Science (BRNS). The issues discussed by these officials were cardinal in making me realize that the concepts of 'Classical Geotechnology' will not be the panacea for addressing their objectives. I feel that this was the time when I started realizing, subconsciously, the importance of laying down the foundations of the theme 'Environmental Geomechanics.'

However, in an era when even the conventional subject was not much appreciated, convincing my students and colleagues about the ideology of this new theme was impeded due to the dogmatic platitudes such as 'it is all about mixing soils with chemicals!', 'We are civil engineers, hence why to learn science?', 'It lacks mathematics!', etc., which prevailed for a long time. It was due to persistent efforts of my research group that we were able to start things from the scratch and evolve this new concept in an otherwise mundane subject. Needless to say, I am thankful to my critics (both in the academia and industry), who maintained faith in my philosophies and some 'excellent' outcomes could be achieved. It proved once again that constraints give birth to creativity.

'Take decisions quickly, your opinion is my opinion,' I always advice my research group, which is my extended family and without whom I would have been nowhere. I have always believed that 'in the academic profession, scholars become a part of your family.' This also helps me in establishing a bond of mutual faith that leads to understanding and respect for each other. I have always felt that I would have been nowhere without 'the mentorship' my students have provided me, and I am extremely thankful to them for the challenges they posed to me, which has given me a lead in my research. I have been fortunate enough to encounter wonderful students; frankly I am yet to come across a 'bad' student. I still remember the first class of Engineering Mechanics at IIT Bombay and the students prompting that I should be doing something new and not merely teaching. That was a great kick start to something innovative which looks like my Environmental Geotechnology Laboratory (www.civil.iitb.ac.in/~dns/ENVGEO/env.html) today! Research cannot be done if one is very reclusive; one needs to cooperate. For me, this laboratory is much beyond the four walls of a RCC structure. Unfortunately, some times the laboratories become more of showrooms 'showcasing variety of instruments and their keepers,' wherein the human touch is missing; if someone happens to break an instrument while working, (s)he is almost certain to be given a harsh word or castigation. I would in fact love such a happening, because it shows that the work is going on! All of us are fundamentally humans, and hence, the requirements of a human behavior must be fulfilled first, and then comes the profession.

I feel I am still a student, being paid for learning new things every day and will strive hard to continue to be, so throughout my life, curiosity has always driven me right from nowhere to somewhere, and I am still exploring new frontiers of my subject by interacting with the renowned researchers from electrical, nano-electronics, chemical, metallurgical, and biotechnology backgrounds. I am not 'scared' of learning from them simply because I believe that one cannot be an expert in every field. I feel my students are my best teachers. I try to catch ideas from the questions thrown at me by them in the classrooms. 'Ideas multiply very soon,' still I find some researchers do not open up. It is important to accept that one cannot know 100 % of everything. I feel that each and every one has taught me something new, for me everybody is a source of knowledge. I try not to antagonize anyone, because for me individuals are not so important, but the personalities are. I may not be meeting the same individual again, but I will definitely come across a similar personality; one should be aggressive at his/her work, for achieving something, but not haughty.

My failures have been various personal constraints; I always wanted to go abroad for higher education, but had to stay back. I remember Dr. A.P.J. Abdul Kalam, the then Principal Scientific Advisor, Govt. of India, appreciating my completely Indian education, during a prolonged interaction with him; in a flight from Agartala to New Delhi. I still believe that had it happened, I would have been a different personality altogether; this is where one has to believe in destiny, take it as a blessing in disguise and turn it into a sequence of opportunities which one could conquer to the best of one's abilities. Be sure that something better than the one you expected will come your way. If one truly deserves, things will happen to him/her, either sooner or later, one just needs to make oneself fit for that. I have never planned for anything in my life except for my 'present.' God has been kind enough to guide me through the best of the paths. To be successful and satisfied in the long run, one must have absolute faith in something external to one's self; it could be as lifeless as a stone to as debatable as God. You have to be religious, but there is no need to go to the temples; the way 'rich people worship.' Indeed, my religion largely lies in building lives of several research scholars for whom their faith rests in me, sitting in the temple of technology—IIT Bombay. It is difficult to estimate how much this institute has given to me, and if I could give back even a fraction, I would be privileged. If IIT Kanpur was my 'Janma Bhoomi,' IIT Bombay became my 'Karma Bhoomi.' I feel that the extent and rigor of industry interaction, and the challenges associated within the 'soil-rock-sea-waste' interaction, which Mumbai provides, is of its own class. Needless to say, I learnt a lot from this city and I owe a lot to it.

I have learnt from my early part of the career that one can sustain even in hostile environment, as long as the person has self-belief. While working in a group, it can be an arduous task to avoid conflicts. Such conflicts are inevitable considering psychologically every 'personality' is unique. But, 'Swantah Sukhaya' is one principle that guides me through. One should give an ear to every surrounding advice but do what pleases the 'self'; since both the physical and mental appetites are diverse for different personalities.

Today, several industries are utilizing outcome of my research, which never was in the realm of the classical Geotechnical Engineering, say about 10 years back. This has broken the common myth that 'academicians and researchers have isolated themselves from the societal needs and live in their own world.' In my opinion, consulting is very important; unless you learn from the field problems, what will you teach in the class? And if you do not teach contemporary things, intelligent minds will not receive them. This also helps in generating the resources, which could help in developing and maintaining day-to-day requirements of 'conducting research and exploring novel ideas.' I could guide several Ph.D. thesis, as I always had industry sponsorship. Industry interaction gives platform both to 'apply your knowledge' and 'generate resources.' This way one can do the research which could help the society, and it is mainly such research that is the need of the hour and would open up further avenues for exploration.

Several colleagues have asked me, 'how do you manage a large research group?' It is fairly simple to manage them since I do not have to! I am one of them and we all manage each other. I make it a point that everyone is satisfied with me, and hence, I enjoy full support whenever I need. This is the sole reason why 'otherwise silent creature like me' could manage to bring more than 105 editorial board members, when he launched an international journal Environmental Geotechnics, ICE (Institution of Civil Engineers) Publishing, London, UK, (www.editorialmanager.com/envgeo). The best researchers from all over the world joined this endeavor happily.

I try to remain 'self-motivated' and 'compare with myself,' I have never ending expectations from myself, and I try to gratify them through hard work and absolute dedication. What keeps me on my toes is that India requires a lot of good human beings and professionals, and I am constantly working toward shaping future, and career, of my students. This country is known due to its 'individuals' and not due to 'the system'; if we want to make any change, it has to be an 'individual effort.'

I am grateful to God that both my 'family' and the 'extended family' are satisfied with me, and I have stirred them together in appropriate proportions such that I get adequate flavors of both in my life. My sister, Dr. Neena Saxena, reminds me of my favorite game during my early child hood that was making a small weighing balance out of the empty cans of Cherry Blossom shoe polish. I have always believed that family support stabilizes you in the worst of the situations, but you should not isolate them. Your family members should be completely aware of what you do. In fact, I make them a part of what I am doing. We hold regular get-togethers at home or outside, Ritu, my dear wife; an artist par excellence knows even favorite dishes of every student and Yashi, my daughter, cherishes these moments well. We wish them on their birthdays and accomplishments; we are all like a big family that travels together for various professional and social activities. I believe that with transparency in ones work, unnecessary misunderstandings can be avoided, which in turn fosters creative thinking and growth.

I would like to conclude with the adage 'remain inquisitive throughout life and have a very different perspective toward nature where most of the things evolve.'

My Life in an Encouraging Educational and Professional Environment

Ganti Prasada Rao

Early Life

I was born in the rainy morning of August 25, 1942, in a small government hospital in Seethanagaram, a little peaceful town, in the present-day state of Andhra Pradesh, at a time when World War II was in full swing. I spent my childhood in our ancestral village Kaluvarayi, where the great scholar Kavyakantha Ganapti Muni, a distant relative of ours and guru to my grandfather, lived many years before. For the first six years, I was a lone child at home until my other siblings began to arrive into our family. My main companions during this period were my parents, especially my father Sri Ganti Venkatappadu who used to carry me wherever he went in the village and into the fields around. He was well versed in Smartham, Vedas, and Astrology and also had usual schooling up to the 9th standard. He taught me many things orally by recitation during my childhood, including 20×20 multiplication tables even before I entered school. These were learnt like songs, and mathematics became a very simple subject for me with this playful preparation before schooling. I recommend to parents to adopt this method of playful teaching of multiplication tables to their kids at home prior to schooling to render their mathematical life easy and enjoyable. There was no nursery education for children in those days, but my time with my father was like nursery experience. Writing was on the surface of fine sand, and calculations were demonstrated with the help tamarind seeds. For my attention and good behavior, I was rewarded with some seasonal fruits and sweets at the end of these sessions.

G.P. Rao (✉)
UNESCO-EOLSS Joint Committee, 2623, Abu Dhabi, UAE
e-mail: gantirao@emirates.net.ae

P. Ghosh and B. Raj (eds.), *The Mind of an Engineer*,
DOI 10.1007/978-981-10-0119-2_20

Schools in Parvatipuram and Bobbili

By the time I began schooling at the time of India's independence in 1947 in Parvatipuram, everything, especially mathematics, became so easy that I could respond even to questions of higher classes, by virtue of my preschool learning. My knowledge of the weekdays, months, names of the years in the cycle of 60 years, seasons, etc. became an added advantage. No wonder, I was called a prodigy and all teachers loved me and appointed me as class monitor whose responsibility included ringing the bell to notify the start and end of the sessions. With the love and encouragement of all teachers and fellow students, this made my school life very enjoyable.

On clear new moon nights as there was no glare of electric lights, my father used to show me the 27 constellations which mark calibration of the sky into 12 sectors of the zodiac. He trained me in certain skills such as weaving baskets with palm leaves, and binding and covering books. We used to work together and make glue with cooked rice flour paste adding copper sulfate to it as a preservative. We also used to collect glue from thorny acacia-type trees (Telugu Tumma). Amber reminds of those clear glue lumps. He kept a wooden roller to draw parallel lines, single or double, by rolling it on plain paper. His attendance registers for his village school were very impressive. He carried pens with black, blue, and red inks and kept a stock of the three inks along with pencil, eraser, and sharpener in his kit, and I follow this practice even today. He advised me to keep a template sheet with bold parallel lines and margin below any plain writing page to see through and be guided while writing. I follow this method even today. His handwriting was beautiful, the lines like garlands of round pearls of uniform size, and he always stressed the importance of good handwriting. On all his books, he wrote a note of appeal to others: "This book belongs to Ganti Venkatappadu. Please do not take it without informing me. In case you take it for a good reason, please return it to me as soon as you finish your work with it. Sincerely, Ganti Venkatappadu." I was also closely associated with my mother from whom I learnt cooking which helped me in many ways in life, and it became one of my hobbies. I went from Bobbili to Doddavari School for classes up to 5 and Samsthanam Branch High School for classes 6 and 7. At the age of 9, I began writing poems in Telugu satisfying classical meter properties—chandassu, well before I began learning chandassu. Poems came spontaneously out of my pen in an organized flow.

Secondary School at Sompeta

As my father obtained a teaching position in a village school at Poornasasanam, about 15 km from Sompeta in the then Srikakulam district of Andhra Pradesh, our family moved to Sompeta in June 1953 for my schooling and my father lived alone in Poornasasanam, visiting us in Sompeta during weekends and holidays.

My teachers in drawing and Hindi also took special interest in me. In the holidays, I was specially taught and trained by them. I became good at free handdrawing, and in 1955, in less than a year, I passed three Hindi examinations up to Rashtrabhasha, without waiting to see Hindi made as a compulsory language. While I was in Class 9, we had a very young and enterprising science teacher Sri Ronanki Krishnamurthy. He was the son of Sri Ronanki Appalaswamy, a famous professor of English and foreign languages at Maharajah's College, Vizianagaram, where I studied PUC later and became close to him. He was a friend philosopher and guide to many students of the college. Some like Arudra (Bhagavathula Sivasankara Sastry) became famous literary figures in Telugu. Krishnamurthy gathered some enthusiastic students to revive the science laboratory of the school which had a good collection of apparatus collecting dust in the shelves due to lack of regular use. A group of students including me agreed to work for a science exhibition, thereby making the apparatus usable for active experimental and demonstration work. For three months, we worked in our spare time cleaning the apparatus, collecting botanical samples, and preparing posters for presentation at the exhibition. I contributed to this effort by making sketches on A3-size drawing sheets. For my part of presentation, I pulled out a very beautiful mechanical model of the solar system, cleaned and lubricated its gear mechanism, and demonstrated in a dark room at the two-day exhibition attracting big crowds in batches. I was very successful in the Board High School at Sompeta and attribute my success, which is still recalled in those areas, to the very encouraging atmosphere of teachers and fellow students. I always topped my class, and in the school final examination of March 1957, I topped all the schools in the district in composite mathematics and received the District Collector's Medal.

At Maharajah's College, Vizianagaram

Any one who knew my performance in the high school would assume that I would routinely proceed to study higher. However, the thought of higher education posed a challenge due to my father's meager earnings. Many spoke encouragingly, but some said that it would be practical and sensible to take up some clerical job and start supporting the family. So I joined a typewriting center to practice typewriting and learn shorthand in preparation to be stenographer. The fee was Rs. 4 per month, and I had 1 h daily for three days a week. This went on for three months, and one day, the turning point in my life came. A friend of my family visited us for a few hours in August 1957 and strongly encouraged my parents to take me to Vizianagaram and admit me in Maharajah's College in the newly started preuniversity course (PUC) indicating the possibility of free boarding in the students' chowltry (free boarding house).

The next day, with a limited amount of available money to meet the fees and other minor expenses, my father took me to Vizianagaram where already the classes were on since the middle of June.

Despite this delay and closure of admissions long before, Principal Sri Vasantarao Venkatarao, a renowned teacher in physics, pacifist, and poet, admitted me in the stream of mathematics, physical sciences, commercial knowledge, and accountancy (MPCA). Sri K.V. Ramanamurthy, Head of the Chemistry Department, who was sitting with the Principal at the time of our first meeting after three days of waiting for the opportunity, also took keen interest in me and gave much encouragement. Unfortunately, free boarding was not possible in the said chowltry, but I and my father went round the town and made arrangements for me to eat in weekly rounds as a guest in seven houses including those of the Principal and the Head of the Chemistry Department. This possibility of free eating known as vaaraalu has been a noble and admirable cultural element in our society. This weekly cycle involved walking around the town both for lunch and for dinner. I became a pet student in the class. The chemistry lecturer Sri C. Madhava Rao once told me that he does not feel enthusiastic to teach the class if I was not present—a comment which I always cherish. In the extracurricular activities, I won the best English Short Story award and best spontaneous Telugu poetry award at the level of the entire college. It was during this period Principal Venkatarao used to encourage my poetry and recite his own poetic works. When eating in one house was stopped, the Principal accommodated me for feeding in his house for that day in the week. I passed the PUC in first class despite all the disturbances and threats, including the dreadful smallpox epidemic, and delayed admission. Now, Maharajah's College added my name to the list of its notable alumni, despite my short association of less than a year. Principal Sri Vasantarao Venkatarao, whom I used to visit regularly thereafter and saw for the last time in 1978, passed away many years ago, but his family fondly remembers me even today. He also encouraged me to go for engineering to ensure a good job and thereby a comfortable life. I applied for admission in the Government Engineering College, Kakinada (GECK), and was selected.

Government Engineering College Kakinada (GECK)

Following WWII, this institution started in the army barracks in the outskirts of Kakinada town as "College of Engineering Vizagapatam at Kakinada" and over the decades evolved into the present-day Jawaharlal Nehru Technological University Kakinada (JNTUK). I studied at GECK from 1958 to 1963 where I had the support of many teachers such as Principals V.V.L. Rao and P.N. Damodaram, and Professors. G. Subrahmanyam, L.G.K. Murthy, C. Sudarsana Rao, G. Mallikarjuna Rao, and C. Tarunayya and received the BE degree in electrical engineering from Andhra University, in 1963 in first class and high honors. Here, I was the Founder Secretary of the Andhra Vignana Samithi, a cultural society of the student body and organized cultural events inviting great Telugu scholars such as Kavi Samrat Viswanatha Sathyanarayana and the mimicry master Nerella Venumadhav. In 1961,

I wrote and directed a short play on the occasion of the birth centenary celebrations of Sri Mokshagundam Visveswaraya, the legendary engineer. Under the auspices of the Electrical Engineering Society, I won best paper award in my final year and bagged many prizes in literary and art competitions. Before the publication of the final results of my BE degree examinations, on the May 16, 1963, I was married to Meenakshi who has also been supportive in my life and in all my academic pursuits. In August 1963, I joined Indian Institute of Technology Kharagpur (IITKGP) for M.Tech. in control systems engineering in the Department of Electrical Engineering. I provided active help in connection with the seminar on electric traction in 1964 organized by the Indian Railways in preparation to its massive electrification program. Several years later, at IITKGP, some among those who came from GECK for postgraduate studies in electrical engineering told me how my notes and my workbook of problems in electrical engineering by Parker Smith, which I left behind, benefited several later batches.

Indian Institute of Technology Kharagpur

I joined the two-year M.Tech. program in control systems engineering which was an emerging discipline. In January 1964, after President Kennedy's assassination, there was an art exhibition in our J.C. Bose Hall of Residence. One wing displayed my works—sketches of Kennedy and Gandhi in pencil and charcoal together and clay models including the busts of Kennedy and Jawaharlal Nehru.

As part of my M.Tech. project work, I built a mechanical contour integrator as a device of mechanizing the computation of the Bromwich contour integral that is related to Fourier series evaluation and conversions between time and frequency domains in Laplace transform. This involved small mechanical components such as guides and gears machined by myself, and this work was later published in the International Journal of Control in 1970. Later an electrical contour integrator based on potentiometer principle was built by my students at PSGCT for their project work, and this too was published in the same journal in 1971. I also participated in a group project in partial fulfillment of the requirements for the award of M.Tech. degree. This was on position control of a massive object. All five students of our class worked hard to build a base on which a huge flywheel was mounted and a gearbox was attached to it to be driven by a motor. I myself did machining of these parts and did welding for gear holders. IITKGP Director Dr. S.R. Sengupta, visiting our departmental workshop while I was welding, mistook me to be a welding mechanic, but Professor K.B. Menon, the Head of our department, whispered into his years and introduced me as a postgraduate student. I still cherish the memory of the consequent facial expression of Dr. Sengupta.

We used thyratron control unit to drive a dc motor and it was a success. For over three decades, this flywheel–gearbox bed was in the control systems laboratory at IITKGP. Different motors and control units were used with it by the subsequent batches of students in their project works. This work strongly united all five of us in the class, and we cherish the memory of our effort on this unit. I received the M.Tech. (control systems engineering) in 1965.

Ph.D. Research

I joined the Department of Electrical Engineering, IIT Kharagpur, as a research scholar and started working for Ph.D. from August 1965. I was asked by my advisor Professor N. Kesavamurthy, to study time-varying parameter systems for my Ph.D. degree. In connection with this, I was advised to study Bethenod's phenomenon of built up and sustained oscillations of pendulum with a ferromagnetic bob suspended above a coil excited by alternating supply of 50 Hz. This phenomenon was observed in the 1930s by the French scientist Bethenod and later explained by N. Minorsky in the framework of parametric excitation using Mathieu equation. I conducted elaborate experiments and found some basic mathematical and physical anomalies in Minorsky's Mathieu equation approach and presented a plausible theory of energy transfer supported by excellent experimental results. This work was published in IEEE Transactions after a few years of invalid arguments by reviewers in favor of Minorsky's theory. I was determined to clear the matter by writing to great experts in magnetism. I sent my work to Professor S. Chandrasekhar whose book on magnetohydrodynamics gave me hope that he would clear some of the very fundamental issues I raised in the context of Bethenod's phenomenon. But Chandrasekhar admitted that he was not competent to judge and assert but hinted the possibility of a good point in my arguments. I lost this letter in the long period of time. Finally, as a result of my persistent effort, the referees saw the truth in my work and asked the editor of IEEE Transactions in Circuit Theory "to encourage the author to lock horns with Minorsky." By then, Minorsky was in his 90s and the editor informed me about his demise in 1970. The paper was published in the March 1972 issue of the journal. This and additional work on time-varying parameter systems fetched me Ph.D. in electrical engineering from IITKGP in 1970. Professors P.K. Rajagopalan and K. Venkataratnam also were often available to review my work and gave helpful suggestions.

During the period of research for Ph.D., I used to repair the electronic equipment in the laboratories and the technicians around used to watch me from a distance since they were not knowledgeable in electronics. I voluntarily coached them at my residence in electrical engineering and electronics including digital devices and systems and made them successfully enhance their qualifications by passing City & Guilds UK examinations, and thereby, they became eligible for higher positions.

They all loved me and became close family friends. I made the electrolytic tank in the control systems laboratory free from the problems of maintenance by etching a grid on the glass by a process of wax coating and application of hydrofluoric acid. This was used in potential analogy problems and simulations of transfer functions of linear dynamic systems; earlier, a graph paper was sandwiched between two transparent plates. When immersed in the tank, the electrolyte used to seep in and make a mess of the apparatus. The analog computer laboratory was maintained by me for a few years. I also used to learn and voluntarily teach computer programming in FORTRAN in my spare time to the interested people, following the arrival of IBM 1620 in the early 1960s.

Professional Life

In 1969 for six months after completing my Ph.D., I remained unemployed and without any source of income. My friends at IITKGP such as K. Venkataratnam, T.R. Padmanbhan, and A.S.R. Murthy stood by my side and financially helped me in travels to attend interviews. Several offers came at once, but I accepted the offer of assistant professor in the Department of Electrical Engineering, PSG College of Technology, Coimbatore (PSGCT), and took the position on the July 4, 1969. I taught several courses in control such as linear systems, nonlinear systems, sampled data systems, and computer programming in FORTRAN. I also took classes in the summer school on control systems and fluidics together with an American expert. I also attended the summer school in systems engineering in this period and actively interacted in the interactive sessions contributing to thoughts on systems science.

During 1969–1971, Professor G.R. Damodaran (GRD) took me into his close circles and gave me the task of compiling and summarizing the results of a National Survey in connection with his Committee's work on Reorganization of Polytechnic Education in India. The method independently developed and used by me to process the linguistic response in the survey reflected some shades of fuzzy sets which were introduced by Professor L.A Zadeh in 1965. I now have the honor of being a member in some prestigious International Advisory Bodies with Prof. Zadeh. The ideas that arose in connection with the Damodaran Committee were published in *International Journal of Systems Science*, in 1974. I participated in the national seminar organized by Govt. of India at IIT Madras on the subject of technical education and also presented three papers with G.R. Damodaran. I delivered the second lecture on January 11, 2014, at PSGCT in the distinguished lecture series organized in connection with GRD Centenary Celebrations; the series was inaugurated with the lecture by Dr. A.P.J Abdul Kalam, the former president of India.

I had the love and respect of all my students both of undergraduate and of postgraduate classes at PSGCT. There was such a grand farewell by hundreds of students and colleagues at Coimbatore railway station when I was leaving by train for IITKGP that I was overwhelmed with mixed feelings.

In October 1969, I moved to IITKGP as assistant professor of electrical engineering and began a long association with my alma mater. I became a professor and served on many administrative positions including the chairmanship of JEE in 1986 and chairmanship of the Central Library. In the latter position, I initiated steps to develop a digital division in the library under the direction of Professor

K.L. Chopra, the director. I declined other higher administrative positions in view of my academic interests.

While I was at IITKGP, all the students of my M.Sc. class at PGGCT came en masse and appeared for interview wishing to work under my supervision for Ph.D. This boosted my image in the department. Some have actually worked and received Ph.D. degrees under my supervision and later occupied distinguished positions in India. Dr. L. Sivakumar, my first doctoral student, retired as General Manager, Research at BHEL Hyderabad after 40 years of brilliant service. Dr. T. Srinivasan served BHU IT in various high positions and retired. Dr. K.R. Palanisamy continued work and submitted a thesis for DSc degree in 1996 to Madras University but unfortunately passed away without receiving the degree because of chronic asthma. Amit Patra and Siddhartha Mukhopadhyay, two toppers of the Electrical Engineering Department of IITKGP who took B.Tech., M.Tech., and Ph.D. degrees all from IITKGP, chose to work under my supervision despite many offers from the USA. They did excellent research in identification of continuous-time systems. They both are now in high positions at IITKGP, and I meet them every year. Dines Chandra Saha, one of my colleagues who did research on the Poisson Moment Functional Method of Identification of Continuous Time Systems, retired after serving as Head of the Department; I must mention here that his handwriting was one of the best I came across in my entire life. A.V.B. Subrahmanyam did brilliant research work and is now in a very high position in the IT sector in the USA. Many students who worked under my supervision at the B.Tech. and M.Tech. levels also rose to very high positions. For example, Amit Bhaya did brilliant research under the Professor Charles Desoer at UC Berkeley and now settled in Brazil as a professor. I feel fortunate to have all these distinguished students of mine still remaining in contact with me.

In the initial years of my research, my choice of continuous-time models in system identification posed a great challenge as I was pitted against the vast global activity with discrete time models, in the “go completely digital” spree of research. In this critical period, the faith my students had in my ideas and effort was a great support to push me forward. The importance of identification of continuous-time models is now well recognized, and applications in various fields such as gene expression mapping are taking place. At the University of Henri Poincare in France, the CONTSID toolbox compatible with MATLAB has been developed on the basis of these methods. There has been extensive following of my work in China and Japan that provided a galaxy of friends and collaborators in these countries.

I visited the USSR Academy of Sciences, Institute of Control Sciences, in 1991 as a one man special delegation from INSA where I interacted with some distinguished Soviet scientists. I was invited in 1990 to give advanced lectures in control to a cosmopolitan team of engineers in the Water and Electricity Department (WED), Govt. of Abu Dhabi, and to guide their research in modeling and simulation of desalination plants. I joined this team on invitation, as scientific advisor to the Directorate of Power and Desalination Plants, WED, and the International Foundation for Water Science and Technology (IFFWASAT) and worked in modeling, simulation, and control of desalination plants. In this period, the team developed mathematical models for large-scale multistage flash desalination plants and came up with adaptive control strategies for the control of these plants under varying conditions. IFFWASAT established the "Systems and Information Laboratory" in the Electrical Engineering Department at the IITKGP in recognition of my helpful role in its activities.

In 2004, Evan Morris, Department of Sociology, University of Regina, Canada, conducted a research project dealing with the development of the theory and applications of Walsh functions during the 1970s and 1980s, chose to interview me as a significant contributor to the field of Walsh functions. My work over these decades unifies all systems of piecewise constant orthogonal functions (PCBF) such as block pulse functions (BPF) (whose completeness was established through my work in 1978), Rademacher functions (RF), Haar functions (HF), and Walsh functions (WF) and finally captures all systems of orthogonal functions into a unified framework of general hybrid orthogonal functions (GHOF) with appropriate applications in the field of systems and control.

I authored/coauthored four books carrying mostly results of my work with my students and coedited one volume of collected works on identification of continuous-time systems. I have been associated with the development from inception of two major encyclopedias: Encyclopedia of Desalination and Water Resources (DESWARE) and Encyclopedia of Life Support Systems (EOLSS), developed under the auspices of the UNESCO. I had the honor of being a delegate of the UNESCO-EOLSS team at the Second World Summit on Sustainable Development in Johannesburg, South Africa in 2002.

I have been giving many plenary/keynote/invited lectures in India and abroad at national and international conferences. I delivered some distinguished lectures such as GRD Centenary Lecture-11 January 2014, PSGCT, Faraday Memorial Lecture organized by IEEE in Hyderabad September 2006, Boral memorial Lecture at BE College Howrah-1990, Karl Reinisch Memorial Lecture at Technische Universitaet Ilmenau, Germany-2007.

Over the last several years, I have been traveling widely upon invitations to give lectures on sustainability concepts and on the Indian heritage to the world of mathematics. My lectures have been on how the Hindu concept of zero and number systems prepared the mind-set of mathematicians worldwide and influenced the development of a host of modern mathematical concepts. These are supported by some ancient texts available in my father's collection. My father introduced me in my childhood to Nakshtranighantu book of synonyms for star constellations and Sankhyanighantu book of synonyms for numbers, from the classical texts of astrology whose importance I realized in this context. My lectures on this subject are the result of my own struggle to understand the hidden connections between many mathematical concepts and the number system. I was once invited by an organization to repeat the lecture which I gave years ago at the same place.

I had the good fortune to be associated with some very distinguished persons in my profession and they include Howard Rosenbrock of the University of Manchester Institute of Science and Technology; Heinz Unbehauen of Ruhr University Bochum, Germany; Achim Sydow of FIRST Berlin, Y.Z. Tzypkin of the

Institute of Control Sciences, Moscow; Liu Bao (son of the famous Chinese painter Liu Haisu) of Tianjin University; Wang Xing-Yu of East China University of Science and Technology, Shanghai; Z. Bubnicki of Technical University of Wroclaw, Poland; Andrew Sage of GMU, USA; and many others.

Conclusion

At home, my parents, Smt. Rajeswaramma and Sri Venkatappadu, provided all support necessary for my education in the face of all economic limitations. I established two scholarships in their memory-one for a female student and another for a male student at Vasavi College of Engineering, Hyderabad. My father is a unique example of robust personality; he had a comfortable childhood but had to struggle to support his family with the uncertain and unsteady income from the work based on his vedic/astrological education. In 1953, he obtained a job as a primary school teacher but had to work in remote areas having no access roads. He rode a bicycle skillfully on pedestrian and bullock cart tracks with considerable loads in all seasons to travel to these remote areas. He always carried a toolkit, a pump, and a drinking water bottle on his bicycle and had two bells on it from which his presence could be conspicuously felt. He was skilled in crafts such as weaving mats and baskets and any other forms of articles from palm reed for which he received a district-level award. He could adapt himself to hard conditions in life and face challenges successfully.

I feel especially proud to have been able to share a very significant time slot in the history of humanity. This time span is marked by events such as WWII at my birth, India's independence at the start of my schooling, and great developments in engineering and technology in the course of my personal development. I witnessed the development of electronics, radio engineering, and electronic computers both analog and digital, since the late 1950s. In fact, I taught these subjects in the early years and I am happy to see some of my past students in India and overseas in top positions in these fields which are now highly advanced, while in school I read about Albert Einstein as a great scientist and remember reading the news about his death in April 1955. I closely followed the news of the first humans landing on the moon. I have seen the reunification of Germany and the end of cold war. I visited the USSR a few days before its collapse in July 1991 which I could sense during my three-week stay in Moscow. I was advised by Yakov Tzypkin to cancel my visits to Tbilisi and Bishkek in view of the turmoil in the areas before the collapse of the USSR.

I believe that the inspiration, love, and encouragement I received from my teachers, colleagues, students, and collaborators worldwide in my entire life as a student and teacher and researcher gave me great strength in facing the many challenges in my professional life. I wrote an invited article in Telugu in an Internet magazine Koumudi a few years ago, gratefully remembering many of my teachers (http://www.koumudi.net/books/koumudi_mamamchimastaru.pdf, pp. 48–52).

Had it not been for the encouraging educational and professional environment enriched by my teachers, students, and colleagues, it would not have been possible for me to write this article. I request the readers to take what seems useful as a message from this writing and ignore all other statements of personal achievement which were made only to make some impact as is required for an article of this kind.

The Journey from a Village as a Farmer's Son to Fashion World as a Scientist in Leather

J. Raghava Rao

I was born in a small town Nellore in Andhra Pradesh on August 30, 1961. My father late Sri J. Subba Naidu was a farmer in a small village with no proper education. My mother Mrs. J. Leelavathi of 90 years is a housewife presently living with me. I have 4 sisters and 1 brother. The village I was born had no proper school. My father's passion to educate his children well made him send all his children to his relative's place with proper schools for seeking quality education. My journey started in my elder sister's house with the first school in Nellore, St. Joseph English Medium School. One of the teachers, Ms. Rita, inspired my young mind for performing well in academics. After class 7, I moved to Vijayawada along with my sister to get admitted in St. Joseph School for my secondary schooling. I was an average student at that time and used to get lots of help from my cousin Dr. Sailaja for subjects such as maths and science. While studying, I used to do lots of house work (which is helping me a lot after my marriage!). Learning to do things yourself always gives you confidence during your lifetime. I qualified in 10th standard and moved to junior intermediate after seeking admission in Andhra Loyola College. I was staying in a hostel for the first time. It is different experience altogether to stay in a hostel and studying in college. I learnt a lot in being independent, and the methodology of teaching and culture really inspired me. One of the maths teachers really inspired me to look for seeking engineering education. I came out successfully with good marks and sought admission in Anna University in Leather Technology. My uncle played an important role at this point of time in shaping me.

The college life at AC College of Technology and the curriculum at Central Leather Research Institute made us proud and fortunate to learn the latest developments in leather research. We had the privilege of being the students of the premier institutions. The academic culture coupled with institutional culture shaped us to think of taking up an academic career. The atmosphere plays an important role in shaping a student's career. I used to have many friends from north and enjoyed and learnt a lot from them. I used to do very well in mathematics that made me to

J.R. Rao (✉)
Chemical Laboratory, Central Leather Research Institute, Sardar Patel Road, Adyar, Chennai 600 020, India
e-mail: jrrao_clri@yahoo.co.in

P. Ghosh and B. Raj (eds.), *The Mind of an Engineer*,
DOI 10.1007/978-981-10-0119-2_21

help the faculty sometimes. This made me to be like a teacher to enjoy certain privileges.

The faculty at both AC College and CLRI made us to correlate the science to life. During my project time, I came in contact with a fair, short, dynamic, and creative personality who had just then returned from LEEDS, UK, my role model and mentor, Dr. T. Ramasami. He inspired me a lot, and I used to seek guidance from him for my project. This contact transformed into a father–son relationship that led to mentorship from him for shaping my future academic career. It is he who shaped me to this stage.

After completing my B.Tech., I joined for M.Tech. without fellowship as there were limited seats with fellowship. Ours was the first batch with 6 students for M. Tech. in leather. My father and brother played a vital role in supporting me to pursue higher studies without fellowship. Dr. D. Ramaswamy, people call him as Vadhiar, was the incharge for the PG courses. He is a very supportive teacher, and I stood first in his subject. I approached Dr. T. Ramasami for his guidance during my M.Tech. project as I was interested in continuing my doctoral program. He tested my seriousness to pursue research by making me wait the whole day. I was so serious to join with him, and I passed his test in commitment. This is the turning point in my academic career to join with him in the chemical laboratory known for its commitment, perfection and culture. I joined the chemical laboratory group and started learning from one and all in the laboratory. The support of Dr. B.U. Nair and Mr. BGS Prasad is enormous that made me to change from timid personality to bold personality. The perfection in analytical techniques was taught by Dr. M. Kanthimathi, whose value does not err even if the instrument errs.

My M.Tech. thesis comprises of both basic as well as applied work on the management of chromium in tanneries. The work was so designed that it bagged first two prizes in the poster session of Tanners Get-together held in 1987. This gave me so much of confidence to do better. The motivation and companionship is something we never get from any place as I got from my workplace, chemical laboratory of CLRI, which I regard as temple. The laboratory was functional round the clock and never used to be closed. It was like our house and used to sleep in the laboratory sometimes doing work late nights. Dr. T. Ramasami, called as Dr. TR, always spent time with the students at bench level. I was caught once from a distance when I was filling the distilled water bottle for keeping the lid down. That is the observation of the person on his students from distance. He always says and practices that the knowledge is for sharing with others and not for keeping it to oneself. I learnt a lot from him both in terms of science and humanity. Chemical laboratory does not practice hierarchy and treats the students as one among the family members. This approach is different from the university approach. The freedom to learn more and do differently is of high order.

After my M.Tech., I continued my Ph.D. under Dr. TR where I continued on the same topic of chrome management as the chemical laboratory was known as 'Chromium Lab' and Dr TR as 'Chromium TR' those days. The foundation laid by him transformed me to improve my knowledge and gain several recognitions in my scientific career. During my journey, this is the best part of my student life I have ever had in my career. Apart from doing research, we were put to several projects, which made us learn several things in doing good science. I still cannot forget that my supervisor travelled by train along with me, while I was a student to work for a consultancy project to M/s Golden Chemicals Ltd. to develop a technology for high exhaust chrome tanning salt. He taught us the discipline, commitment and sincerity to do things made us climb the ladder. He is instrumental in transforming me to a young dynamic scientist with several rewards. The knowledge which he imparted made us strong to face any difficult situation. He was a hard task master. One incidence I recall is my methodology examination which was so tough with grilling from my guide that the doctoral committee members came in support of me. We were four doing Ph.D. at that time: N.K. Chandrababu, B. Chandrasekaran, and V. Subramanian. Dr. Nair was instrumental in molding us to the tune of Dr. TR through his presence in supporting us to perform better. The members of the chemical laboratory are always supportive of me to stand taller at difficult times. I used to be so organized that Dr. TR used to use my workplace for any experiment. I used to maintain my workbook with perfection so as to easily identify any mistake that happens later.

On June 25, 1993, I joined CLRI as a scientist in chemical laboratory. I was put to several projects both national and international, and my commitment to do things with sincerity helped me to perform better. I became the teacher for the B.Tech. students in 1994, and I was very eager in sharing the knowledge through the fundamental principles learned from my mentor, Dr. T. Ramasami. The passion to learn and do research with commitment made me a well-organized researcher. Like in normal life, we move from childhood to adolescent to adulthood, now I moved from school kid to researcher learning things and now turning to a scientist to be a teacher. The teachings of Dr. TR fascinated me to practice teaching as well as scientific job with perfection.

I started working in the new areas such as changing the fundamentals of leather processing through crazy ideas, and it worked out well that we published in the high-rated ACS journals and brought laurels to the group as well as to CLRI in leather research. My mentor was impressed with my presentation at one of the Research Council meetings and sent an emotional letter to keep my tempo high and look further to perform better. His yardstick always changes to higher side to his group people. He admires his students when they perform well, and he always treats us as his children. This is the special feature of the chemical laboratory that it operates since then as a family without any hierarchy.

We gained global prominence in terms of developing new pathways for leather making. This resulted in several awards from Govt. of Tamilnadu, DBT, NRDC, Burhani-NEERI foundation, TANSA, WIPO, and the prestigious Indira Gandhi Paryavaran Purashkar Award from the Ministry of Environment and Forests. All this was possible because of hard work and having committed students in my career. Getting a committed students makes the supervisors life simple and easy to perform better. I was so fortunate that I was associated with several young and bright students who delivered in the same lines as I did to my mentor. Now, the role changed from being mentored to mentoring others. I still practice his principles in mentoring with a human touch.

I still remember and follow the teachings of my mentor, while he moved to take a responsible position as secretary, Department of Science and Technology, GoI, New Delhi. Never get disturbed or perturbed by looking down to others while moving up in an inclined platform in your career. Dr. A.B. Mandal, a person of simplicity in the Director's Chair, is very supportive for me in doing science of my choice. Dr. Nair's personal and professional support made me to perform well in new areas of leather science and technology.

At times of frustration in research career, Dr. Aruna used to say that hard work always pays you Rao, please be cool and go ahead. The company of young colleagues Dr. K.J. Sreeram and Dr. Nishad Fathima gave me enormous energy to take challenges with ease to perform better. There are times we outreached our fellow scientists in setting goals. Our group received several national and international recognitions for the scientific work done in the area of leather science and technology.

I was associated with several students at B.Tech./M.Tech./M.Sc./Ph.D. and diploma as a supervisor or faculty. The students were many starting from K.J. Sreeram, P. Thanikaivelan, B. Madhan, N.N. Fathima, R. Aravindhan, S. Saravanabhavan, A. Musa, Swarna V. Kanth, V. Punitha, Jayakumar, Ivy, and several others. Some of them after their doctoral program joined CLRI and some went to industry. I take the pride to say that our group generated 3 CSIR young scientists, 1 INSA young fellow, and 1 young ISCA award with quality research. Our group has published maximum number of papers in leather and non-leather journals globally. My success in this phase is due to the people associated with me, and the collective efforts of all made me successful. I was elected Fellow of Indian National Academy of Engineering (FNAE). We need to take people along with and make them grow independently as they 3 move forward. This concept helped me to attain popularity in attracting several students for their postgraduate and doctoral programs. We need to teach the students in a way that they understand you to be a part of the program with commitment and ownership.

I am happy to share that once I was asked to coordinate an international project in Ethiopia when it was faced with several challenges, and this opportunity helped me to display my leadership quality by making this project successful. Once an opportunity comes, you need to grab and perform well to get yourself recognized. An impact-making presentation was made to the Minister of Industries during the Steering Committee Meeting held in Ethiopia. I have been member of several international projects and visited several countries in collaboration with research activities. I have really enjoyed my research career, and it took time to understand the importance of doing research to meet the social demands. The leather research is always associated with societal needs through technology transfer for the needs and sustenance of the leather sector.

As a farmer's son, I have seen cattle in my village and took a career as a scientist in leather, which is made from the outer skin to meet the global needs. My wife Dr. K.V. Rajeswari, son Mr. J.V. Sai Bharat, and daughter Ms. J. Sravani played a very vital role in bearing the pain in supporting me to perform well to reach this position. At time of difficulty and frustration, they are always there to provide me an ambience to be cool and gain confidence. They encouraged me to reach taller in excelling science. Once after marriage, I remember my mentor telling my wife not to expect me to be back home by 6 p.m. Research career is always associated with peaks and dips, and you need to be careful in having patience to face the challenges. My father's and brother's dream to see me as a person of highest education from our village was fulfilled, and this I would like to carry forward in giving a supportive hand to the school in my village as well as to the needy students of our school to pursue higher education.

My success is due to my family members at one stage (childhood) and then due to my teachers and mentors (researcher) and finally due to my students, colleagues, and family (scientist). The passion to do science in meeting the societal needs to be looked at as a service rendered by a doctor to a patient in saving his life. Our life needs to be fulfilled to do service to the mankind through scientific inventions in pursuing science through a research career. Hard work always provides you happiness and joy. Practice what you preach and take people along with you and share the success and allow them to grow has been my mantra for success. The success of your student is your success. Research career makes you happy when you perform science with passion.

Forty Years of Academics and Administration: Experiences and Lessons

K.G. Ranga Raju

'Facebook' did not exist during most of my career, and 'face-to-face' contacts were what my generation yearned for—particularly with the greats—in the hope of enriching one's professional and personal experiences. I also belong to a generation which adhered to certain traditions of the time and was thus lot less informal than the generation of today. For instance, Prof. Ram. J. Garde with whom I worked very closely for over thirty years at the University of Roorkee (now I.I.T) was always 'Sir' to me and never 'Ram', because he had been my teacher and mentor! Considering that most of the readers of this article would at least be one generation younger than me, I, therefore, had doubts about the relevance of my experiences and the lessons I learnt from them to the younger generations, who are certainly different from my generation in many ways. Even though the incidents I talk about may be rather mundane, the value systems they led me to believe in are still probably important and relevant, and it is this belief which emboldens me to write this article.

My entire career of over 40 years was spent at the University of Roorkee (now I.I.T.) where I was involved in teaching, research, consultancy and administration. It is only natural, therefore, that the experiences I write about are all those from Roorkee. One of the greatest strengths of Roorkee was recognising right from the time of inception of the institution over 160 years ago that while academics are important, it is not everything. The emphasis was on the overall development of the personality of the student. The stress on sports, hobbies and cultural activities helped instil the qualities of leadership in the students, and it is no wonder that many of their alumni occupied important positions, particularly on the national scene. My own association with River water Disputes—domestic and international—in recent years has convinced me that you need many more attributes than a strong academic background to make a significant contribution in the resolution of these disputes.

The University of Roorkee always laid great stress on discipline and had a system of imposing fine—in cash or in terms of marks—for misdemeanours of students.

K.G. Ranga Raju (✉)
102, Rajalakshmi Apartments, 24th Main, JP Nagar (6th Phase),
Bangalore 560 078, India
e-mail: kgrangaraju@gmail.com

P. Ghosh and B. Raj (eds.), *The Mind of an Engineer*,
DOI 10.1007/978-981-10-0119-2_22

As Dean of Students Welfare, I also imposed such fines in a few cases. An incident I remember in this connection was of a student who was shattered when a fine was imposed on him and kept on pleading with me that it should be waived. I explained to him about his guilt and that the punishment was commensurate with his wrongful actions. I also advised him to put it all behind him and turn a new leaf. A few months later, he showed exemplary courage in coming to the rescue of some of the girl students of the campus who were being teased by some outsiders. I wrote him a personal letter sincerely appreciating his actions and thanking him on behalf of the university. When it was time for him to leave the university after graduation a few months later, he came to me to seek my blessings. About five years later, I was thrilled when this student came running to me when he sighted me at an airport, introduced himself, enquired after me and told me how well he was doing in his career. After all, if he bore any ill will against me, he need not have come to talk to me after he saw me. This incident convinced me that the student community, in general, is willing to accept punishment gracefully if they are convinced that the administrator in question is fair and applies the rules and regulations uniformly without bias or prejudice. By and large, they bear no rancour against those who punish them if they are certain that there was nothing personal and there was fairness overall.

One of the problems we faced at the University of Roorkee for a while was the delay in the submission of marks by some teachers, which often delayed the declaration of results. As Dean of Academic Affairs, I had evolved a system by which a rather stern letter would go to the defaulting teacher from my side after the due date of submission of marks, pointing out to him that he had not yet sent the marks and that he should do so immediately. It so happened once that such a letter wrongly went to one of our sincere and very senior teacher, because the office had included his name my mistake in the list of defaulters, and he was naturally furious. I called the concerned clerk to my office and told him that I would take the blame and apologise in writing to the teacher concerned without ever mentioning the clerical error which had occurred, but that he should take care to ensure that our office is not so embarrassed ever again in the future. This action gave the entire office the confidence that I would protect my staff in the event of unintended mistakes without any adverse entries on the file and, as a consequence, I was assured of their loyalty and commitment.

On another occasion, I felt that the assistant superintendent in my office ought to have shown a little more intelligence and initiative than he actually did and told him so. His response brought a smile to my lips: He said, ‘Sir, were I so competent, would I have remained an Assistant Superintendent and not become an officer?!’ I then realised how we often have expectations from individuals in a group, beyond their training and competence. By giving vent to our frustration at those expectations not being met, we are probably being unfair to the individual, undermining his confidence and in the process making him less efficient. Most of us would never have the luxury of picking each member of our team; we would be required to work with a set of people of varying intelligence and competence, and getting the best of the team needs an understanding and tolerance of the shortcomings of the individuals as much as an appreciation of their strengths.

Prof. R.J. Garde and I were once having discussions with another professor from a sister institute, who was visiting Roorkee. The visitor was talking at length about some research work done recently in another premier institute in the country. Having heard him, Prof. Garde merely said, 'I am aware of that work, we have a copy of that thesis in our library'. I was truly amazed at his modesty, because I knew that it was Prof. Garde, who had examined that Ph.D. thesis and then passed on the copy of the thesis to our library! This incident happened in the 1960s when not many doctoral theses were submitted in the country, and examining a Ph.D. thesis was certainly considered an honour and most of us would have found it hard to resist mentioning the fact of having examined the thesis! Prof. Garde was the acknowledged expert in the country in the area of fluvial hydraulics at that time, and he was probably trying to stress, in a subtle way, the value of modesty to me, his disciple.

Prof. Garde's doctoral supervisor, Prof. M.L. Albertson of the Colorado State University (USA), spent a couple of hours with Prof. Garde and me sometime in the 1960s. He had some interesting observations to make on how to ensure pleasant and harmonious relations amongst the members of a research team. He said that once the pecking order is established, meaning when the peers know from track record who the prime mover of the research project is, it is immaterial whether his name appears as the first author or as the third; he will always receive due recognition. Unfortunately, there are often heartburns about the sequencing of names in the academic world. Once your reputation is made, you could always opt to be the last of the authors even though you may feel that your contribution is no less than that of the other members of the team. Such magnanimity often spurs the younger members (who are looking for their own places in the sun) to greater heights and certainly fosters team work.

The last of my experiences I talk about is one concerning Prof. H.A. Einstein, a legend in the field of Fluvial Hydraulics and the son of the more famous Einstein of Relativity Theory. We were running a summer course on fluvial hydraulics in 1969 at Roorkee for which Prof. Einstein was one of the resource persons. He desired that we the faculty of Roorkee should go ahead and complete the planned material on our own and that his lectures should be deemed to be supplementary. Interestingly, he chose to sit in the lectures of all of us! I was required to deliver a lecture on Einstein's Bed Load Theory and that too with Einstein sitting in the audience! Wanting to be sure of not making any mistakes in my presentation, I went to Prof. Einstein the previous evening to understand from the great man himself about the physical significance of one of the correction factors introduced by him in his classic paper. Smilingly, he told me, 'Young man, that diagram is the most illogical thing in my paper. On the one hand, I had my Mathematical Model based on the Theory of Probability and, on the other, the experimental data of Gilbert; the two did not agree. I had to introduce that empirical correction factor you mention to get a match between the two!' I left his office wondering whether such honesty and candour comes from greatness or people rise to great heights because they dare to be frank and honest under all circumstances.

Creativity—A Personal Challenge

M.L. Munjal

Born to nearly illiterate parents on the wrong side of the border, and starting my life from a refugee camp, I rose to receive awards from three different prime ministers! Here, I share with the reader my life story with the hope that it will inspire the younger generation to challenge themselves at every stage and convert adversity into advantage, instead of cribbing about inadequacy of resources, facilities, and opportunities.

As a three-year-old child, living in a refugee camp in Hoshiarpur, I overheard my father thanking God. I asked him what he was thanking God for when we did not even know where the next meal would come from. He answered simply, 'Son, we are alive, isn't that sufficient?'. That bottom line has stayed with me all my life. I realized that God gave us life and the rest is in our hands; no excuses!

Luckily, the Government gave us some refugee grant, and the Sanatan Dharam Higher Secondary School waived all tuition fees. I never mind if I had to walk several kilometers daily to the school without proper shoes. One day, during morning prayers in the school, a guest lecturer told us that as physical strength can be increased by means of regular physical exercises, intelligence could be increased by intellectual exercise; nobody is born intelligent. This appealed to me instantly. I started challenging myself intellectually. Playing with numbers, I developed 'magic tables' to entertain myself as well as my classmates. I turned adversity into an advantage. Having no entertainment at home, I engaged myself with mathematical challenges that made me excel in not only mathematics but also everything logical.

At the end of the 7th grade, during vacation time, when challenged by my brother who would not let me play, I took up the 8th grade mathematics book, taught myself from the text, and solved all exercises in two weeks' time. This not only took my brother off my back (he could no longer stop me from playing) but also led to self-discovery. I realized that I did not need a teacher. With the resultant self-confidence, despite my spending evening hours to play cricket for my school, I secured 5th rank in the Punjab University in my Higher Secondary Examination.

M.L. Munjal (✉)
Department of Mechanical Engineering, Indian Institute of Science, Bangalore 560 012, India
e-mail: munjal@mecheng.iisc.ernet.in

P. Ghosh and B. Raj (eds.), *The Mind of an Engineer*,
DOI 10.1007/978-981-10-0119-2_23

However, the future was bleak because my father could not afford to send me to college. But then, as the saying goes, God helps those who help themselves. The Government introduced the 'Government of India Scholarship' for life for the first ten rank holders of every university in the country (in 1961, there were very few universities in the country). So suddenly, my higher education right up to Ph.D. was taken care of.

I wanted to pursue my career in mathematics and physics. However, I was persuaded to take up engineering for its lucre. I joined Punjab Engineering College, Chandigarh, in mechanical engineering, but my mind was always in research. During the summer vacation after 3rd year, we visited Bangalore on an educational tour. That is when I discovered the Indian Institute of Science (IISc). It was the place I had been always looking for. Fortunately, I got selected to do my M.E. degree in the Department of Internal Combustion Engineering without any formal interview or test. During my stay at IISc, I developed great interest in vibrations as well as mathematics.

In the 4th semester, I was asked to work on analysis and design of mufflers for the engine exhaust noise control. However, acoustics of noise control had not been taught in the class. In fact, it was not a part of the curriculum those days. I had the option to request for a change in the project. However, I decided to take it up as a challenge. I went to the library, selected a book on 'Fundamentals of Acoustics,' and gave myself a crash course in acoustics. I read the book by Kinstler and Frey cover to cover and solved most of the problems. Then, I did a literature survey of the journal papers available in the area of muffler acoustics. I realized that very little had been done on rational synthesis of exhaust mufflers. The normal practice those days was one of trial and error. I felt intuitively that I could do better than that. Over the next three months, making use of mathematical induction, electroacoustic analogies, combinatorics, and heuristics of matrix multiplication (all self-taught), I developed an algebraic algorithm by virtue of which I was able to write out the expression for insertion loss of a given linear dynamical filter, without having to write and solve the governing equations simultaneously. This algebraic algorithm was a breakthrough not only for analysis of one-dimensional filters but also for rational synthesis of vibration isolators as well as exhaust mufflers. Had I not challenged myself, I would have ended up making, at best, a small incremental contribution to the field. Instead, I developed a niche for myself.

I was offered a lectureship in the Department without any interview. In fact, a supernumary post was created for me, waiving off the requirement of a Ph.D. for the lecturer's post. Based on extensions of my algorithm, I got my Ph.D. degree and was awarded the Science Academy Medal for Young Scientists in Engineering Sciences for the year 1975. I received it at the hands of Mrs. Indira Gandhi, the then Honorable Prime Minister of India.

During the next seven years, I worked on analysis of commercial mufflers, most of which make use of perforated elements. This posed a formidable challenge. In association with one of my Ph.D. students, I developed a distributed parameter approach along with eigen analysis of perforated element mufflers. A paper based on this work was adjudged the best paper in the world in Muffler Acoustics by

Nelson Industries in USA, and we got the Nelson Acoustical Paper Award (first prize) in 1984. Based on this breakthrough, I got two projects from the Volkswagen Foundation, Germany, and published a few papers that have been cited widely.

I had my first sabbatical at the Institute for Technical Acoustics, Technical University of Berlin, during 1979–1980, where I gave a course on a muffler acoustics in English. I did not know much of German, and my audience were not proficient in English. So, I used to prepare and hand over xerox copies of the same one week in advance in English. These sets of notes became the first draft of a monograph. Thus, I converted a problem into an opportunity. I improved the notes over the next few years, and finally, it was published by John Wiley, New York, in 1987. This monograph has remained till today the only book on this subject and has been cited in nearly all papers that have appeared in journals during the last 25 years. (Incidentally, its second edition is under preparation and may be published by John Wiley, Chichester, UK, in 2014).

Around the same time, I received the 'Shanti Swarup Bhatnagar Prize in Engineering Sciences' for the year 1986 at the hands of Shri Rajiv Gandhi, the then Honorable Prime Minister of India, Fellowship of the Indian National Science Academy (1987), Indian Academy of Sciences (1987), and the newly established Indian National Academy of Engineering (1987).

In 1988, I was approached by DRDO to work on acoustic stealth for underwater vehicles. This was an entirely new field, but then I took up the challenge and eventually developed design guidelines for the stealth linings for submarines. The graphical user interface (GUI) and the codes for analysis of these resonator linings were passed on to the Indian Navy. This and some other associated pieces of research got me the coveted DRDO Academic Excellence Award for the year 2009 at the hands of Honorable Prime Minister of India, Dr. Manmohan Singh.

Incidentally, I have always worked on practical problems. It is more challenging, yet it is more satisfying to an engineering scientist at the end of the day. I have carried out more than 100 consultancy projects during the last four decades, apart from helping DRDO and the defense forces.

Concurrently, I have been active in the environmental noise control. In fact, I have been Chairman of the National Committee for Noise Pollution Control, which advises the Central Pollution Control Board, the executive wing of the Ministry of Environment and Forests. Since its inception in 1998, based on this committee's recommendations, the Ministry of Environment and Forests has issued gazette notifications for the control of noise from diesel generator sets, portable gensets, automobiles, firecrackers, public address systems, etc. This work has been recognized recently by the Government of Madhya Pradesh with the Pt. Jawaharlal Nehru National Award in Engineering and Technology for the year 2010. I received this prestigious award recently from Mr. Vijay Vergeeya, Minister of Science and Technology of the Government of Madhya Pradesh.

Arising out of all these personal experiences, there are a few observations that I would like to share with the reader:

1. Be resourceful. The best of science was not created by scientists with great resources. In fact, I often turned the lack of resources into a challenge, and the result was amazing. A fellow scientist from USA once remarked, 'But then, you had the advantage of adversity!'.
2. Everybody wants to work at, or be associated with, a world-class Institute like Indian Institute of Science. Well, IISc is what it is today because of its outstanding alumni. Why don't we aim at excelling ourselves, so that the Institute's name shines further? Let us rise to a level that the Institute feels proud of us.
3. If you enjoy your work, you will never feel tired. I proved this to myself again and again in my life. As per the Science of Psychology, a man can work 24 h a day, provided that he is not doing what he is obliged to do. Enjoy the process of research; do not make it a job. Work does not kill, stress does; and stress is a state of mind, an attitude.
4. Creativity increases as we think deeper and deeper, and this depth can be achieved by thinking about a problem undisturbed for long hours at a stretch (no coffee breaks). It was this kind of long, uninterrupted sittings in the library that resulted in the development of the algebraic algorithm for analysis of one-dimensional acoustic filters and vibration isolators at the start of my career.
5. When you listen to a teacher in the class, you may not grasp or retain everything. However, if the teacher tells you to read yourself what he is going to teach the next day, you will be alert in the class next day to clarify your doubts, and when teacher touches upon those points, you will never forget the answers. This is what I do in my classes. My student, incidentally, discovers his potential and gets ready for research by the time he finishes the course.
6. Often students complain about a particular teacher being too harsh in marking. However, I told myself in similar circumstances that the teacher may deduct more marks for a mistake, but if I did not make any mistake, what could he do? This led me to strive for perfection, which resulted in self-confidence, self-esteem, and creativity.
7. In the long run, clarity about basic principles, laws, and concepts is more important than cramming for the examinations. Concepts remain with you for a long time and, during spells of deep extended thinking, result in creativity.

Life—A Journey of Learning

N. Raghavan

It has often been said by several wise men of various faiths that life on Earth (school) is meant for learning. At the end of the day, it is good reminiscing to see what lessons came one's way, what one learnt and how one applied the learnings; also, to see how one passed on the best of one's learnings to others. Life, they say, is a series of lessons to be learnt and taught. I find that this has been the case with me also and that I have been given many chances of learning various lessons over time. Thus, learning never stops and continues all the time in our life. This learning can be in terms of one's chosen career, in personal life, in interactions with family, friends and the society at large, in interactions at the workplace and so on. Also, while we have been given free will and we are supposed to do our actions/karma on our own, the hand of Providence steps in now and then to give new twists to the directions of life or to give a helping touch. I believe that I have been working quite hard in my career, as well as on the personal front, and when one gives all that one could for the task at hand, the guiding/helping hand is not too far away; that is to say, I would not feel that I can take much personal credit for the various milestones that I have achieved. I feel that the above principles have played a significant part in my life.

In the early part of my life till about the age of 10, I was living with my grandparents due to certain family circumstances. Thereafter, in the course of my studies, I had spent seven years in hostel life in IIT's and had also spent four years in various sites, till the time of marriage. Such a life had given me *certain independence, as well as* a willingness to accept any situation (and quality of food.) that came my way. While I tell my wife that she should be happy with my acceptance of whatever food that comes on the table, she is unhappy that one is not able to appreciate the wonderful cooking that she is capable of.

After passing my B.Tech., not being able to afford higher education then, I was looking for a job, but jobs were not looking my way for quite some time, it being a recession period for civil engineers. I got a break with a construction company later, after trying out with a sales job and in PWD. The first time I went to work at a

N. Raghavan (✉)
2A, Krishna Mandir, 27, Thiruvenkadam Street, Mandaveli, Chennai 600 028, India
e-mail: nraghavan01@gmail.com

P. Ghosh and B. Raj (eds.), *The Mind of an Engineer*,
DOI 10.1007/978-981-10-0119-2_24

construction site as part of a contracting firm was the first time I had ever gone to North India and, that too, to a remote site—pitch-forked from the port construction works at Cuddalore to the hydro-project in Chambal ravines—the start of a career always intertwined with water. And I went with hardly any Hindi speaking skills. Controlling the labour or dealing with working levels of the client or just managing one's daily life (such as ordering food), all without knowing the local language well were all full of learnings. I learnt the fascinating skills of rowing a boat on Chambal alone in the night (the boatman used to go away in the nights) on the way home from the site late, walking on a simple steel joist without hand railing strung across 20-m-tall piers over a dam spillway, learning to eat north Indian village food, etc. Once in a village in a northern State, my Jeep was impounded and I had to wend my weary way alone on foot late in the night on a hungry stomach, for over 12 km along a mountain path through a nallah rumoured to have been haunted. Hard lessons indeed. Fortunately, I did not know the local language well at that time, and local ghosts would have had a tough time scaring me in their native language convincingly. Being in charge at remote sites does teach a lot of independence and adaptability. An interesting interlude was a project for strengthening of the foundation of the famous Qutb Minar tower in New Delhi, working day and night. I am glad that I could do my mite for flying India's colours very high. The rigours of site life are often well compensated by fun and learnings and happy camaraderie.

There is a lot of debate on the subject of the actual usefulness in life of all theoretical studies which one has to do in the course of education. It has also been said that what one learns in college is only the ability or art of managing life, career, et cetera, and also the art of interactions with our fellow human beings. The ability to find solutions for problems for which a solution is not that readily apparent is a skill which one should pick up in college. I guess that in this respect, seven years of IIT education has been of great help to me. I would also say that if one is going to continue a technical career, then some of the foundations laid in the course of learning the subjects in college would definitely come in handy in the course of a long career.

A lot of different viewpoints have come out over the ages on why IIT's are successful or why they make the students do things differently. Apart from the academic learnings, one also learns a lot in the hostel life, from friends and peers, one gets broad exposure to think out one's approach to life, etc. Looking from the importance of the environment shaping one's life, my association with Ramakrishna Mission schools and IIT's was of great help. Good teachers do provide a lot of inspiration, and I could do some original work for my M.Tech. project only with the inspiration I derived from my excellent top-notch guides.

As an instance of serendipity in life, I got posted by chance—after a long period in sites—in Mumbai for a short period, came across an ad for M.Tech. course in IIT Bombay, applied and wrote the exam, got selected by IITB, and resigned and got relieved from the previous employment—all in a matter of days. The M.Tech. course was a "life-changer" for me. It was during this period that I did a lot of introspection and engaged myself in many self-improvement measures—developing more self-confidence, improving my physique, postures and manner of speaking, engaging in extra-curricular activities, becoming more extrovertish, etc. Doing well in my M.Tech. course, I got back into the technical stream and immediately got an assignment in a top design company.

I owe a lot to my engineering career in STUP Consultants, where I picked up skills of problem-solving, finding innovative solutions, perseverance and hard work. I also learned that an atmosphere of excellence breeds further excellence (as they say, for the life in IIT's also). I could progress considerably after the initial learnings from several of my seniors—who had great intelligence, structural design capability and intuitions, skills for developing construction methods, innovative out-of-the-box thinking and engineering values—a very good team built up by Mr Alimchandani. Very fortunately, opportunities came my way to attempt the design of various types of structures covering practically the entire gamut—the tallest structure in India (a 323-m-tall TV tower—again keeping India's colours flying very *high.*), 67-m-long cantilever tie-back roof, a marine project involving launching and positioning of precast concrete caissons, several hyperbolic paraboloid natural draught cooling towers including thermal and structural designs evolved practically from first principles, parts of nuclear containment structures, long-span continuous bridges, several flyovers, innovative schemes of design and construction, etc. As mentioned earlier, the inspiration received from the seniors, the conducive environment, the able back-up available for construction engineering, availability of good juniors, etc., contributed to steering through uncharted waters. Another basic value that I learnt was *integrity to the client* in ensuring the best possible design, that tender-time quantities are not to be exceeded during the detailed engineering, keeping the interest of the client in mind at all times, etc. I had to get many designs approved by third parties, and this was often quite hard—due to different perceptions, different agendas, difficulties of others in understanding the structural intricacies involved, etc. Everyone would like to be conservative and play safe, but the project cost goes up. One has to persevere and apply all of one's technical knowledge and arguments as well as persuasive skills to get one's design approved as submitted. A great challenge and full of learnings. *A good skill learnt was* how to explain complex matters in simpler ways and the art of convincing others. I also learned *the importance of integrating design engineering with construction engineering/construction methods* as the two always go hand in hand for most structures. The highly innovative technique developed by my colleagues there of launching 650 t bridge girders using only tidal power was a great revelation. I learnt to do safe but economical designs, which meet the requirements adequately but without any fat. But this often meant putting in more work to convince the approving authority by looking up for more references and doing more computations.

At that time, I got an excellent opportunity for working with computational mechanics. I had the responsibility for developing computer systems as well as technical software. Software development or analytical abilities have been some of my very strong points, right from college. For the engineering of the Second Thane

Creek Bridge, which was at that time a high technology landmark project, we developed a great extent of software for completing the designs in time, with innovativeness and keeping the cost within limits. It was a challenge getting design approvals for this project successfully from an international consultant, RPT, and it was a pleasure dealing with professionals with whom one could see eye-to-eye. I also recall the very able support from bright younger colleagues. *Apart from one's own abilities, the environment, the seniors and the supporting juniors also matter for success in one's projects.* I was able to develop and use software for repeated optimisation to ensure efficient designs for many structures, and to make up for any lack of intuitive design outlook. I could also do some pioneering work in the country at that time: implementing the SAP IV software, managing to do with only available computational resources—which were quite scarce, of much limited capability compared to today's versions and also expensive at that time—for a variety of problems. It had 14,000+ lines of code and was too big to be run as a single unit. The coding as well as the theory—developed by leading international professionals—had to be understood, split up into smaller self-supporting modules and run for actual cases after validation with trial problems. Some of the problems which had to be tackled included complex problems for nuclear containment structures, which were being analysed in the country for the first time. The project was being handled by a senior with tremendous insight into structures, and it was a pleasure being of assistance with the software support. It was a challenging task which was successfully accomplished with a lot of determined efforts. *The lessons here were tenacity, adaptability and determination.*

These attributes were also useful for unravelling the mysteries of thermal and structural designs of hyperbolic paraboloid natural draught cooling towers. At that time, indigenous design know-how was not freely available for these aspects, and in a period of limited computational resources, our methods had to be developed from first principles. I recall—now with horror!—the enormous time invested by myself and an associate and heart-rending trial-and-error explorations without the availability of any learned counsel, before coming up with workable solutions. These projects were all design-and-build (D&B) projects, and the challenge was keeping to the quantities committed at the time of tendering even after detailed engineering and proof-checking by third parties. I believe *the D&B type of project realisation helps integrate the design and construction schemes well, with innovative outcomes and well suiting the capabilities of the constructor, paving way for fast and smooth completion.* For two tenders, we had to fight tooth and nail to get the projects even with being the lowest in price, for various reasons. From these "battles", *I learnt from my seniors the principle of not giving up and fighting hard for one's rights.*

Another landmark project was a railway bridge, which was designed and built by the precast segmental construction method in the 1980s (still in use), probably for the first time in the country. Again, it was a challenge to master the basics of this (then very daring) concept (particularly for a railway bridge) and get the design approved, keeping one's commitments to one's client duly fulfilled. Today, practically all elevated metro construction uses this concept, but I must salute Er. E. Sreedharan, who pioneered this concept with tremendous engineering courage. I have been fortunate to work with this great engineer and administrator on a couple of other projects as well. To me he was a person with his head high above in the clouds with tremendous vision, but with feet firmly planted on the ground to ensure realistic implementation of the vision. On the railways side, I was also fortunate to design the first leg of the elevated MRTS superstructure in Chennai with full spans precast and launched into position.

In this context, I have to emphasise *the need for young engineers to keep studying and learning to improve their technical abilities*. I owe a great deal to a great book by Podolny and Muller on bridge design and construction, and every time I went through the book, I would mine some golden nuggets of ideas, concepts and knowledge. Such books are indeed a real treasure trove. How I would love to write a book like that. In STUP, I also got some opportunities for writing in the house magazine, as I was supposed to have a flair for a "breezy style".

In L&T, Dr. A. Ramakrishna was a great inspiration. His passion for innovation and good technology, efficient and quality construction, strong customer orientation and genial outlook were key factors for me to desire an entry into L&T. He had done pioneering work in precast concrete construction, particularly with shells and folded plates. Though I would have liked to work with innovative design and construction of buildings or bridges, I was assigned to work with power projects, mainly in a developmental mode. I was part of L&T's thrust into thermal and hydro-power plant construction, and the hydro-connection also took me to underground construction. This association with L&T came about due to a chance meeting I had with AR, as he was affectionately called, in Mumbai airport when he persuaded me to join him. This was another instance of Providence taking the lead.

There were many lessons to be learnt in my L&T career, where again a culture of excellence was flourishing. Joining an established company with somewhat conservative orientation and joining laterally at somewhat senior levels have their own challenges, and *adaptability and humility are two important requirements, coupled with lots of patience*. Changing over from a life of design engineering to construction management, learning the ropes of site work without being able to spend much time in sites, etc., were some of the challenges. I also received a lot of managerial training to supplement my technical knowledge. An "EPC fever" had gripped the nation in the early 1990s when India seemingly opened the doors for international private sector participation in power generation. Many opportunities came along for associating with many international agencies—developers, EPC contractors, consultants and so on, and one had to learn the new ways of these foreign agencies, including understanding the nuances of the jargon, mainly from the US-based companies. The contracting formats and many intricate contract conditions—one can recall the Dhabol project with consternation!—were new to us, and it was an interesting time learning and mastering these. The EPC bidding format involved working with different equipment manufacturers, engineering consultants and project management associates, and it was *very instructive and interesting to be part of such multidisciplinary, multinational and multiskilled teams*.

This gave me *a realisation that the life of an individual gets shaped not only by his own endeavours but also to a major extent, by the aspirations of the employer and the developments in the workplace milieu.* Flowing with the times, I tried to weave into whatever developments were taking place and which were taking me along. *In any case, L&T had many good people from whom I could learn many things.* Two interesting projects for which I could add significant value were a 330 MW gas-based power project and a 54 MW cogeneration project with a petroleum coke calcination adjunct. It was quite fascinating to deal with associates from many countries ranging from US, Europe, Korea and Japan and building friendships with many diverse people. As a creative aside, I helped make a film, including developing the script, showcasing our capabilities for power projects.

My stint in the field of hydro-electric projects was quite fascinating, for most of these projects were in the Himalayas. A slight diversion here. *I believe that the universe does work towards giving everyone what one desires deep inside, in the long run, consistent with the greater good of all.* Rhonda Byrne's best-selling book, "The Secret" enshrines this principle. In my life whether it was working with L&T, living in Bangalore or being in the Himalayas, all came out of this deep—though unexpressed outright—desire. One caution is that *what seems to work is not what one professes outside but what lurks deep inside.*

Regarding being in the Himalayas, my jurisdiction in hydro- and nuclear projects included projects right from J&K to Arunachal Pradesh, covering in between HP, Uttaranchal, Nepal, Sikkim and Bhutan, and apart from them, others in WB, Karnataka, Chhattisgarh, AP and Kerala. The country's largest hydro-electric project—2000 MW Subansiri project—was another landmark project for me. It had many problems and was very slow to take off. Difficult access, tough terrains, poor geotechnical conditions, inappropriate designs, hostile locals, interstate border problems, etc., conspired to make our life very difficult. Many of these hydro-projects involved considerable road travel, mostly through winding narrow roads with a hill on one side and a precipice on the other. Some of them were in areas with inimical environments. Apart from these, there were also projects in other parts of the country, thus enlarging my scope of travel to practically all corners of the country, at uncomfortably frequent intervals. Incidentally, in the course of my stays in various parts of the country, I picked up smattering of a few Indian languages, supplementing a couple of foreign languages which I had studied, which all led me to realise that I had a liking for languages. Though the travels took a considerable toll on my health and physique (particularly the back), it was an interesting experience dealing with many different and diverse projects. I liked the hills very much and incidentally, and I believe (though probably somewhat irrationally.) that the hills "stoked up" my spiritual aspirations to a good extent to make me "let go" at a somewhat early stage. At that time, we were mulling setting up a joint venture engineering company with a leading international engineering company and though it did not fructify, the experience was quite interesting. This reminds me of another learning which I had come across that *people come into one's life either to teach one something or to learn something from one, and most of them have only a limited window in one's life, suited to that purpose.* When one ponders over this, it is wondrous to recollect how some people who were so important to oneself at a given time did not matter at all later or simply vanished from one's life; strained attempts to retain such contacts often prove to be infructuous.

The project for an underground unlined rock cavern for storing LPG was a very challenging project with many lessons. An underground rock cavern of 1.25 lakh cubic metre capacity had to be built 200 m below ground using only two small-diameter shafts of 4 and 6.5 m diameter, adopting the drilling and blasting technique, under stringent specifications. The contracting format was very interesting, and the key factor was *the stringent safety requirements* which had mainly come from the French part of the owner team, TotalFinaElf—HP. Their overriding concern for safety during construction and the processes and procedures adapted to ensure the same were real eye-openers. Working totally underground, taking in and bringing out all the large equipment in dismantled condition, evacuating all the blasted rock pieces only through the 200-m-deep small-diameter shafts were awesome requirements. What took the project through were the *excellent enthusiasm and man management of the knowledgeable site team, its coherence and the never-say-die positive attitude.* Though the project had many hiccups in the earlier days, it picked up very well afterwards and it was very satisfying to see the positive surge of energy in the site and the excellent camaraderie, not only within the site but also with the clients. The whole work took a team approach, and it was very humbling to be part of such a great team. The continuous emphasis on safety ensured that the entire project was completed without a single fatal accident underground. *I would say that the cavern project was one of the excellent landmark projects of L&T.*

Another interesting short-term stint in my life was my stewardship of a new L&T company, L&T-Ramboll Consulting Engineers Limited (LTR) for about four years. To fulfil a need for a strong engineering company to service infrastructure construction projects, L&T set up this company in collaboration with a leading Scandinavian engineering company. Given L&T's Danish origin, association with another Danish company was somewhat natural. Ramboll had many wonderful people who gave their support unstintingly. Setting up a brand new company, evolving and organising all the procedures and processes, recruiting all the skilled staff (a very precious commodity, indeed, at all times.), getting business and executing the same and satisfying the two different parent companies were all big challenges. This engineering company had to be totally separate and independent from the parent construction companies but with the strong ethical precepts of the parent companies, and hence, all the formative work had to be done independently and with a strong vision. A wonder was we could make a profit right from the first year of inception. It was very invigourating to work with an excellent team of expert, committed team of highly enthusiastic people in the nascent organisation. It is satisfying to see that the association had lasted more than 15 years. In this assignment, I could further broaden my horizons in going into development of projects as well as trying out some pioneering concepts. The greenfield Gangavaram Port and the MIHAN airport at Nagpur are some examples. We also seized a good opportunity for designing a cable-stayed bridge, the first for our group of companies—a rather simple application for a pipe bridge, but nevertheless a cable-stayed bridge (again probably the fulfilment of a deep-lurking desire of mine).

One of the projects which I can never forget came across my life at this juncture. We landed up with a project for designing nine flyovers at predesignated locations—including all surveys for topography, traffic and geotechnology—and detailed engineering including good-for-construction drawings and bar bending schedules along with draft tender documents for construction contracts—all to be completed in just four months. This was LTR's first project, when the company staff strength was just building up. This project was a tremendous challenge. We were able to deliver the job quite well with very cost-advantageous, innovative, aesthetic solutions with high functionality and which were also built within the short time periods envisaged by the client. This was possible since we had integrated the design scheme and construction methods (which were very simple in nature, suiting most constructors), and these methods were spelt out in detail in the tender document itself, ensuring that the bidders could minimise their risks and quote on a common platform. *The learnings was the importance of construction methods, integrated engineering and well-made contract documents.*

Coming to the concept of giving back, I was fortunate—with a lot of serendipity again, thanks to INAE's Distinguished Professor scheme and an earlier referral by Mr K.V. Rangaswami—to join IIT Madras to share with the students something of what I had learned during my career. I was also elected to a senior position in the IITM Alumni Association and was fortunate to play a leading part in the setting up of the first-ever IIT Alumni Club, in Chennai. I must say that these developments took place *due to chance encounters with certain others and paying heed to certain promptings which could have been attended to, or left to float away*. Incidentally, again as a creative aside, I produced two videos for IITMAA showcasing the alumni's involvements with charitable work.

In the current phase of my career, I am engaged at IIT Madras, covering some teaching, guiding some research, participating in consultancy assignments and driving new technologies such as Lean construction management. The latter is a recent paradigm, which has caught on quite well in the developed world, but is yet to find strong practice in India. In the campaign undertaken by IIT Madras and ILCE, I have been playing a lead role in popularising Lean, as well as trying to get Lean implemented at actual construction sites. *It is never too late to learn*. In this regard in a recent programme carried out at nine sites of five major construction organisations, Lean was implemented over a period of eight months in a fairly successful manner with significant gains. I had also presented a paper from IIT Madras on this programme at an international conference held in Norway recently and as a follow-up, and I was the chair for the first national conference on Lean construction, which took place in February this year at Mumbai. In the civil engineering, Department of IITM, I have come across highly accomplished and knowledgeable colleagues with international standing and with their support discovered a few subjects in which I could contribute with my industry background to supplement the academic teachings; the range covers construction methods, project management, contract management, underground structures, construction safety, Lean construction management, etc., and even ethics in construction.

When I look back on my long career, I find that too often I had placed significant focus on work, sometimes at the cost of personal life or family life. Not that I was too much of a workaholic, because I also had other interests such as reading, music, films and walking. I do realise now that one has to evolve a healthy balance between work and family life, and also "personal life" and it has been a regret. The tricky thing is that the children are born and the family is developing, especially at the time when one's career is also at a key stage demanding a lot of attention. In my case, I console myself that I was doing a lot of pioneering work at that time, which demanded a considerable amount of time for evolving solutions for difficult problems without much external guidance. A few awards and recognitions have come my way, but fortunately, I learnt not to chase them. However, with things turning out not so bad at the end of the day, one can only thank God for setting things right in his own way. Getting back to the concept of giving back, I am trying as much as possible to give back to the society in various ways. I still need to write some books—a long-pending goal to share some of my technical learnings as well as some spiritual learnings (which I am yet to understand properly). Also, I still have not learnt how to say "No" or how to stay non-judgmental.

Looking back, I have to admit that I could not have led the life described above without the active and commendable support from my very able and understanding wife, who has been a great pillar of strength, all through—steering not only myself but also our children successfully through the streams of life. My father—who rose to good heights in his career from very humble beginnings through personal brilliance and sheer hard work, but still supporting the family through many difficult times—was also a good inspiration.

Knowledge is an important part in the lives of engineers and contacts with other professionals, and organisations play an important part for realising one's objectives and those of one's organisation. I realised at an early stage that seminars and conferences help considerably towards the above. I have delivered about 170 articles and talks in various technical forums in India and abroad. I have also seen that *more than reading books, interactions with other professionals and visiting actual project sites help us learn a lot more and see many angles which one would totally miss by having a merely theoretical approach.* Of course, one has to find time for these in addition to doing proper justice to the main work at hand, and this also contributes to longer hours of working and work pressure in the shorter run.

Finally, a pet subject somewhat dear to my heart, at this stage of my life. Of late, I have taken to reading many books on subjects which could be called spiritual or New Age or theosophical. Based on these readings, interactions with certain people who have come across my life in the recent years and a good amount of introspection, I have formed certain new outlooks on life. Though this is not the forum for expounding on the above, I would still like to summarise the above essentially as following: *life on this earth is for learning in the Earth School; people we come across and events that happen to us are all meant to further us in our learning or help others in their learnings; all religions lead to the same God, and all men are but a part of the one, giving rise to an all-important requirement of universal love. Seek and ye shall definitely find.*

Summing up, one owes a lot to one's seniors and other professionals from whom one has learnt various things, as well as several others who had helped one in so many ways and my heartfelt gratitude to all of them. I have not mentioned any names of my numerous benefactors and precious friends, because there were so many. "Endaro Mahanubhavulu" (salutations to all those great men in this world who have come before") comes to my mind. I realise that on many occasions in life *sincere prayers have worked to get the blessings of the* one. *Apart from playing the cards, one has been dealt with according to one's best ability and as shaped by the then-prevailing environment; the providential hand is always benevolent; the present moment is always the best offering by the universe.*

Writing this article has really helped me to introspect and discover more aspects of my life than I would have imagined otherwise. I wish all the readers the best of everything in their lives—work, family and personal.

Some Views and Experience on the Mind of an Engineer

C.G. Krishnadas Nair

I remembered the long-delayed promise made by me to reflect on my career as an engineer, which could be of some value to the readers, particularly young and the future generation. Being an INAE Fellow for over two decades and an INAE Lifetime Achievement Awardee (2007), I owe it to our young engineers. The question which came to my mind is "Who is an engineer?" A person with a diploma/degree in engineering? or one who practices the profession of an engineer with or without diploma/degree? The dictionary definition of an engineer is "one who designs/builds/maintains tools, instruments, equipments, systems, machines, vehicles and the like; buildings, roads, railways, bridges and the like".

Engineers use knowledge to produce technology and products of utility to the world; human beings and other living beings. There were engineers among humans even in the prehistoric age, when humans were illiterate and cave dwellers or nomads. In the Stone Age, these creative-minded humans imagined (designed) and fabricated tools and other articles of utility from stones in varying complexity of shapes and sizes. In fact, it is this creative mind which helped humans in the ascent of human civilisation right through Stone Age, Copper Age, Iron Age and the Modern Age of Information Technology and Intelligent Materials. An engineer's mind must observe, dream, imagine and be creative.

When great engineers contributed by dreaming/imagining and by creative work, what else was in their mind? Bharat Ratna, Sir M. Visveswaraiah, designed and directed the construction of the Osman Sagar (under the governance of the Nizam VII) in Hyderabad which saved Hyderabad from annual flooding during the monsoons and also created a reservoir with sustained supply of water to the growing needs of the city of Hyderabad. He designed and built Krishnarajasagar dam across the River Cauveri in Mysore and the beautiful Brindavan gardens with dancing fountains using the out flow of the water from the dam. When engineers design and build such things of beauty and value to humans and environment, the

C.G. Krishnadas Nair (✉)
Society of Indian Aerospace Technologies and Industries (SIATI), Aeronautical Society Building, Suranjandas Road (Off Old Madras Road), Bangalore 560 075, India
e-mail: krishnadnair@hotmail.com

P. Ghosh and B. Raj (eds.), *The Mind of an Engineer*,
DOI 10.1007/978-981-10-0119-2_25

mind is committed to "selfless services to society and environment". This must be another important feature of the engineer's mind.

Engineers must be responsible environmentalists. An engineer's mind should be responsible for environmental protection and sustainable development. While engineer's work is essential for progress and prosperity, it can also lead to injuries to life and environment. For example, mining of coal, minerals and metals, building dams for irrigation and hydro-electric projects will disturb the local ecosystem and destroy forests. Industries produce chemicals and fertilisers, and metals and alloys, petroleum products and such others regularly pollute air, water and ground. Automobile and aircraft exhausts poison the air. But we need minerals, metals, chemicals and fertilizers, and dams for irrigation and electricity, automobiles and aircrafts. Engineers must have an understanding and concern of the environmental problems and reduce or eliminate the same using technology and other appropriate measures. For example, pollution from the industries can be neutralised by appropriate treatments. Forests can be regenerated, and automobile and aircraft emissions can be controlled by improving the engine design and also change of fuel. Engineer's mind must be committed, right from the design stage for designing and executing projects, processes and products in such a way as to reduce/eliminate injury to health and environment.

Engineers are not magicians; they create real things out of resources provided by nature. Whatever the products be the metals and alloys and other materials, buildings and habitat, machines, ships, automobiles, aeroplanes and the alike or energy from thermal hydro-electric projects or nuclear, solar or wind power all these are from resources provided to humans by the Almighty who also gifted humans the power to observe, imagine and create. So let all of us engineers remain humble and remind ourselves that it is God's work, we are doing. "Work is worship" as Swamy Vivekananda exhorted to the World. Humility is another important aspect of a good engineer's mind. "*Vidya Vineyena Shobitham*"—Knowledge Shines with humility.

I had my basic engineering education at IIT Chennai and graduated with B.Tech. in Metallurgical Engineering in 1964. It was a 5-year course of which the first three years curricula gave the students a broad-based knowledge in Maths, Physics and Chemistry and also on all basic engineering subjects covering civil, mechanical, electrical, electronic metallurgy and materials. This included laboratory work as well as workshops in all these areas. Specialised subjects for each branch of specialisation started in the 3rd year and identified in the 4th and 5th years. I think that the present semester system with 4 years does not give adequate time to the students and teachers for such a broad-based foundation in engineering with some amount of multidisciplinary knowledge and skills. This is important to develop the inherent Power of the Mind to observe and imagine to design and build products of utility and blossom the creative part of the engineers mind. The gap in the present 4-year semester system could be partly compensated by careful planning and execution of the project work which is a part of the curricula. For example, we can assign the project work to a multidisciplinary team of engineering students and guide them to design and build a product jointly.

I will narrate a small story from my IIT student life. During the summer vacation, most of the students went to their native towns/villages but a few from far north stayed in the hostel with special permission. Poornachandra Maji was an electronic engineering student who was staying in the hostel. During the third-year-summer holiday, Maji told me that he is going to build a Transistor Radio during the vacation and invited me to join him. I readily agreed as a partner of this project. Maji and I went to an electronic and electrical component shop in the city and bought several small items such as transistors, capacitors, resistors, micro- and mini-switches, thin wire, printed circuit boards, solder and soldering iron. We bought two sets of components to assemble two transistor radios. We assembled the components over the printed circuit boards and connected through soldering. We worked throughout day and night and completed assembly and connected to a set of torch batteries and switched on our two sets of transistor radios. We tuned the radios and received Vividh Bharati and it was a great achievement and excitement for us, the third-year engineering students, when a melodious music was heard through our transistors. Next day, I left the hostel and took the train from Chennai to Aluwe carrying my newly made transistor radio and I ran all the way to my native village from the railway station and announced that I have built a transistor radio and brought it along with me. Soon curious neighbours joined my mother and sister as they were all inquisitive, as there was no radio in our village. I hooked up the naked radio and tuned not only to Vividh Bharati but also received All India Radio, Trivandrum, and even Radio-Sri Lanka. Soon we had a crowd from the village listening to the radio. At that time, we did not have a radio even in our panchayat office. All were excited and congratulated on my achievement. I was already accepted as an engineer. It was a lesson for me and should for all engineers. An engineer should not be just a theoretician, but should be creative and build things of utility.

An **engineer's mind** should have a major "heart component", to be a good human being and to contribute to society and environment. Our Director at IIT Chennai was a nature lover and he along with professors imbibed in us love for nature. IIT curricula included a compulsory subject in humanities, covering literature, history of human civilisation, ethics and values. IIT students coming from various parts of India formed a multicultural group and campus life was vibrant with cultural and sports activities. Such activities will make an **engineer's mind** and personality multidimensional and enhance the EQ very much required for team work and success in life. We have several examples of great scientists and engineers who distinguished not only in scientific/engineering achievements but also were exponents/appreciators of music, art and literature with compassion and empathy for fellow humans, society and environment.

After my graduation from IIT Chennai in metallurgical engineering, I joined the university of Saskatchewan, Canada, for a Master's degree in Mechanical Engineering, instead of continuing in metallurgical engineering as I thought that a combination of metallurgical and mechanical engineering will be useful for me to become a design and development engineer. Later, I specialised during my Ph.D. in fatigue crack propagation and structural failure in aircraft aluminium alloys. This led to my joining Hindustan Aeronautics Ltd., as a Materials Specialist/design and development Manager in 1971, after a brief period of teaching and research at the Regional Engineering College at Surathkal and at the University Sheffield, UK.

At that time, HAL was importing almost all materials and several structures in composite (FRP) materials. My task was to develop these items indigenously. A major hurdle was airworthiness certification. Whatever was imported were accepted as the same came with a certificate from the foreign supplier. Materials developed in India even if it is one-to-one exact equivalent in chemical composition and properties, it had to undergo elaborate quality assurance testing and certification. There was a bias (as is there even now) that quality of imported items is better than those produced in India. There were many interesting and humoros incidents with respect to indigenous development of materials and certification. I will narrate briefly two examples. "Decapex" was an imported solvent used to soften adhesive layer bonding to facilitate stripping of the sheet bonded to the aluminium spar and plastic foam core of the helicopter blade to facilitate repair. The solvent was not costly but had to be brought by ship only as it being a chemical was forbidden to be despatched by aircraft. Often there was delay in supplies affecting the execution of repair in time. So, indigenous development was a strategic requirement. I got the imported sample Decapex analysed through fractional distillation in our chemical laboratory, identified the ingredients, bought these solvent ingredients and made a mixture exactly matching the composition of Decapex and same was offered for shop trials. Shop technicians conducted trials and opined that the shop trials were satisfactory but imported solvent is better. Our team in the laboratory and I knew it cannot be so, as both were identical. The verdict from the shop was surely based on our mind set that imported product is of better quality. So we decided to play a trick on the shop in charge and technicians. We collected another tin of Decapex from the store and informed that we needed a two weeks' time to improve and develop a better substitute. After two weeks, we emptied the contents of the cans to a bottle marked laboratory sample and filled the empty can bearing the label Decapex with substitute made in the laboratory. Both were offered for shop trials, the same certificate came from the shop, the sample in bottle labelled as Lab substitute has marginally improved but imported Decapex in the tin is better! They did not know that tin labelled Decapex actually contained the laboratory sample and the bottle contained the imported Decapex. We disclosed the secret and all had their laughs and blamed it on our Indian mind set with a bias for foreign quality.

In another case, we took up indigenous development of fibre-reinforced composite structural parts for a helicopter. Development of these FRP structures was a great challenge as these were made out of woven glass cloth and special polyester resins. It was necessary to develop these first before developing the composite structures. With a lot of efforts and time, the basic materials were developed and also the process technology for the FRP structures. The product met all the physical and mechanical test requirements. However, this time the certifying agencies insisted that the indigenously made structures must withstand 1.5 times more stress than that stipulated for the imported structures as it is the first time we are using such structures made in India and a higher safety factor (1.5) is required. With further efforts, we developed higher strength FRP structures. Then a new requirement was put up Quality Assurance Agencies demanded that the structures must be fire proof. We argued that there was no such requirement and the imported FRP structures were not fire proof. We took specimen from the imported and indigenously made FRP structures and subjected them to fire test; both samples burnt vigorously. But even after this demonstration, Quality Assurance Agencies were not willing to give certification unless we develop fire-proof structures for greater safety. The engineer in me became thoroughly frustrated and it appeared to me that only I was interested in indigenous development and manufacture and these certifying agencies were interested in import and not bothered about saving valuable foreign exchange and creating employment in India by stopping import and taking up manufacture of these structures indigenously. I complained, but our General Manager advised that as an engineer I should accept the challenge and do it as it is for a better product. So I took the challenge and after a number of additional research and experiments using additives for fire proofing, succeeded in coming up with a composition, which was fully fire proof. All were delighted including me and the Quality Assurance Agencies as we developed a product better than what was imported. HAL set up an FRP shop to manufacture these structures for our production requirement creating more employment, wealth and welfare.

It was a lesson in "triumph over challenge". For me personally, it was a remarkable achievement boosting my self-esteem, and also winning friends and admirers from among my colleagues and employees. All these would not have been achieved, if I had given up the development out of frustration arising out of the increasing quality demands from the airworthiness certifying agencies and blaming it on their lack of patriotism.

An engineer's mind should be tough and positive while facing criticism and should never allow frustration to creep in leading to despair. Constructive criticism for improvement must be responded with positive action. A product is not developed until it is accepted by the customer.

During this period, I was advised by the General Manager to give a series of lectures on aircraft materials and process technologies to the newly inducted design and management engineering trainees. Although I had some basic knowledge of aerospace materials, I needed to know more and make in-depth studies. So I gathered books on design of aircraft, helicopters, engines and materials, special properties to meet the design requirements and international aerospace materials specifications. As HAL was importing materials from UK, USA, France, Germany and Russia, I was amazed at the large number of material specifications. I realised that many of these apparently different alloy specifications were one and the same and there was scope for rationalising and standardising these to develop and manufacture equivalents in the country rather than importing small quantities from several countries at high costs. I prepared my lectures illustrating the design requirements of various aircraft, helicopter and engine components and structures and relating to the materials and their properties to meet the design requirements. My classes were well appreciated because of the elaborate preparation made by me. I used my lecture notes to author a book on "Aircraft Materials" and later with additional research and rationalising and standardising guided a team of metallurgical and design engineers to prepare a Reference Data book in three volumes on "Aircraft International Materials specifications properties and equivalents". Had I taken the suggestion of the General Manager as a boost to my ego and just taught what I knew, I would not have achieved this. For an engineer, learning should not stop with the end of academic career receiving a degree and learning should be continuous. An engineer's mind must have a sustained thirst for knowledge.

The news of my early success in indigenous development of materials and FRP structures reached the Chairman of HAL and one day I was transferred to the newly created development department in the Corporate Office. The low point was change of my designation to an insignificant Import Substitution Officer. I was thoroughly disappointed as I felt I was reduced from position of Design and Development Manager to an Import Substitution Officer. My dream at that time was to become the Senior Development Engineer/Manager and then Chief Metallurgist and eventually become the Chief of the Central materials laboratory of HAL. This, I thought will be the best career for me and of the maximum I can reach in an aircraft industry predominated by Aeronautical Engineers and Senior Air force personnel. I poured out my heart to the new boss in the Corporate Office and I informed him of my considering to resign and go back to Canada, if I am not put back in the helicopter design division or in the Central laboratory. My new boss, a very senior Air force officer and an accomplished electronics engineer, made me think differently. He advised me not to look at the current change of designation but look at the opportunity to work in the corporate office. I am part of the Corporate R&D and reporting to the chief of development who reports to the Chairman of the company. Hence, I am only two levels below the Chairman. If I continued in the helicopter division or as Senior Manager in the laboratory, I am several levels below the Divisional Head/General Manager, who is three levels below the Chairman. Whenever I go and discuss with Directors and Senior Scientists of National Research laboratories, senior officials of industries, GMs of other HAL divisions for indigenous development, I will be representing the Chairman and not a divisional head of HAL and therefore I will be working with greater authority. If I am interested and focussed on my objective of enhancing research and self-reliance, then the change is indeed a positive one and should forget about the temporary unhappiness of immediate promotion and lack of a classy designation. I accepted my mentor's view and this was a great turning point in my career. Had I insisted on going back to the helicopter division as Design Engineer/Manager and perhaps get a promotion as Senior Manager in a few years, I would have perhaps retired from HAL as the Chief of the Laboratory, instead of growing up in the Corporate World and finally becoming the Chairman of the company. An engineer's mind should respond to change and adversities with a positive attitude.

After some years in the Corporate Office, I was assigned to the Foundry and Forge factory in Bangalore, which was a loss making unit. It was also the centre of aggressive union activities. I was given a mandate for closing down the Foundry and Forge division or reviving it as a profitable unit. Both options were challenging but reviving a loss-making unit rather than closing it down was more challenging as well as spiritually more rewarding. In that case, I will not have to make people lose their jobs. I may even provide opportunities for creating more jobs and enhancing production. I found that the motivation level of employees was very low. Foundry and Forge employees and officers were not considered as part of the main stream as compared to those in aircraft, helicopter and engine factories. F&F was making low-cost items and not the high-quality precision forgings and castings required for aerospace. As a team, we outsourced the low-price commercial quality castings and forgings to industries with lower infrastructure cost. We retrained the technicians and engineers and took up the manufacture of aircraft quality precision forgings and castings which were then imported at high cost. We arranged for our technicians and supervisors to visit the aircraft, helicopter and engine factories and see the parts made by them fitted on the aircraft, helicopter and engines and explained to them the function and importance of these parts. This made them feel proud of their job and enhanced self-esteem and commitment. From mere component manufacturers, they became partners in the manufacture of aircraft, helicopters and engines for enhancing Nation's self-reliance. This is a great management lesson for all. Communication of the vision/mission to all team members make them emotionally committed, proud partners in the fulfilment of the objective of the organisation. An *engineer's mind* must focus on team building and inspire the team members.

While engineers should continue learning to enhance their own performance and contribution, they must also be good teachers. Teaching is a noble profession as teachers impart knowledge and skills to the students without any inhibition and bless them to become great achievers. In industry, some engineer–managers hold their knowledge and information from imparting to their team members as they fear of the subordinates performing better than them and possibility of their going higher up in the organisation. This is not at all a good practice, as the more you train and impart knowledge to the team members, the team performance will excel and hence a good engineer/manager must not only be a continuous learner but also be a willing and a committed teacher.

After an interesting and eventful career as General Manager of HAL's F&F and Engine Divisions, I became the Managing Director of HAL's sprawling Bangalore complex consisting of aircraft, engine, helicopter and other divisions. I do not wish to give even a brief account of the engineering and management challenges during this period and how these were solved. These are covered in several other publications and in a recent book "Anuyathra"—*Story of an Aerospace Personality of India*[1]. However, I shall attempt one case as an example of engineer's duty to convert science into technology.

HAL was manufacturing helicopters, military aircraft and engines under license with transfer of technology agreement with the licensor. However, perusal of TOT agreements revealed that a large number of critical components in each case were excluded and technology for these was denied, even if we were ready to pay for the

[1]"Anuyathra"—Story of an Aerospace Personality of India, K.A. Muraleedharan and Thankachan, Poorna Printing and Publishing House, Muttappalam PO., Varkala–695 145.

same. Hence, these critical technologies were to be developed indigenously for enhancing our design and manufacturing capability. I was sure of existence of several scientific research papers in these areas. A perusal of these will be a starting point. Our team searched for published literature with the help of CSIR & DRDO laboratories and obtained considerable literature. We set up laboratory scale plants (mini pilot plants) for each type of processes and carried out experiments, analysed results and developed the technologies. Some examples are Co-deposition through electro-plating of cobalt and chromium carbide particles, Chrome-aluminsing and platinum aluminsing, Detonation gun coating and the like for high-temperature erosion and corrosion resistance. Similarly, industrial processes were designed based on published scientific work and followed by actual trials and evaluation using production facilities. Some examples are the development of precision investment cast turbine blades, precision forged titanium compressor blades and isothermal forging for compressor discs for jet engines. An engineer's mind must scan the scientific literature and convert the acquired knowledge into technology and processes and products of utility.

Let us consider another area, say building science. Civil and electrical/electronic engineers and architects have converted scientific knowledge on materials, energy and conservation and developed the green technologies for designing and building of eco-friendly buildings with platinum/gold rating. Likewise engineers have used scientific knowledge of semi-conductivity, dielectric and piezo-electric properties, shape memory alloys, conductive polymers and other sensors and developed Micro-Electro Mechanical Devices (MEMs), intelligent materials and smart structures. Engineering mind must have a passion to search for scientific knowledge in their own and related fields.

I took over the reins of HAL as its Corporate Chairman in July 1997. It was an opportunity for me to transform the cost plus low-profit organisation to a globally competent and highly profitable aerospace industry with a paradigm shift to R&D diversification and growth through partnership. This I accomplished with the committed involvement of "Team –HAL". "Flight 99-HAL"[2] covers the story of transformation of HAL as an interesting case study for all engineer–managers. I shall not attempt to cover the same in this article.

Let me, but add that all engineers in their career growth whether in an industry or R&D/Academic Institutions will necessarily increase the management content in their job as they climb up the career ladder. At the top as CMD, Director, Dean, etc., management and administrative content will be substantial and engineer's mind needs to expand to accommodate the growing corporate and social responsibilities. Responsibility to society, environment and sustainable development, in any case, is engineer's paramount responsibility.

[2]Flight 99 HAL—Story of Transformation of HAL, CG Krishnadas Nair, Navbharat Enterprises, Bangalore–560 020 in association with Society for Aerospace Studies, New Delhi–110 003, First Edition–2003/Second Edition–2007.

Some engineers choose a management career from the start or after sometime in their chosen engineering career. For example, some may join Commerce and Business Management, Civil Services, Armed forces, NGOS, Public Governance and Administration and such others and even Politics. Why not? Engineers by virtue of their knowledge and analytical skills can take up any job and be successful. They should acquire additional knowledge and skills in policies and governance so that they can influence the establishments to develop good policies and governance and use technology for good and ethical use. Engineers must develop worthy goals and missions in life. Professional knowledge and skills must be complemented with good character and ethical values. Moral excellence and technical (professional) competence will lead to success.

During the last century, development and applications of technology produced spectacular developments and human achievements in various fields including in our Space Missions. Science and engineering enabled mankind to make exponential progress surpassing the material progress in the last 2000 years. Science and technology are growing at an exponential rate, but there is considerable deterioration in morality, ethical values, social science and disciplines with devastating effects in management and governance. Hence, some engineers ought to become leaders of society so that science and technology does not become an instrument of destruction. Engineers are blessed as destiny has given them a chance to be an engineer. So live the life fully as an engineer in whatever career path is chosen: a hard core engineer, scientist, technocrat, teacher, researcher, manager, administrator, politician, etc and be leaders. *An* engineer's mind empowered with character and competence, courage and wisdom, humility and selfless service, toughness and positive response to hurdles, dreaming, creating and leading enable one to reach one's goal of contributing to the growth of human civilisation. "*Loka Samasta Sukhino Bhavant*" (let there be well-being of the entire World).

A Tribute to Some of My "Teachers" Who Mattered the Most

Nikhil R. Pal

Finally, when I decided to write, I thought it would primarily be a tribute to some of those who made me what I am today. I was born in Tribeni, a little village in the district of Hooghly located about 55 km north of Calcutta. My father was a primary school teacher—a true bookish "school master", who was very smart and a symbol of honesty. He was able to do almost anything that I could imagine—from repairing watches to plumbing jobs. My mother was a homemaker, completely dedicated to the family. Both of my parents were happy to work, more than 18 h a day, not only to make us happy but also our neighbours and relatives. My father taught me never to be ashamed of any work as long as I believed that it was right to do so. He used to say that the primary goal in the school should not be to get good grades but to learn as much as you can and that would help in the long run. In numerous occasions, my parents would say, "whatever you do in life, do it with full dedication, but first and foremost try to be a good human being". I cannot assess where I stand with respect to their expectation, but these were the guiding principles injected by my primary mentors.

One of the important lessons given by my father was about way to "success". He said that there were three important attributes (I shall use the Bengali words and within parentheses the English translation) of a person: Saadh (volition, love for what you do/learn), Saadhya (ability), and Saadhana (devotion, pursuit). If someone has all three attributes, then (s)he is a genius. But any two of these are enough to be successful in the usual sense of the word "successful". He inspired by saying that I might not have the ability, but I could grow love for the subject/job and I could train myself to be dedicated. I always considered myself a mediocre person and still believe so and hence continue to bank on growing love for what I do and try to be dedicated to extent that I can. Whatever I am today is a consequence of this training and a set of events connected by AND gates—had any of these gates failed, I would have lost. I mention below a few of these events.

N.R. Pal (✉)
Electronics and Communication—Sciences Unit, Indian Statistical Institute, 203, Barrackpore Trunk Road, Kolkata 700 108, India
e-mail: nikhil@isical.ac.in

P. Ghosh and B. Raj (eds.), *The Mind of an Engineer*,
DOI 10.1007/978-981-10-0119-2_26

I studied up to higher secondary in village schools. My father was not solvent enough to support private tutors except for a short while in class eleven. But I was extremely lucky to get wonderful teachers at my schools, for example, our headmaster, Mr. Satya Bhanu Sur, Sir, who would be waiting for me in his office after office hours to help me with English. He wanted me to go for a research career and that is why he was very upset and expressed his unhappiness when he came to know that I joined industry after completion of M.Tech. This incidence was about nine years after I left school! What more can one expect in terms of love and motivation from a teacher? At that time, although I was working for my Ph.D. thesis as a part-timer, things were not progressing the way I expected. However, after meeting Sur Sir, I was determined to finish work for my Ph.D. and started dreaming of an academic career.

While in school, I had a dream of becoming a scientist. But my immature mind visualized a scientist as a person with dirty clothes, dirty long beard, who does not sleep and eat; does not have time for a bath; and is always in the laboratory doing something dangerous to know the unknown! So I thought of studying physics and got selected for admission to a few well-known colleges including St. Xavier's college, Calcutta. The very first impression of St. Xavier's College was so strong compared to the other colleges, I decided to join there. However, at some point of time during this process, because of comparatively higher tuition fees and other expenses, I thought of getting admitted to some other college. But, there was someone to rescue me. Mr. Satya Narayan Banerjee, a professor from a local college, when learnt this, came forward to help me through Rotary club. Finally, this was not needed because I got a tuition waiver after a few "class tests"; not only that, but also I got all the required books from a few professors of the college. They were kind enough to allow me to keep the books for the entire duration of the course. This is also an event in that serial circuit.

While in the college, at the beginning, I was too stupid to think that poverty is a "shame" (well, today I believe, it is indeed a shame, but not for the poor but for those who run the show!), and I used to sit on the last row to eat my tiffin (some chapatti and vegetables that my mother used to prepare for me) so that others, who were eating "fancy" tiffin, could not see it. But with time, my college gave me courage and confidence. Although I was a shy boy in the class, in three years I wrote three articles, "Views from a train", "A modified Poynting's method for computation of G", the gravitational constant, and "A correction to Poiseuille's method for finding viscosity of water". These were published in college scientific magazines in different years. The magnitude of the correction to Poiseuille's method that I had found was much larger than the other corrections that were taught to physics students. Unfortunately, the text books still do not include this correction. Those were my first "research", if you allow me to say so and they gave me a lot of satisfaction and confidence. I got my first taste of research!

After this, I went for a Master of Business Management (MBM) course with specialization in Operations research. During the three years of MBM, I was teaching a few students at home to provide some financial support to my family as well as to cover my own expenses. After MBM, I was about to join a job because it was a need for the family. But my parents and others in the family were happy to go through hardships and wanted me to continue study, if I wanted to. Very reluctantly I sat for the entrance test for the M.Tech. (Computer Science) program of the Indian Statistical Institute (ISI). To my surprise, I got selected, but provisionally because my MBM result was not announced. The institute wanted me to submit my final results by some specified date, but that was almost impossible. Somehow, one of my MBM teachers, Prof. Anil K. Sarkar, came to know of it (I still do not know, how Sarkar Sir came to know it). One day Prof. Sarkar came to ISI from his residence near Hindusthan Park (a place quite far from ISI), talked to our course incharge, and found a way to rescue me. Prof. Sarker made all necessary arrangements so that the viva voce examination of my MBM dissertation could be completed and confidential results could be sent to ISI. I would not be writing this article, if Prof. Sarker would not have had voluntarily took so much of trouble for one of his students. Another major event in that is serial circuit.

Although my family never asked, I decided to go for a job after passing M.Tech. I joined Hindustan Motors, as it was near ISI. My intention was to work for the Ph.D. degree as a part-timer. I could not enjoy my work at Hindustan Motors. So after about eight months, from the car manufacturer I moved to work with a tyre manufacturer, Dunlop, India Ltd, which was near my hometown. Simultaneously, I was doing some research work under the supervision of Prof. Sankar K. Pal. Although I had published some papers during that time, I was not happy with my research progress. After working for a period of two and a half years in Dunlop, India, I joined ISI in 1987 as a programmer and became a full professor there in 1995, and I am still there. My journey to complete the Ph.D. thesis was rather smooth, but took more time than one would be happy with. My supervisor gave me enough freedom to work on whatever I wanted and never imposed any constraint—possibly he understood my basic characteristics. My Ph.D. thesis is primarily on image processing. I love modelling and quantification of uncertainty irrespective of its form, and in my thesis, there are two chapters on new measures for probabilistic uncertainty and fuzzy uncertainty, and of course, they are connected/applied to image processing. I obtained my Ph.D. degree in 1991 from ISI. My love for dealing with uncertainty (although I hate gambling in any form) is still there, and I have been continuing to work in that area. Immediately after submitting my thesis, I attended a workshop on neural networks at the Indian Institute of Science, Bangalore. This made me interested in neural networks.

After completion of Ph.D., I got an offer for a regular academic job from a university in Australia and a postdoctoral fellowship from USA to work with Prof. James C. Bezdek (subsequently I shall refer to him as Jim) who at that time was (and still is) one of the most renowned fuzzy researchers. Of course, my ISI job was there. I was in a dilemma. Part of the postdoctoral fellowship was for working on self-organizing maps, and this was the major driving factor for me to opt for the USA. Some senior professors including my advisor also suggested me to join Prof. Bezdek. It was another major step in my life. This is the time when I took fuzzy sets and pattern recognition seriously. The postdoctoral period was probably the most enjoyable and academically most productive period in my career. Since 1991, I have visited Jim many times. I have done research collaborations with many national and international researchers, but I have co-authored the maximum number of papers with Jim. Prof. Bezdek is known to be an excellent researcher, but many think that he is a "rough and tough" person. In reality, he has a very soft heart and is an extremely nice person who prefers to call a spade a spade irrespective of the person in front of him. He was a great mentor, both academically and otherwise. He tried to make me adapt so that I could be more acceptable by the so-called civilized society, but probably he failed there! I have worked with him on different interesting problems. One of the problems that I enjoyed very much was quantification of total uncertainty in evidential framework. We have written many papers on clustering and cluster validation. Fuzzy clustering was introduced to eliminate some problems of crisp clustering, but in the case of fuzzy c-means (popularly known as FCM) points equidistant from the cluster centres get equal memberships irrespective of the magnitudes of their distances. To eliminate this problem, possiblistic clustering was introduced, but that suffers from the "coincident-clusters" problem. In one occasion, I visited Jim along with my wife, Dr. Kuhu Pal, who was a postdoctoral fellow. Incidentally, at the same time, Prof. James M. Keller, one of the originators of the possibilistic clustering, also visited Prof. Bezdek. During that enjoyable trip, we all four worked together and found a solution to both problems in a single framework leading to what is known as possibilistic fuzzy c-means clustering algorithms.

Things will remain incomplete, unless I share at least one story here. During one of the visits, I became quite sick because of problems with my cervical discs. Jim took me to a neurologist who among other things suggested me to sleep on a firm bed. While coming back, Jim asked me about my bed. I told him that it was very soft and I would be doing something, I was not sure what though. The following weekend, in the morning, I heard my door bell ringing. When I opened the door, I was surprised. Jim brought a piece of plywood of size about seven feet by four feet in his pickup truck and some tools for carpentry work. He carried that heavy plywood all by himself, as I could not help due to my illness, and cut it into the right size to fit my bed. I suppose, only a few lucky persons like me can get such a caring friend and a great mentor. It is interesting to note here that Jim is a great woodworker.

My health has never been very cooperative with me. However, like in academics, here also there is someone to look after me. He is Prof. Subrata Banerjee, a renowned neurologist, who has been treating me for almost all illnesses that I have suffered from for about 30 years. More importantly, in my crisis, he is always there with suggestions, courage, and strength. He has been like my elder brother, a mentor, and a guide.

Since 1991, I have been primarily working on two facets of computational intelligence—neural networks and fuzzy logic, and I occasionally do some work in evolutionary computing. In 2002, I became interested in various problems in the area of bioinformatics and still continuing to work in this area. I am happy with what I am today. My work has given me some international recognition. I became a fellow of the IEEE, USA, and a fellow of the International Fuzzy Systems

Association (IFSA). I have served as the Editor-in-Chief of the IEEE Transactions on Fuzzy Systems, possibly the most prestigious journal in Fuzzy Systems, for a period of six years (2005–2010)—the maximum period that one can serve. At present, I am serving as the vice-president for Publications of the IEEE Computational Intelligence Society. I have obtained the *2015* Fuzzy Systems Pioneer Award of the IEEE Computational Intelligence Society.

Looking ahead, I want to continue my work in computational intelligence/ machine learning. One of my favourite topics in pattern recognition is dimensionality reduction where in my group we have made a substantial amount of contribution. I would like to continue working in this area. Recently, I have started spending quite a bit of time in the area of brain–computer interface, and I hope to continue the same because this area is quite challenging and it can offer many things to the society.

In addition to the few great people that I have mentioned above, I gratefully acknowledge many others including my students and numerous national and international collaborators who have helped me in different ways in different contexts. Of course, things will remain incomplete if I do not acknowledge my family who continues to sacrifice the most.

The world is still so beautiful because "they" are there to help.

A Journey of Self-realization

Rajaram Bojji

My father Dr. Bojji Subbannachar, a gold medallist of Stanley Medical College (Madras those days), for lack of resources for further education, joined Auxiliary Med Corp in 1940s, lived in some makeshift hut with wife (my mother) Mrs. Manjula Bai in a village, somewhere close to Yellamanchili, a small town, in Visakhapatnam, Andhra Pradesh. Daily he attended patients in nearby villages riding in his bike with medical kit. My grandfather, not a well to do man, kept praying at an ancient temple in agraharam, a collection of Brahmin households, near Kadapa, anxiously for safety of his only son as Japan was threatening to bombard those areas! Well in such a hut as habitat, they begot a son in 1945, that too after a few miscarriages. That happened to be I and received unstinting love and care from the parents. I never had to shed a tear in my living memory. I never heard a harsh word as a child or as a kid.

My father's services automatically got transferred to Composite Madras and later to Andhra Pradesh State Medical Services. So my education followed his postings. I did my elementary school in Onipenta village, somewhere close to Kadapa, later on Kandukuru, close to Nellore, followed by Vijayawada, and then Yellamanchili where I finished my high school from 5th to SSLC—all government or municipal public schools.

Simple friends, generally poor background, status-wise our family better placed, studied with kerosene lanterns, upgraded to Petromax light in later years. It was exciting to use old battery-powered table radio, with 20-m stretched copper wire as antenna outside home, to tune into shortwave stations and listen to VOA and BBC in those days in villages! With little English knowledge and help from father, I slowly started to follow these transmissions, starting from my class 4, which did influence my thinking. Programmes on science, the worldwide coverage of events, redefined my own borders of world within my mind. Much time was left alone to handle myself to find something to do. Home generally being a little far away in most of the postings for my father, friends of school came not too frequently and that too only in evenings. Alone, I used to sit outside in verandah of reasonably

R. Bojji (✉)
Rajaram Bojji Inc., 834, Spring Knoll Drive, Herndon, VA 20170, USA
e-mail: rajaram.bojji@gmail.com

P. Ghosh and B. Raj (eds.), *The Mind of an Engineer*,
DOI 10.1007/978-981-10-0119-2_27

spacious old bungalow attached to hospital, watched birds, and ants! Yes, ants fascinated me. So much so that, when I found they had to crawl over a portion of stone-paved areas around my house, I used to cut a few rubber balls into pieces to place them as shelter against hot burning sun, and their tiny "feet" would get shade to crawl. In a manner of speaking, ants remained my close friends and amazed me the way they communicate and build up their living habitats and care for each other. So tiny but such a grit to stand up to world with giants like me, I used to think.

Hindu editorials, my dad told me, are good for my English and the practice of reading kept up my touch with the language even though I was in Telugu medium schools throughout. High school gave me the opportunity of social service through Scouts and Guides Organisation as well as Auxiliary Cadet Corps.

My parents encouraged. Without any stress of studies, I found myself getting first rank and always wondered why the more deserving hardworking friends are not getting that number one tag. I used to think they are better actually than me. I used to tell them the same. But in a friendly manner, they used to laugh it off. Those were truly wonderful days! One problem with me was I could not study too much, after getting a gist of textbook and used to find easier to write whatever I felt like in my own way. My friends used to practically reproduce the textbooks! That used to surprise me.

Days were truly beautiful. School fun. Algebra took my interest and used to finish the textbook in first quarter and to clear doubt I used to go to maths teacher's home and disturb the good lady to clear my doubt, portion being much ahead of the class. Never had tuition, but what a wonderful loving teachers they were! Every single game I played—hockey, football, cricket, and badminton—I enjoyed, suffered injuries, went back again to play and enjoyed the free kid's life fully!

For preuniversity course, I moved out of home to Loyola College, Vijayawada. I stood first in school winning some medals at high school, but by that time, my father moved on transfer to Jaggayyapet, near Vijayawada. I never bothered to collect medals. For some odd reason, thought to myself, let them remain with the social studies teacher who collected them in my absence. I used to think the honour if any belongs to the teachers, not me.

Loyola college studies for one year, now totally in English medium proved to be different experience and a challenge. Actually, when taking admission, the Principal Fr. Gordon, a very gentle graceful tall man in robes, summoned me and my father to his office. Our application read that I had opted for biological sciences. No doubt I had excellent qualifying marks both for maths/physics stream and for biological science stream. Those days, the choice has to be made. My mother and dad wanted me to be a doctor, like father. Well, I used to be so scared of even touching a cockroach (true even today), and I was not sure it was a great idea. But then, I listened to mom because they never asked or forced me to do anything in life. Father Gordon told my father looking at the marks in maths in my SSLC certificate that, with algebra as main, the boy should actually take maths, physics, chemistry, and logic. He told us that I would be much better off as an engineer, naturally. My father looked at me. I was happy within, but then I said that if dad agrees, I would change option to become an engineer. But what to tell mother on returning home?

My father smiled, told me not to worry, agreed for the change, thus, my fate line changed to become engineer instead of a doctor! How sensitive my father had been —a mere look at my face he could read me and did not lose one instant to change what we had decided after days of struggle and debate at home.

Loyola hostel life was memorable; compulsory study hours, compulsory sleep, timely lunches and dinners. Either you finish food within allotted 15 min or lose your plantain leaf and get out of the mess. At the end of the year without my knowing, I was announced to have won the Best Conduct Prize by the tough sounding and looking but actually kind-hearted Warden Father Miranda! I received Tagore's Gitanjali on stage, to loud applause of a thousand students and faculty in 1960. I never knew they advised my parents to come for the function. How can I describe the joy of that moment! I owe quite a lot to the dear fathers in Loyola who in one year moulded me substantially. That also was my first year away from my parents in my life, a kind of traumatic but life changing.

With PUC marks, one enters university those days. With double centum in maths and pretty high standing in other subjects, I practically received invites from universities without tensions. The seashore of Visakhapatnam made me choose Andhra University. Since my father also was in his first postings in war times worked there, enthusiastically I concurred.

The university life, started with stay at those circular-shaped round hostels, may be a unique feature! But Loyola Hostel too was oval-shaped and closed and so a sense of continuity for me.

After the usual rather toughening welcome games of seniors we settled down, with a resolve that we won't do such things as seniors. And we didn't too, breaking a tradition. We welcomed juniors in a more civil and loving manner, I believe. The period marked activity in literary circles as Literary Secretary, usual Hostel Secretary, then General Secretary of Students' Union followed finally by being President in final year. The NCC, their camps gave ample opportunities to be trained in fire arms too! Cleaning of guns, polishing shoes to perfection was fun in its way. By choice, though merit wise eligible to take any branch, I became a Civil Engineer, a strong desire of parents to see me in Andhra Pradesh working as Superintending Engineer in PWD! In the University days, I became a good player of Tennis and table tennis. I was a stage performer of music with Banjo, and sparring partner in Chess for the University.

Without much effort, I was bestowed first rank in university with distinction in 1966, which landed open invitation from IIT Kharagpur to join M.Tech. programme in structural engineering.

That was the time of dilemma for me. Like everyone else, I too filed papers for admission to US universities, and University of Maryland (I recollect, but could be another one) offered admission to MS programme with the promise of assistantship too.

A strong sense of nationalism I used to advocate as student leader forced my student body to forego rice in hostel mess during Chinese war, to conserve rice, a staple diet, replaced with rotis, saved money foregoing usual long distance study tour donating the funds for jawan welfare. Well, when education of the same level as in USA is available in our country, why go abroad? With that thought, I parted with my friends who left for USA without assistantship, but I went to IIT Kharagpur, foregoing my assistantship. My friends did call me a little crazy, as usual.

University taught the usual subjects as per syllabus, but I was fortunate those days with availability large number of Russian (translated into English) and US aid material on scientific matters at substantially subsidized cost. I took tremendous pleasure in reading and understanding tensor calculus, advanced physical sciences, space exploits, and principles covering the space travel and emerging computers. Truly a rewarding period!

IIT Kharagpur opened up a totally different work culture and study atmosphere, so different from that of university. The Professors, lecture notes more as pointers and much to be gleaned from the extensive Library, that too from latest current Journals publishing papers of the most recent advances in each field, brought me to the frontiers of knowledge, where my mind can fly. The easy relationships, the eager expectations of discovering something different and new is what appealed to me. I was never a man who could be straitjacketed even in high school curriculum. I used to write my own things, discussing and debating. Many of my friends used to say lucky, and some crazy teachers seem to have evaluated allowing me to get even higher rank. So IIT was a fresh breeze for me, to my liking. Shell structure, advanced theory of elasticity, and philosophical approach of von Mises tickled my brains. Kani's work attracted me. My own forays into true ultimate strength assessment for concrete by detecting different stages of cracking using ultrasonic waves, formed thesis to finally earn my master's degree, and went on to become a research scholar. I wanted to take a Ph.D. and then seek positions abroad for further research. But having married young, the girl I loved, in my second year of master's programme, I found it tough to balance life with stipend I received as research scholar. Finances being what they were forced me to write UPSC examination seeking to join the Indian Railway Service of Engineers, in 1970.

I entered altogether a different world. Training at various levels, the codes of practice, technical inputs very special to railways, etc. Hands on real life experiences too were included in the two years of probation. It took me all over India, followed by tests to certify me fit to take a working position. Perhaps the best programmes to go through, provided you are serious. As an officer, there was always scope for taking things easy, which actually some seniors used to suggest. But I enjoyed working with gangmen in harsh sun and going through use of the tools myself, feeling at ease with them. The rides on steam locomotives with drivers and the inspection methodology all add up to tremendous confidence by the time, 1972, I was posted as Asst. Engineer Shahabad, near Wadi, in Gulbarga Dist. Karnataka.

A huge impressive bungalow of old style with garden and a saloon too added to my status with almost a thousand working those days with lots of temporary gangs on rolls. The 24/7 work schedule hardly allowed me to savour the privileges! Gheraos, unionized labour test one's leadership qualities. Missing sleep riding locomotives boarding past midnight and getting off at 2 a.m. at some far away station was normal. Daytime push trolley over the entire stretch at walking speeds, every month kept a hectic work schedule. But the relaxation for me was to be the change I want to be for the system. SO following my dictum that a solution can have any amount of complexity, but for working it should be simple, invented a few gadgets for the largely illiterate gangmen to make curve realignment simple. A lady's hairpin with a locally cut GI sheet triangle with 6 in. × 2 in. sides, providing 1:3 ratio, used with a 20 m string as tools allowed the gangmen to do realignment of railway curves all by themselves. The principle was simple. The versine at mid chord is 3 times the versine at half chord for a perfect circular curve. The hairpin used as ref marker for versine at mid chord holding 6 in. side perpendicular to rail, then the device held vertically at quarter point of the chord string. If it was matching with hairpin location, nothing to be done. Else slew applied by gangmen at mid point to move the chord to match. This progressively goes from one end to another and then a reverse pass given to correct for accumulated errors on variable curved transitions. The quality of work done recorded by taking routine versine measurements is broadly the method. The job otherwise involved laborious measurements of versines, over several kilometres, and then send them to Hqrs CE office, for working out on a computing device the slews required to correct, and on filed fix reference points and then slew under supervision of an Inspector. Called it a Standard Curve rectifier, which got a National level award at a conference organized in Delhi. First year of service this was produced. Then over next two years, a new washable apron which can be executed without sanction of any work by Hqrs, being too economical, packing tube for measured shovel packing steel trough sleepers of those days, were created. All these works recognized, peer reviewed, and honoured by publications in the railways. I used to enjoy the feeling that we do things which no one produced earlier in the world, and we were the first. This doing something different kept high motivation and excitement within staff, making every day exciting.

The normal heavy workload of managing personnel files, the inspections, 24/7 routine on tracks, contracts, and execution of works were being managed to the best satisfaction of superiors. In addition, these more exciting things were done making the job a joy!

We worked together and eliminated the practice of manual scavenging in railway colony, involving 250 toilets. Government was not sanctioning the budget. So we organized the scrap tar drums, two of them forming septic tank, buried underground, used cement pans locally cast and called entire work as drainage repairs and finished off all the 250 units in 3 months. Staff felt elated. It is simple but touches lives. I had to face some administrative grumblings, but unions stood by me. Well, I had more than normal fill of brushes with administrative machinery normally working to a rote. Change begets it.

Three years over, I promoted and posted as Divisional Engineer of Sholapur. From then on within short span moved to Secunderabad, then to Research Designs and Standards Organisation, Lucknow, because by that time already had a few papers in our technical journals and some awards at All India level. Posted to research Directorate in track wing, in 1976 remained there, as Joint Director/Research till 1987.

Perhaps, this suited me the best.

The first task was to find a method of manufacture for the mass imported elastic clip called Pandrol clip, as RDSO was trying to indigenize its production. A complex spatial curved one was defying the die-making process in those days. That is the time I within 2 months could envisage a new concept die constant, by plotting the variable spring constant with rotation of clip and optimizing the correct orientation of the main spatial curve. This was used to correct the die, and the first Indian clips manufactured within three months with our own technique. The reward is that Pandrol Company heads visited my laboratories, to understand under permission from Railway Board and, of course, then copied, to produce a new variation which they promptly patented! Our government has little respect for Indian intellectual property I realized.

Anyway, stoically, worked further and making the lab technicians my co-partners in spirit, we recommissioned our own furnace scrapped years earlier, and we produced a brand new clip much economical and better performing, which as RD3 got international patents including USA. The reward I could give my staff was to make them co-patentees! Govt was lukewarm. But we enjoyed the journey. Then I realized the reward is not to be looked for from the government. We will make life more interesting and rewarding by our own work and fight the system, by getting patents to prove we are doing innovations and thus followed further innovations from the laboratory and patents. Yes, under pressure, government did conduct field trials, etc., but had a way of favouring foreign origin items only. There are many ways for blocking intelligently implementation. But this did not affect our spirit. What did not happen in RDSO record happened in my time with the same staff and lack of resources. Those dilapidated sheds certainly made a shining example to all others in RDSO. We refused to be unhappy—enjoyed the forays to frontiers of knowledge.

This is the period of high productivity with 300 plus research test reports filed, and I personally getting international recognition because of my papers on New Theory of Rail Wheel Interaction, Simple Theory of Track Vehicle Interaction, and two more original papers. It was gratifying that in Europe students worked on the basis of my papers for their Ph.D. and sent me thanking communications. But of course not a mention within peer groups nor a debate!!

Fulfilment is that I could do what I wanted to do. Rest whether, the system benefited or not depended on many other factors, beyond my control. I learnt to accept success norm as my work should remain of world class and recognized by third-party international bodies and peer reviewed at world level. Local biases notwithstanding, I succeeded by my norm. Since I did not look for any returns, by donating all intellectual property to government, I fortunately did not face any disappointments. I knew that I, as a man of organized services, should be mere manager type and not supposed to add research inputs of such serious degree. So I remained an odd man in.

I had an opportunity to go to Jordan in 1983 or 1984 as consultant to investigate the cracking of aluminium freight stock made by French for Aqaba Railway Corporation—a three-month assignment. With simple accelerometer, analogue one, and one assistant I landed there. Manually and laboriously, we worked frequency responses and found the cause was ill-designed suspension. This report of analysis got accepted by World Bank which granted relief to the Aqaba Railway Corporation. Work being finished, earlier than the allotted time, we came back, foregoing the remaining daily allowances in foreign country…It is great that my assistant too followed me without demur!

At the same time, I was in Jordan, and I had to make a trip to Vienna to assist on field trials for one of my inventions, microprocessor-based rational analysis for track maintenance (RAMTRAK), which aims to correct long-wave defects in tracks for very high speed (>250 kmph). It was my first trip to Europe and I was amazed at the respect I received there and the privileges granted for my stay as their guest. As one accustomed to be mostly ignored or treated indifferently, this was unforgettable experience and realization why we Indians are technologically a resounding failure, while Europe puts on pedestal technology and innovation. I was a nuisance back home, but here it was different. I came back with a firm resolve we should make some structural changes in RDSO.

This experience forced me to pen the restructuring the RDSO report for changing the attitude of the RDSO from being a system of just a doormat for blindly stamping whatever executives wanted to import, without own effort of building original knowledge base. Well, it took lot of effort, but it was adopted by the progressive minister at that time, and RDSO reform was affected with budget grants for setting up new laboratories and relaxed norms for international conferences. But the fear is that finally the relaxed norms will be used by the mighty to make ego trips out of the country but really not to help the country always lingered, as mentioned wryly, by Minister himself, while agreeing to my report. Rankwise, I was too junior to deal with him. But he was gracious. Then, I did not stay to see the results. I had to move on.

But the period of 7 years spent in Research Directorate gave unique rewarding experience of hands on working on real-time systems of microprocessors, measuring wheels, track recording cum research vehicle and finally resulted in my own conclusion that instead of power spectral density, one can as well simplify using appropriate chord measured versines as in–out data and the standard deviation defined as standard definition for track geometry assessment. Similarly, working opportunity for Planning Group for Railways at Planning Commission level and assisting the likes of Mr. Menezes, Dr. Jagjit Singh inspired me to develop computer simulation models for complex railway operations and assess effectiveness of alternate options for improving railway performance and safety. Every department of railways I covered for what should be the future path to be adopted for railways. The knowledge gained and contributions from my thinking getting accepted at such high level tremendously boosted my sense of fulfilment.

Landing in Lusaka for a stint of three years with Zambia Railways in 1987 with my wife opened up a starkly different kind of experience. We had a 2-h drive to reach a motel called masiye to stay along with our team for a few weeks before regular accommodations were organized. We Indians are great as a team working abroad. Practically, our ladies took over the hotel kitchen and trained the cooks to prepare a few Indian dishes!! Well, some fun.

Once settled, as the design Engineer my duties covered anything and everything connected with cerebral work in regard to civil engineering starting from simplest residential home construction along with electrical wiring plans to track health monitoring and replacing programmes, not leaving out procurement activity preparing needed documentation and tender proceedings for Directors' approvals. But the entire workload was not too taxing really, used to take only 15–20 % of working time.

So difficult to remain idle compared to 14-h work a day, I was used to in RDSO. But the spare time became an opportunity to seek my dream to create a new railway system which has all the advantages of coned wheel with single flange on rail, providing high-speed transportation, but risk of derailment or capsizing killing people be totally eliminated.

I started to work from the basics and building up a system radically different from the known system. To prevent the bogie with wheel sets escaping the guiding tracks, I enclosed the same in a restraining box, but the coach taken out suspended from the travelling bogie overhead. Though similar to an overhead crane, it was modified to have high speeds by using the same railway bogie sets with coned wheel sets and single flange; behaviour and physics same as conventional ones, but with the coach containing the people, travelling below the tracks not above the tracks!

In Bologna University, as part of their 900 years of completion celebrations, this technical paper was presented in international conference and it was a memorable experience, in 1989.

But on return to India, in 2004 as MD Konkan Railway could really try the concept out!

Lucky that with industry support I could put up a full-scale prototype within 3 months, costing Rs. 7 Crore, and that put Railway Ministry to shame for refusing me earlier the funds.

Out of the sanction of Rs. 50 Crore, tested the live system over standard 1.5 km track to international standards. BARC scientists too chipped in. It actually speaks volumes of the prowess of our scientists and engineers that within 90 days such momentous work was done! International technical bodies like American Society of Civil Engineers, and even channels like National Geographic and Discovery covered this new Metro system, SkyBus. Prime Minister of Malaysia offered a grant of Rs. 2000 Crore to build first 14 km route for Hyderabad. But the Govt of India for its own reasons did not allow. Now the test track is proposed to be sold as scrap.

Story of technology not denied from abroad, so we deny our own technology.

On return from Zambia in 1990, I was posted as Chief Engineer, Goa fixing the railway alignment in highly aware Goa! I could use remote sensing satellites to prove that railway does not hurt environment. This work on remote sensing issues became a technical paper in an International Journal, London along with scientist from IIT Mumbai.

Finally, political pressures resulted in appointment of Justice Ozha commission who upheld the work done by us.

Then, I moved on as board member of Konkan Railway in charge of the project and later became MD in 1997. The construction experience, challenges faced, and resistance of higher authorities and seniors to geotechnical science are something unbelievable. Administrators rely on outdated own experience rather than advancement in engineering and technology.

In Konkan Railway, I followed my own norm that technology should be used to serve people. New innovations resulted in like anti-collision device, the SkyBus metro, maintenance of free tracks, and self-stabilizing tracks, to mention a few.

The ACD, anti-collision device, looks simple, but has very many features of artificial intelligence built in to handle uncertain data and information gaps, and the assurance of safety. For the typical railway signal engineer, who is conditioned on go no go kind of thinking it was too difficult to absorb. But the innovation met all the demands made by them of even implausible situations. That was confounding. But this development forced French Consulate to come all the way to dissuade us from pursuing, because it adversely affects their industry. That is the difference between us and western countries. But we received Parliament approval after due tests, and implemented over 2000 km. But lobbies still worked, and after my retirement, the RDSO started another project renaming it. Technology gets easily defeated by administrative actions!

But innovative Roll on Roll off service, for trucks bridging the distance from Mangalore to Mumbai, almost 600 km, by rail, saving fuel and decongesting the NH17, still continues successfully earning hundreds of crores over the last decade and half. Similarly the wide area networked IT based railway operation and management, which I took pains to build from scratch, handling even real-time decision making to run trains apart from the standard ticketing, accounting and three levels of decision making capacity with built in knowledge base to train station masters too through on line refresher courses has proved itself to be sturdy, reliable, delivering over last 15 years 24/7 service, with no breakdown. Truly is something India can be proud of.

I am fortunate to have achieved many a dream in India. On retirement, I moved to USA as resident to stay with children. The association with Empowered Board for delivery of ITER, for India, is a rewarding experience over 5 years. Then, I continued research for using gravity as alternate perennial source of energy for mankind. The paper Gravity Powered Road, Rail, Water and Runway transport systems got peer approval in international conference APM2011 Paris in 2011. I feel happy it has become part of proceedings of ASCE.

Currently, I am working to put up experimental Gravity Power House to generate energy from sea waves and transport systems using gravity power towers. We can stop using fossil fuels on the planet, practically if we succeed here.

That is my dream.

A Sketch of My Life

Rabindranath Ghosh

It was beyond my imagination that I may have to write a short sketch of my life. It is indeed a privilege to be a fellow of an academy that encourages doing so. I come from a family of teachers. My grandfather used to teach mathematics at colleges at Patna and Muzaffarpur in Bihar. He died in 1949 a few months after I was born. My father had a distinguished academic career. He was a captain in the Army Medical Corps during the World War II. Later, he came back to teaching. He retired as principal, Darbhanga Medical College in North Bihar.

My early education was through private tutors at home. This was primarily because my father had to spend nearly 3 years in Edinburgh during the early 1950s for his doctoral degree. My mother along with me and my sister had to shuttle between our ancestral house at Rajmahal and Patna, the then state of Bihar, where my grandmothers lived. On his return, my father joined Darbhanga Medical College as a professor, and we moved to Laheriasarai where the college was located. I started going to school as a regular student when I was 9 year old (1957). I was admitted directly to class VI in M L Academy (also known as Saraswati School possibly its initial name), a Hindi medium secondary school. It was established with assistance from the Maharaja of Darbhanga. Although there was a Zila school in the town, my father insisted that I should go to this school because of the reputation of Shri Jhingur Kunwar its headmaster and Shri Nani Gopal Chakraborty its assistant head master. Both of them were highly respected teachers. I was an average student until 1961. It was possibly because of my poor eyesight. Once I started using glasses, my performance started improving. This was also the year our school was upgraded to higher secondary. It meant on completion of school one could get a direct entry into engineering or medical colleges. Mathematics was my favorite subject. Ours was the first batch of this school to appear for the Bihar Higher Secondary Board examination held in March 1964. Our teachers in the school had great hopes, but only six of us in a class of around 80 students passed in the first division. A few of my classmates appeared for the Joint Entrance Examination for

R. Ghosh (✉)
Department of Metallurgical and Materials Engineering, Indian Institute of Technology, Kharagpur 721 302, India
e-mail: rnghosh@metal.iitkgp.ernet.in

P. Ghosh and B. Raj (eds.), *The Mind of an Engineer*,
DOI 10.1007/978-981-10-0119-2_28

admission to IIT held during the first weekend of May 1964. When I went to appear for the examination held at Bihar Engineering College, Patna, it was a great surprise to find that Arun Pradip Sinha the topper of our class was sitting next to me. We had four papers Mathematics, English, Physics, and Chemistry. Questions were of much higher standard than those we were familiar with so far. I still remember that the Mathematics paper had 25 short questions of 1 mark each and there were negative marks for every wrong answer. Luckily, I concentrated more on the remaining fifteen questions of 5 marks each. We were certainly not happy with our performance in the examination. It was a great surprise to find that both of us had qualified. Those days successful candidates had to appear for counseling for allotment of seats in various disciplines and face medical examination in respective zonal institutes. We had to come to IIT Kharagpur, and we were impressed with its sprawling campus and the facility. Students were called in batches of 60. Arun's counseling was held on the first day he was admitted to Mechanical engineering, whereas my counseling was on the following day. I had to settle for metallurgical engineering. In those days, IIT Kharagpur used to be the first preference for most students. For a little-known Hindi medium school in a small town in North Bihar, it was indeed an achievement to have two of its students qualify in such a competitive examination and get admitted to IIT Kharagpur.

My association with IIT Kharagpur formally began from the July 1, 1964, as a first-year B.Tech. student of metallurgical engineering. Hostels in the campus were named after our national leaders. I was a resident of Nehru Hall during my undergraduate days. Barring a few months, most us got single-seated rooms. I preferred to stay in a double-seated room during my first year along with Pobitra Majumdar a student of naval architecture. Those days only the double-seated rooms used to have ceiling fans. For a sixteen-year-old boy, it was tough to be away from home. It took some time to get adjusted with the new environment. The tight academic schedule and examinations held at regular intervals kept us busy. The first two years were mostly devoted to basic science and engineering courses. Every year we had to face three examinations held during September, January, and April. The end-term examination used to have questions from the entire course. There was also a system of open-book examination on a few subjects where students were allowed to consult books during examination. Interaction with the Department of Metallurgical Engineering was minimal during the first two years. We had occasional orientation classes where we had opportunity to listen to a few faculties of our department. This is when I met Prof. E.H. Bucknall. He gave us a general overview of metallurgy. On his suggestion, most us purchased a book titled Metals in the Service of Man which I still posses. It gives an excellent exposure on the extraction, processing, properties, and uses of all common metals and alloys. The first three years were mainly devoted to basic courses in mathematics, physics, chemistry, and engineering, whereas subjects related to metallurgical engineering were taught in our fourth and fifth years. My favorite subjects were metallurgical thermodynamics, X-ray diffraction and metal physics. These were taught by Professor P.R. Dhar, Professor P.G. Mukunda, and Prof. S.K. Mitra, respectively. I still remember having solved all the numerical problems given in the book of

Metallurgical Thermodynamics by Darken and Gurry. We were also fortunate to have access to most of the laboratory facilities that were available in our department.

After obtaining B.Tech. degree, I joined the M.Tech. program in physical metallurgy offered by the department. The first year was devoted primarily to lecture and laboratory classes. The courses were a little more quantitative. Therefore, I started liking the course, and my performance in the class improved significantly. I was assigned a project on tempering characteristics of steel having Si as one of the intentionally added elements. This was done under the guidance of Dr. R.A. Tewari and Professor P.R. Dhar. Dr. Tewari around the time had just completed his Ph.D. His thesis was on the tempering behavior of Si steel. Professor Dhar wanted me to investigate the effect of Ni addition on transformation characteristics of this steel. One of the objectives of Ni addition was to suppress the effect of auto-tempering. The amount of Ni to be added was found to be close to 18 %. The alloy was made in an air induction furnace. The alloy was amenable to both hot and cold working. It had to be quenched in liquid nitrogen for hardening. I had to design and fabricate a simple setup to measure magnetic permeability after quenching the sample to subzero temperatures so that the temperature at which the soft austenite in this steel transforms into hard martensite could be determined. On tempering the steel at 400 C after quenching in liquid nitrogen, it was found that the hardness has increased from 400 VHN to 500 VHN. Hardness versus aging time plot was very much similar to that shown by age hardenable alloys. Attempts to identify the precipitate by recording X-ray diffraction pattern using Debye-Scherer camera or selected area diffraction patterns obtained during the examination of extraction, carbon replica or thin foils under EM6 transmission electron microscope only showed reflections that were coincident with those expected from austenite. It was found that diffraction peaks from Ni_3Si could be coincident with those from austenite. In the absence of access to SEM or TEM having EDS facility, its presence could not be proved conclusively. Nevertheless, it fetched me master's degree with excellent grade. I also received the Alumni Association Gold Medal on being adjudged the best postgraduate student for the session 1969–1971.

I subsequently got selected for the post of lecturer at IIT Kharagpur. This gave me an opportunity to serve my alma mater from October 1971. I took up the profession of teaching seriously. This helped me develop a clear concept on the evolution of structures in materials during processing and its effects on its performance. While going through the literatures on strengthening mechanism in metals and alloys, I came across two papers by Sandip K. Mitra one of my teachers; one was with Donald McLean of NPL Teddington, UK and other with J.E. Dorn of University of Berkley, California. Both of these were widely referred in technical papers and books. Therefore, I decided to work for my Ph.D. under his guidance. Amitabh Biswas who was two-year senior to me was also working for Ph.D. under his supervision. He had an excellent skill to setup experimental facility. He had built a unit which could grow single crystal of aluminum and a constant stress creep testing machine. We started working together as a team. Development of dispersion strengthened alloys for high-temperature service was an area of considerable interest in the early seventies. At high temperature, all metals undergo time-dependent deformation even if the stress is lower than its yield stress. This phenomenon, called creep, is associated with the movement of dislocation (crystal defects). The presence of stable dispersed particles inhibits such movement. This results in the improvement of its creep resistance. I used co-precipitation technique to produce silver-coated silica particles so that these could be incorporated in the metal to improve its creep resistance. It was possible to control the silver-to-silica ratio. However, it could be melted and cast only if the amount of silica was very less. In cases where the amount of silica was very high, it did not melt at all. It was possibly because the coating was too thin. There was marginal increase in hardness wherever melting and casting were possible yet the tensile strengths in bulk of the cases were found to be lower than that of pure silver. Meanwhile, I also picked up FORTRAN language to write computer programs to solve numerical problems. Those days the only facility that was available was the mainframe computer. We had to write the code and submit the job in the form of a pack of punched cards. Often, we had to wait for several days to get an output in the form of a print out. It used to take several months to debug and run a couple of hundred lines program. In spite of this constraint, I was able to develop a model that could simulate the effect of sudden change in stress on the shape of the creep strain time plot. This technique was used by Mitra and McLean to estimate creep recovery rate. The measurement showed that at a constant temperature, the recovery rate (r) is proportional to stress raised to the power of 3. The work hardening (h) was known to be inversely proportional to stress. Thus, it was possible to explain why steady-state creep rate, which is given by the ratio r/h, is of the order of 4. Although the recovery–work hardening model could explain the creep behavior of most metals and single phase alloys, it failed to account for the unusually high stress exponent found in the case of precipitation strengthened alloys. In order to explain this, the concept of effective stress which is defined as the difference between the applied stress and the internal stress that develops within the metal during deformation had to be introduced. I used this concept for numerical simulation of the creep strain transient tests. The results showed that such tests give correct estimate of recovery rate only if the internal stress is equal to the applied stress. It was also experimentally validated. This formed the main part of my Ph.D. thesis. Unfortunately, Professor Mitra passed away in 1979 when I was still in the midst of compiling my work. Nevertheless, it resulted in several quick publications and I got my doctorate degree in 1981.

One of the major constraints for research those days was the access to precision testing machines even in premiere institutes like IITs. In order to generate creep strain time plots to estimate work hardening and recovery rate, I had to use a unit that was fabricated by us. We also had to build an emergency power supply unit to provide power to the creep tester in case of power failure. This is when I got an opportunity to visit National Metallurgical Laboratory (NML) in November 1975. I was to present a paper at the Annual Technical Meeting of Indian Institute of Metal which was being held in its premises. This is when I saw the newly built Creep Testing Facility at NML. It had over 200 creep test points and a centralized data collection station housed in an air-conditioned hall. I was so impressed with the setup that when I got an opportunity, I readily accepted the offer to join NML.

When I came here as a scientist in December 1982, I was placed in the Creep Laboratory. This was a part of the Material Science Division. Dr. Rajendra Kumar was the head of this group. Professor Altekar was the director of the laboratory. Those days, it was extremely difficult to get family accommodation in NML campus. The problem was so acute that many had to wait for 10–15 years. I too had to suffer. The situation might have been different if I would have asked for lien terms which allowed consideration of past service for the allotment of a quarter. I missed this opportunity as I followed the advice of Dr. Kumar and resigned from a permanent position at IIT on the false belief that I may be allotted a quarter out of turn. This did not happen. I had to live in a rented house in Sonari, Jamshedpur, with my family for over six years. In those days, getting a good private accommodation in Jamshedpur was extremely difficult. Nevertheless, I learnt a lesson that apart from being a scientist, one has to be more aware of rules and regulation. When I look back, now I do not repent the decision I took. If I would not have done so, I might have gone back to the peaceful and relaxed academic environment of IIT possibly within a year.

Initially, I was associated with the evaluation of creep resistance of indigenously produced steel so that these could be accepted for use in boilers of power plant. So far only graphical comparison of stress rupture plots of the steels under evaluation with those of the established grades was being used as the acceptance criterion. This was quite subjective in the presence of the scatter normally associated with such data. The absence of any computational aids was a major bottleneck for the proper analysis of the huge volume of creep data that were generated in the laboratory. I had to present the work that has been done so far on creep resistant steel before a meeting of Indian Creep Panel having members from Steel Plants, BHEL, Atomic Energy, and Boiler Board in early 1984. Luckily, this is when Professor O.N. Mohanty my former colleague at IIT Kharagpur had joined NML as its deputy director. He had a programmable calculator. This helped me analyze the data quickly and present these in a way that was appreciated by all. This is when I came to know that attempts to setup computational facility did not materialize because of stiff resistance from the employees union. Subsequently before Professor Altekar retired, it was possible to buy and install two word processors which could also be used as a computer. I was given the responsibility to install, run, and train those who wanted to use the system. We also developed a set of programs that could be used to predict remaining lives of service exposed high-temperature components such as steam pipes from accelerated tests performed in the laboratory. Apart from creep testing, I was also involved in failure analysis of oil refinery and power plant components. This gave me an opportunity to visit several power plants, oil refineries, and fabrication units. The interaction with those involved in running these helped me appreciate and learn a lot about the material-related problems being faced by them.

Professor S. Banerjee joined NML as its director in December 1984. Before taking over this responsibility, he did visit us on several occasions. I met him possibly in March 1983 when he came to NML on a short visit and delivered a couple of lectures on the application of fracture mechanics. He again came to NML during summer vacation for a little longer stay. During this period, I did interact with him much more closely. We were able to initiate fatigue pre-crack in 3 point bend specimen using the then existing electromagnetic fatigue tester (Vibrophore). The pre-cracked specimen was finally tested on 25 ton screw-driven Instron testing system in creep laboratory. Apart from creep laboratory, Professor Banerjee visited various other divisions and sections. As a consequence, he had a fairly good idea on the strength and weakness of this organization much before he got an offer to take over as its director. He came with a clear vision and a broad outline of a plan to transform its work culture. This was also the time when it was extremely difficult to get enough plan grants to upgrade the basic infrastructure so crucial for the survival of aging laboratory. He was primarily responsible for major reorganization of the divisional structures, relocation of facility, and introduction of documented procedures for project initiation, monitoring, and closure. Some of the new concepts particularly those associated with project monitoring introduced in this laboratory are now being adopted at other places as well. Open-door meeting involves all scientists, be it in-house committee or Research Council (RC) meeting originated from here.

During this restructuring, I was inducted as one of the members in House Allotment Committee possibly because I was the only one who could use a computer. Dr. Dhanjayan was the chairman. One of the targets set by us was to display the eligibility list for allotment of quarters by the first week of January 1986. Preparation of this list was the main bottleneck for quick allotment of vacant quarters. The criteria of allotment were also very complex. There was no alternative but to take the help of computer. Only facility available then was two Merlin computers procured in 1984 March/April for word processing. It had only two 8-inch 128 KB disk drives. We started entering relevant data for all the employees in floppy drive in a tabular format using word processor. Most of these were done on weekends. The number of records and fields necessary to prepare the list was too large to be accommodated in one floppy. It had to be saved under different file names in several floppies. The final eligibility list was prepared by merging records from several files and running the sorting program. In spite of this, it was possible to display the eligibility list for the allotment of quarters as per our target. The database was also used for the allotment of individual identity number for all employees. This was done by sorting the data in an ascending order of the date of joining NML. This helped us in demonstrating how use of computer could help in administrative work as well and make it more fair and transparent.

Hereafter, there was no resistance in setting up computational facility in the laboratory. I was given the responsibility to set up new computational facility and to promote its use in scientific as well as administrative work. We got generous help from several institutes and organizations having prior experience. The support we got from Professor D.B. Phatak of CSE Department IIT Bombay is worth mentioning. He helped us prepare appropriate specs of a distributed computer network with several terminals and accessories. We got overwhelming response from several suppliers of computers as soon as the query was floated. Professor Phatak was also involved in the selection of the final configuration and the vendors. A site (the present location of servers) was identified for setting up a new computer application division. It had to be thoroughly renovated to accommodate the proposed system. None of us had any formal training on running a computer network. We got cooperation from all. It took us nearly a year to refurbish the location and install the computer. I was fortunate to have a few dedicated colleagues (Dr. K.M. Godiwalla and Dr. Subroto Chattopadhyay) whose untiring efforts lead to the establishment of a new division under my leadership. This was formally inaugurated in the month of November 1987 by Dr. V.S. Arunachalam the chairman of Research Council of NML. Soon we realized how difficult it was to generate enough work to keep the system busy and in running order. We were always on the lookout for a properly trained system administrator. It was extremely difficult to find one. Most of our time was spent in providing training and maintenance support. There were several eager learners from other divisions as well. Some of them helped us develop software for payroll preparation. This was being used by the bill section of NML until the new accounting package from CSIR was installed. Some used the drawing packages to prepare layout drawings of the entire laboratory. This proved useful for planning relocation of divisions and infrastructural facility. Soon after we realized that we have stepped into an area which has a very high rate of obsolescence. We were badly in need of additional funds to upgrade and also persons for technical support. Around 1990/1991, it was possible to recruit a person having zeal and dedication to pick up operation and maintenance of computer system. With his untiring efforts and the generous support from major projects such as Component Integrity Evaluation Program (CIEP), Steel Development Fund project on Blast Furnace, and planned grants for modernization, it was possible to upgrade the facility periodically with more modern and rugged system. It was possible to keep pace with the unprecedented growth in ICT during the subsequent decades. NML was amongst the first few laboratories under CSIR to provide access to internet and email through WAN with terminals located at every desk of scientists and administrative staff. The introduction of computers in 1987 was a major turning point in changing the face of NML.

Another major turning point was launching of Component Integrity Evaluation Program (CIEP) at NML, in August 1991. Initial plan started sometime around 1986/87. It was visualized as a multi-sponsored project. In view of our interactions with thermal power plants and activities connected with Indian Creep Panel, it was identified as one of the promising areas where concerted efforts could be initiated to bring a large externally funded project. The initial proposal was prepared by Dr. O.N. Mohanty around 1985/86. It was also the time when a few initiatives were taken to involve petroleum industries. This included investigations on the health of Haldia—Mourigram product pipeline segments having type III defects and cracking of Horton sphere of Gujarat Refinery of Indian Oil. Our proposal was sent to various funding agencies and potential customers such as power plants and petrochemical industries. This was also the time when World Bank (WB) showed interest to support R&D activities by providing low interest loans. Our proposal was short listed by CSIR. We were invited by the WB team to present our case. It was possibly around June/July 1987.

Professor S. Banerjee gave the responsibility to me to present our case. The meeting was arranged by Dr. H.R. Bhojwani at CSIR headquarters. Our linkage with petroleum and power sector proved helpful. It was incidentally amongst the first few proposals accepted by World Bank for funding. It was going to be in the form of a long-term interest-free loan. The total project cost was worked out to be 6 crore. Detailed proposal in the format received from WB was subsequently prepared by Dr. Amitav Roy who joined NML around May/June 1987. The idea of getting 50 % of the total cost from the potential beneficiary was floated by us although WB could have given the full funding. The final proposal was presented before RC during 1988/89. Dr. J.J. Irani was present as RC member. Several important dignitaries of Tata Steel were also present as invitees. In fact, the first encouraging response came from Shri P.N. Roy, General Manager Tata Steel. He said this could be of considerable importance to steel plants. He cited as an example the importance of integrity assessment of critical components such as crane hooks used to carry molten steel ladle. As far as I recollect, they were the first sponsor of this program. The first installment of payment (Rs. 25 lakh 50 % of the total commitment) was released soon after. With this initial encouragement, it was possible to get 4 more sponsors, namely Atomic Energy (AEC), Bharat Petroleum (BPCL), Indian Oil Corporation (IOCL), and Steel Authority of India (SAIL).

The project was unique in several respects. It not only focused on detailed plan of activities but also suggested a unique monitoring system. Program Management Board (PMB) was the apex body with members from each of the sponsor, WB representative, and Director NML. Group leader (GL) of the project was the convener. The next level was Project Technical Committee (PTC) with GL as the chairman and representatives from sponsors as members. The overall technical activities were divided into 5 broad areas: high-temperature creep, fatigue and fracture, corrosion base assessment, nondestructive evaluation, and modeling and simulation. The broad R&D activities were decided based on the feedback and needs of the sponsor. Apart from this, there was provision to take up one sponsor specific work from every sponsor. The first meeting was held on August 18, 1991. Dr. O.N. Mohanty, the first group leader of the project and convener of the PMB ably conducted the meeting as scheduled even though the same morning his father passed away in USA. Many of us came to know about it after the meeting. This was also the time when we were able to recruit some of the best young scientists to take up the challenging problems identified for this activity. Two of them now happen to be the key persons who could take the laboratory to greater heights. In fact, some of them preferred to join NML even though they had offers from Tata Steel and other well-known R&D establishment of the country. The PMB used to meet once a year and PTC twice a year.

The program was so well planned and structured that when Professor P. Ramachandra Rao took over as the director of NML in September 1992 it continued as per schedule. When Professor O.N. Mohanty left NML to join Tata Steel, I became the group leader. However by then, steps for procurement of new equipments were already initiated. A separate account was opened at Canara Bank, Bistupur, at the instance of ICICI Bank through which the World Bank fund was routed. The purchase procedure was also suggested by WB. A workshop was organized. This was attended by Dr. Roy and me. It was indeed very helpful. We were able to insist staggered payment terms within 100 % LC (Letter of Credit) and extended warranty clause. We also had to insure all the equipment purchased through WB fund. All these proved extremely useful as we faced minimum problem in subsequent years in their upkeep and maintenance. Apart from fulfilling the commitments made to the sponsor, the facility was being increasing used for other organizations to generate funds to pay back the loan. During the course of the program, a number of site work related to remaining life assessment of power plants were also taken up. In spite of the difficulties in reimbursing actual expenses one used to incur during site work, we did get voluntary support from a few young colleagues. In one particular year, total earning from such work did exceed 1 crore. Although there were immense possibilities to increase ECF (external cash flow) through such effort, it could not be pursued because of the stringent TA/DA rules admissible for such work. Nevertheless, we were able to attract many laboratory-based investigations under this umbrella project. It was possible to pay back the loan as scheduled from the external cash flow of this program. The last installment was paid in the year 2002.

A number of workshops were organized to disseminate the knowledge. Proceedings were also brought out. A few of these had excellent circulation. Professor P. Ramachandra Rao also took a special initiative to bring out a special volume on Life and Integrity Assessment. This was brought out by Indian Academy of Science. This also had several articles from well-known experts from abroad with whom we were able to establish close contacts during these years.

With better infrastructure and expertise, it was subsequently possible to attract many challenging jobs. NML was also indentified as one of the laboratories where failure and life assessment of aircraft components could be taken up. Innumerable failure investigations thereafter have been carried out at NML. Many of these have been highly appreciated particularly by Indian Air Force (IAF). The concept of network projects involving several laboratories under CSIR started during the 10th five-year plan. This is when Professor S.P. Mehrotra joined as our director. NML was identified as the nodal laboratory for a project involving life and integrity assessment of critical engineering components. New state-of-the-art testing and characterizing facility needed for such activities were established. On the whole, CIEP was indeed a landmark. This did help us build close linkages with industries and focus our attention toward the development of tools and techniques needed for the life extension of aging infrastructure.

Although I had to leave Creep Testing group after the establishment of Computer Application Division, I did have close association with activities associated with life and integrity assessment of engineering components. This was possible because of my involvement in several projects apart from CIEP. In 2003, nearly a year after Prof. Mehrotra took over as the director of NML, there was a major reorganization of the divisional structure of the laboratory. I became the head of the Material Science and Technology Division (MST) having around 50 scientists. The division had four major groups: material evaluation (where the focus on characterization of mechanical properties), microstructural characterization, and nondestructive testing and advanced materials (where the focus was on synthesis, processing, and characterization using non-conventional routes). I was fortunate to have active cooperation from every members of MST division. During this period, we got liberal funding from several sources. Bulk of it came from Planning Commission Grants for Network Projects and Department of Science and Technology (DST). The fund from DST was primarily for the establishment of nondestructive testing (NDT) facility. This was initiated at the behest of Professor P. Ramachandra Rao. Most of the ground work was done by Dr. D.K. Bhattacharya. Unfortunately, he decided to leave NML to join Central Glass and Ceramic Research Institute (CGCRI), Calcutta, soon after the formal approval was obtained. The responsibility to fulfill the commitments fell on me. I got whole-hearted support from Dr. Amitav Mitra, Dr. Narayan Parida, Dr. Sarmistha Sagar, Dr. Gautam Das, and their colleagues. Today, it happens to be one of the most active groups on NDT in India. The group has helped several industries all over the country in the evaluation of service induced damage assessment of critical components. A number of portable instruments have been developed for the integrity assessment of service exposed high-temperature components such as steam pipes in power plants or heater tubes in petrochemical units. These are currently being used in several such units all over the country.

Professor Mehrotra nominated me to be a member of a multi-disciplinary task force set up by Ministry of Power, Government of India, to look into the problem faced by the sudden shutdown of all the six units of Nathpa Jhakri Hydel Power Plant in Himachal Pradesh within a few months of its inauguration and suggest remedial measures. This is when I got an opportunity to visit several hydroelectric plants located in the Himalayan region. Severe erosion of several underwater parts of turbines was the major problem. This was primarily due to the presence of high amount of silt in the water. Parts made of hardened and tempered 13Cr4Ni or 16Cr5Ni steel were getting eroded within a few months. Therefore, the plant had to be shutdown as and when the silt content in the water goes beyond a threshold. The amount of silt in water found to be much higher than the permissible limit during April to September, when the potential for power generation is the highest. One of the suggestions of the committee was to develop silt erosion resistant steel for such parts. The responsibility fell on NML. The project started under my supervision in 2007. It involved several members from different groups at NML. Examination of damaged components suggested that there are specific locations that are more prone to erosion. We thought of developing steel whose ability to resist erosion would increase, while it is in use. Initial trial looked encouraging. However, I had to leave in February 2009 NML to join IIT Kharagpur, but my colleagues under the supervision of Dr. S.R. Singh and Dr. Sandip Ghosh Chowdhury did a commendable job. I am told that a component made of the new steel made at NML has given at least two and half times longer life than that of the conventional grade. If this could be up-scaled to produce Francis turbine blades and guide vanes, it would indeed be a major milestone in combating the erosion problem being faced by the hydel power plants in the Himalayan region of our country.

During the eighties at NML, opportunity to interact with a foreign university or laboratory was limited. Professor Banerjee asked me to explore the possibility of interacting with Professor Mike Ashby of University of Cambridge. Since my interest was in the area of modeling of high-temperature creep behavior of engineering materials, he advised me to approach National Physical Laboratory (NPL), Teddington, UK. This is also the place where my Ph.D. supervisor Dr. Mitra spent a useful part of his career. In October 1988, I got an opportunity to visit NPL under TCTP program of British Council. Dr. Malcolm McLean was my supervisor. I got associated with CRISPEN program developed by NPL and University of Cambridge. This was a menu-driven software for **CR**reep **S**train **P**rediction of **En**gineering Materials. It was based on continuum damage mechanics describing the evolution of creep strain at a given stress and temperature by a set of state variables representing different forms of structural changes that occur in engineering materials during service. The exact forms of the coupled differential equations were derived on the basis of the dominant mechanisms of damage accumulation. Experimental studies had shown that apart from nucleation of creep cavities, the creep of superalloy was determined by dislocation (strain) softening, whereas that in steel was due to coarsening of precipitates (time softening). The solver was developed using Turbo Pascal so that it could run on a desktop computer running on MSDOS. My initial task was to incorporate a solver that could estimate the material parameters in situations where both time and strain softening were important. My experience in setting up computer facility at NML proved handy. Within a short time, I could pick up PASCAL language and incorporate the additional module that was assigned to me. Malcolm was extremely happy and I was asked to demonstrate the program to Professor Mike Ashby during one of his visits to NPL. Soon after this, Malcolm asked whether I could help them develop a model that could predict the effect of crystal orientation on the creep behavior of single crystal superalloy SSR99. This was developed by Rolls Royce, and it was being used as blades in their aero-gas turbines. The cube <100> orientation was found to have the best creep resistance. The bulk of the creep strain time data was available for this direction only. Could these be used to estimate the material parameters that could simulate the creep behavior of an arbitrary orientation? Luckily, a few dataset were provided to us by erstwhile Royal Aerospace Establishment (RAE) for <111>, and creep data on a couple of complex orientations were also available for the validation of the predictive model.

This is where my experience in teaching a course on dislocation theory at IIT Kharagpur proved useful. There was a part that dealt with plastic deformation of single crystal. I used to explain the method of converting shear strain on specific slip systems into tensile strain using tensor notations. Creep too occurs due to dislocation glide on a set of specific slip system. Therefore, the model used in CRISPEN had to be reformulated in the form of shear stress and shear strain. The accumulated strain in each of these had to be compiled and transformed into tensile creep strain. The number of operating slip system was at least 12. There were doubts if this could be implemented on a desktop computer. Nevertheless, very soon I was able to develop a solver that could predict orientation dependence of creep behavior of single crystal using a desktop computer. The stay was so productive that the interaction with Malcolm McLean continued for over 15 fifteen years even though he moved to Imperial College in 1990. He as well as some of his colleagues (Brian Dyson) did visit NML on several occasions. We were also successful in getting EPSRC UK supported project on modeling and simulation of creep of superalloy. One of our colleagues got an opportunity to work at Imperial College for a couple of years. The collaboration culminated in organizing 2-day workshop on Physically Base Constitutive Models for high-temperature materials at National Aerospace Laboratory, Bangalore, in January 2004. Apart from Dr. Sanjay Sondhi and me, there were lectures by Prof. McLean, Prof. Brian Dyson, Dr. Hector Basoalto from Imperial College London, and Professor David Hayhurst of UMIST, UK. It was attended by over 30 participants from various industries and academic institutions. Unfortunately, the following year Professor McLean passed away. My association with him was indeed a turning point of my career.

In 2009, I came back to teaching after a gap of 26 years. I joined the Department of Metallurgical and Materials Engineering at IIT Kharagpur as a chair professor. I am currently offering two courses one on design and selection of materials and the other on metallurgical failure analysis to undergraduate students. This gives me a chance to share with them my long experience in helping several industries solve their problems related to processing and performance of a wide range of materials. When we were student in the late sixties, the choice of materials for design and fabrication of engineering components was limited. Today, the options are too many. Even in the case of steel, the most common material of construction the number of new grades introduced after 1980 far exceeds the total number of grades that were available till then. While I was in UK, I got chance to spend a week at the University of Cambridge to interact with Professor Ashby. This was the time he was involved in the compilation of material database in an organized fashion. This was being used to generate material selection charts for various applications. He had given me several of these which I shared with my colleagues at NML. This is when I realized that the way we introduce the wide range of materials of construction to engineering students is far removed from real-life application. The approach is likely to change soon. Today, it happens to be one of the main objectives of granta.com, an organization established by Professor Ashby. I do often visit their Web site and ask my students to do so while offering these courses. With the growing availability of wide range of information on the net and the ease with which these can be accessed, the way these courses are being taught in the class is going to change. I got an opportunity to offer a course on principles of physical metallurgy under the NPTEL program of the Ministry of Human Resources of our country. The video version is available on the net, and the Web version has recently been uploaded. It may be too early to judge if it is useful to students interested to learn about structure and properties of metals and alloys. However, students occasionally raise interesting questions. I enjoy answering these.

All this has been possible because of my long association with IIT Kharagpur and NML Jamshedpur, establishments that recently celebrated their Diamond Jubilee.

I shall fail in my duty if I do not acknowledge the support that I have received from my family particularly my wife Jayasree. We were married in 1976. She was brought up in Pune where her father worked initially as a scientist in Explosive Research and Development Laboratory and later as its director until 1979. In spite of her good educational background, she preferred to be a homemaker and managed it efficiently. We have a son and a daughter. Both are married and settled. We have a grandson too. He has just started going to a nursery school. This makes me think what kind of future lies ahead of the present school going children in the face of stiff competition and growing peer pressure. We were possibly lucky to have gone through this stage when there was little peer pressure and relatively relaxed curriculum. It is true that knowledge is going to expand exponentially. However, human capacity to assimilate and use it is limited. We have to depend on new tools and techniques that help us exploit it fully. The problem can never be solved by expanding the syllabus and increasing the work load in schools and colleges. Learning does not end with graduation. One has to keep learning all through his career. Therefore, one of the main objectives of every academic institute should be to promote self-learning and provide an environment that promotes it. Physical insight and experimental skill are often not enough to solve a challenging problem in any discipline of engineering you must have the ability to combine these with mathematical rigor. A flexible curriculum that suits the aspiration of students and the social needs may help this happen.

A Humble Journey

Soumitro Banerjee

When I received the mail to write an article for the INAE Newsletter related to my childhood, education, mentorship, challenges met, etc., it got me thinking: Am I really the right kind of person to project before the upcoming generation? There are many people out there who are far more intelligent than I am, and who have shown much greater promise during their formative years than I have. In contrast, I never was a first ranker in school or college. True, I had a reasonably good education in the Ramakrishna Mission School at Narendrapur, West Bengal, and in the Bengal Engineering College where I did my bachelor's degree in electrical engineering. But I was never considered an "ace" student. I was more interested in things other than studies: I painted, I played many games, and I was deeply involved in students' activities. Friends did not see me studying much and were pleasantly surprised when I managed to sail through the exams. The secret lies in a peculiar aspect of electrical engineering: it is possible to derive things from the first principles sitting in the exam hall, even though you may not have any clue what the right answer could be.

At IIT Delhi, where I did my master's and Ph.D. degrees, it was the same story. I got involved in all sorts of activities. At the IITs, the intensely competitive atmosphere of interhostel championships extracts the best out of anybody who has some prior exposure to a cultural or sports activity. I was in a postgraduate hostel, where the number of such people was very small. So I had to take part in almost every activity. I was into music, arts, dramatics, and most of the sports. After starting to do Ph.D. work, I added two: I started to learn the violin and Judo (what a combination!). So, again, I had very little time to do research.

I framed a strategy: I requested my supervisor Prof. S.C. Tripathy to allow me to choose my own topic of research. He consented. So I went to the library and spent hours looking up what new is happening in the horizon. I chose a topic on which my supervisor had no prior exposure. In the beginning, he was not willing to let me work on that. But I somehow managed to convince him that it was a "doable" topic. So we reached an understanding: I would do the work independently and would

S. Banerjee (✉)
IISER Kolkata, Mohanpur Campus, Nadia 741 252, West Bengal, India
e-mail: soumitro.banerjee@gmail.com

P. Ghosh and B. Raj (eds.), *The Mind of an Engineer*,
DOI 10.1007/978-981-10-0119-2_29

periodically report to him the progress. In case the work fails, it will be entirely my responsibility and I would not blame him for the failure. Since the planned work was in the twilight zone between power systems and power electronics, he only advised me to get a cosupervisor from the area of power electronics. So I requested Prof. J.K. Chatterjee to be a cosupervisor, with the same understanding.

Thus, freed from the problem of being micromanaged, I took responsibility of my own time management. I was in the thick of students' activities and would schedule my research work as per my convenient timings, balancing the two priorities. This gave me very valuable training and experience in time management. I consider it an important factor behind the success that I have achieved. Most of my colleagues at IIT or IISER complain that they do not get adequate time to do research due to "teaching load", and they are so very busy doing these that they get no time to think about the society at large. I have never had any problem in being involved in social work while doing science of the highest standard—thanks to the training in time management I acquired at an early stage of my life. My involvement in extra-curricular activities also gave me exposure in handling people and situations.

The risk I took in choosing a new field somehow worked out. After the initial groping in the dark, I began to obtain new and promising results.

After two-and-a-half years of work, an advertisement for Lecturer's post at IIT Kharagpur was published in the newspapers. I applied casually (since an M.Tech. degree was specified as the minimum qualification), without ever hoping to get it. I had no publication and no Ph.D. degree. They had to judge on the basis of only the interview. Miraculously, the interview went very well (they asked just the questions whose answers I happened to know), and I got the job. That was 1986.

I still remember the first day I entered the campus of IIT Kharagpur to join service. I stood silently before the large façade, overtaken by emotion, wondering if I really deserved this honour. I had spent my student life playfully, never spending much time on studies, and now, I was faced with the challenge of teaching the best pupils and doing research of the highest standard. I was not sure that I was up to the mark. That day I made a pledge to myself. I gave myself a period of 10 years to prove my worth. At the end of the 10 years, I would make an honest assessment: if I proved myself fit to be a faculty member of an IIT, I would remain there. If I fail to make a mark, I would quit IIT and will move to some less demanding institution.

Thus, I started my journey at IIT Kharagpur. I felt indebted to IIT because it had allowed me to use its name as my place of affiliation. We did not have much else, and for me, it did not matter. I just had a table and a chair in a room shared with another faculty member, and no laboratory of my own. At that time, IIT had shortage of faculty accommodation and I got just a room in a guest house with a shared toilet. I and the colleagues who joined around that time were fine with it. Nowadays, I see the younger colleagues behave as if they have done a great favour to the Institute by joining it, and now, it was the duty of the Institute to satisfy their requirements. They keep talking about what the Institute has given them and what it should give them. Such thoughts never crossed my mind. IIT had given me a foothold to prove myself and that was enough. I was more concerned about what I was giving back in return, rather than what the IIT was giving me.

I had that concern about my role in the society also. The poor country was giving me a fat salary, and all the time, I was asking myself whether I served enough to justify it. So I tried to do something extra in the free time that I had in the weekends, by involving myself in science popularization work and such social service.

The first two years were tough, as I had to study a lot in order to teach. In addition, I had to work for my Ph.D. Fortunately, I was not much dependent on guidance and could work on my own. Every vacation I visited Delhi to report the progress to my supervisors. Finally, I wrote up the thesis and submitted in 1987.

After finishing the Ph.D. and communicating a two-part paper out of this work, I decided to change the area of work. My reasoning was as follows: if I continued to work in the same area, others would think that I have failed to come out of the cradle of my supervisors. So I started looking for a new area of research. The criteria of choice were as follows. Firstly, it should be a problem in which little work has been done so far. Secondly, it should have the promise of being important in future. Thirdly, it should be doable within the resource available at IIT Kharagpur at that time. Those days we did not get much of a starting grant and had to set up experiments at a corner of a teaching laboratory. So I could not choose a topic that places great demand on experimental or computational resources.

With such constraints, it was difficult to home on to a subject of research. I kept looking at various problems, doing a bit of work on this and that to keep the paper clock ticking. Finally, around 1990–1991, I found what I was looking for.

By then, only two papers had reported peculiar nonlinear phenomena including the occurrence of chaos in power electronic circuits. The problem satisfied all the criteria I had set. Around that time, I got my first Ph.D. student. So I decided to explore the area. By the time the student submitted the Ph.D. thesis in 1995, we managed to publish a couple of IEEE papers out of the work.

To me that was not enough. My self-imposed deadline of 10 years was nearing, and as yet I had not achieved anything that would satisfy my conscience. I was not yet worthy of the IIT.

But there was something strange in the results we had obtained. By then, I had read up much of bifurcation and chaos theory, but still I failed to explain the instabilities (or bifurcations) we observed in power electronic circuits. These bifurcations did not fall into any of the categories known till then. In the papers, I had just reported what we observed and left it at that. I had a gut feeling that there is something big in it and decided to explore it myself.

At that time, I got the BOYSCAST Fellowship of the DST and went to work with the physicists and mathematicians of the University of Maryland for four months. There I got to learn the tools and techniques used by the insiders of the field. I used these to explore the problem I had in mind. I discovered that the bifurcations I observed in power electronic circuits belonged to a new class called border collision bifurcation. Prof. Yorke's group at Maryland had investigated piecewise smooth maps as a mathematical possibility and had observed similar type of bifurcations. At that time, no physical example of such bifurcations was known.

The bifurcations we had observed in power electronic circuits were in fact the first examples reported in physical systems.

This discovery posed a great many new questions. What is the class of physical systems where this kind of bifurcations can occur? Can the observed bifurcations be properly explained on the basis of the available theory? These questions demanded answers, but the period of my leave came to an end and I returned to the IIT. I continued to work on the problem alone, working feverishly as my self-imposed deadline was nearing. I had married in December 1995 and had to leave my wife at home during my trip to Maryland. After the return also, I was not being able to give her much time as I was returning home past midnight almost everyday. I am indebted to her for the understanding and cooperation at that important juncture of my life.

Over the next few months, I managed to prove that all switching systems would give rise to piecewise smooth maps on discrete-time modelling, and hence, the new type of bifurcations should be observable in all such systems. Now, systems which have some kind of switching action are quite common in science and engineering. This meant that the theory of border collision bifurcation will have very widespread application. But the problem was that the mathematical theory developed until that time was in a very premature stage, and much work had to be done in order to give it a shape that can be applied to physical problems.

Prof. Yorke invited me the next summer to visit Maryland to work on this problem. During that two-month period, we initiated the work, but could not progress much within that short period. I completed the work after coming back to India. The papers coming out of this effort proved to be of enormous impact. One paper was published in Physical Review Letters, one in Physical Review E, and three in the IEEE Transactions on Circuits and Systems.

My conscience was cleared. I had proved myself worthy of being in the faculty of IIT and decided to stay on.

Many people have expressed surprise why I did not continue the collaboration after these pieces of work. The problem of working with famous people is that others tend to assume that the basic idea came from the most famous person among the authors. At some stage, it is important to come out of the shadow, to establish oneself as an independent researcher. That can be done only by publishing papers of similar quality on your own, or with your students.

When I established myself as a leader in the field, younger researchers of different foreign universities started seeking my help in solving problems, and that resulted in further collaborative works. But I made sure that in these collaborations I am the senior person involved, and the others benefit from my ideas, not the other way round.

Recognition took some time to come by. Until 1998, I was ignored in the successive promotion exercises and remained in the lowest cadre for more than 12 years. I did not mind that at all, because in my own assessment, I was yet to prove my worth. True, it hurts when you see contemporaries promoted while you are left behind. But I had confidence in the system: if you are good, it is impossible to ignore you for long. So I kept working silently. Finally, when my work was recognized, I did not have to look back. The Bhatnagar prize and the fellowships of the INAE, IASc, and the INSA came the year my name was proposed, and I never had to wait a year.

I feel it is important to make an objective and critical assessment of oneself all the time and to keep alive a sense of dissatisfaction with oneself. The moment you develop a high opinion about yourself, you are likely to become a victim of your own ego and would not be able to produce much in life. It is important to be your greatest critic.

It is this dissatisfaction with me that drove me to leave IIT Kharagpur in 2009. By then, I had won most of the awards and laurels that one covets in life, and there was very little to achieve further. Life was becoming too cosy, devoid of any challenge. In this situation, most people would stop being productive and would spend the rest of life basking in the past glory. I found that unacceptable. So I moved to the newly established IISER to force myself into the challenge of building a new institution, learning, and teaching subjects I never studied as a student, where I will have to take up new research problems. I am finding it challenging, and I am finding enjoyment in the challenge.

My Experiences

S. Kalyanaraman

Challenges Due to Birth Place

Being born in a village and studying there a few classes before moving to a nearby town/city is no disadvantage. Actually, it gives an exposure to the entire spectrum of people who have their own wisdom. A fully city-based 'urbanite' need not be considered as a typical person to 'know all and do all.' I was fortunate to be born in a village—called Mudikondan in Thanjavur district, Tamil Nadu, in 1942 and had my early education in that village up to 8th standard and then moved over to a nearby town called Mayiladuthurai and got my secondary school education in Municipal High School.

My teachers during the primary and secondary schools gave me a reasonably good education. My higher education was in Madras University, Madras Institute of Technology, and University of Florida. This spectrum of places and cultures gave me the realization that that this world is diverse and we have to accept them initially as they are and work toward improvements.

My Inspiration

A couple of very good teachers during my early childhood could inspire in me the importance of education and kindle my interest in studies.

When Soviet Union launched its first satellite (Sputnik) during 1957, newspapers flashed that an artificial moon called Sputnik has been launched. I used to go out in the nights with the fond hope of seeing it. But I could not see it with the naked eye.

S. Kalyanaraman (✉)
Department of Electronics and Communication, Dayananda Sagar College of Engineering, Bangalore 500 078, India
e-mail: kalyanarin@yahoo.co.in

P. Ghosh and B. Raj (eds.), *The Mind of an Engineer*,
DOI 10.1007/978-981-10-0119-2_30

But I was inspired by that effort and I felt that I should also do one day something similar to that.

Fortunately, I could join the courses which I desired to, in reputed institutions, and I always had some curiosity to know and thus could do well in studies.

Later after completing my studies, again, I was fortunate to join ISRO during 1972 and work with renowned space science experts such as (i) Dr. Abdul Kalam for ISRO's first satellite launch vehicle (SLV) project by designing and delivering communication systems of Rohini satellite which it carried, (ii) with Prof. U.R. Rao, for the first Indian satellite project Aryabhata, for design and development of communication system of the satellite, and (iii) Dr. Kasturirangan, as his deputy for first operational Indian Remote Sensing Satellite Project.

All those with whom I worked, whether they were my superiors or subordinates, could motivate me with their approaches to work and desire to succeed. In turn, I also did my best to reciprocate.

My Work and Technology Challenges

After my graduation, initially I worked in LRDE (DRDO) on Radar signal processing projects. Later, I joined Indian Space Research Organization as a scientist/engineer and worked on the first Indian satellite project Aryabhata as a system engineer for developing communication systems. We took a little help from Russians, as they were advanced in that field in understanding the complexities involved in the design, fabrication, and testing of satellites and the technologies involved. This satellite launched during 1975 paved the way for future growth of our space program and many successes. Always, starting is a problem. Once we start, it grows.

Space will not forgive us if we make a mistake. As everyone knows that satellite systems flown in space cannot be repaired unlike the earth-based systems, the reliability of the space systems over its designed life must be very high close to 100 %. A reliable design, analysis, use of right type of materials and components, right type of fabrication processes backed by quality control and full complement of test, and evaluation to ensure that the system will withstand the harsh environment of space during its lifetime are all required. The basic satellite technologies were developed during that project in India.

Engineering and Management Challenges

When we start working in an organization or a company, we meet with some challenges. The area of work can be anything—say, design and development, analysis, fabrication, testing, management, or some service. Generally, the objective in terms of outcome of the work may be specific to that organization or company. In order to meet or exceed the objective, one has to be creative/innovative. In satellite projects, where I worked, the objective is different for each project. And no two projects are same. Each satellite must be engineered to meet its final objective. Even if two or three satellites are duplicated, some improvements from the previous version of the product are always required making the room for innovative thinking. Many technologies are unavailable in this area. There could be many impediments for achieving success in one's effort, and one should consciously find other options of achieving the goal. We did exactly the same in ISRO and found alternative workable options to achieve the set goal.

There are management challenges as satellite projects are interdisciplinary. A manager's responsibility includes

1. Human resources: It is necessary to put right person on the right job and it starts with the recruitment. And obviously, proper incentive and recognition are required to be given for such persons. Then only success is guaranteed.
2. Other resources: The minimum required infrastructure must be provided for carrying out the work.
3. Inputs: To make a product, if raw materials to the required quality are not available to make the final product, due to some reason or other (sometimes, due to geopolitical reasons, sanctions may be imposed and were once placed on India by some countries after Pokaran nuclear test), they were to be faced and alternatives are to be thought of. For example, if it is an important component to be used in a system, either we develop it on our own or find alternative sources or find an alternative route to develop the product by avoiding that component. There are many ways of realizing a product meeting the end user's requirements.

In ISRO, we faced such problems and solved by one of the above methods.

1. While engineering a product, it must be ensured that conscious efforts are put to optimize in every direction without diluting the end goal. This makes the product acceptable and competitive.
2. A proper guidance and a review system must be in place. Regular reviews and need-based reviews of work must be conducted frequently and concluded for possible action.
3. Reliability and quality assurance: This world is a competitive one. Only companies that provide quality products or service at competitive price can survive.

What I Learnt

I retired as Program Director of Indian Remote Sensing Satellite Program looking after the satellite developments. We achieved many successes. In all these projects, maximum has been achieved with minimum budget compared to advanced countries in the world. For achieving success, the points learned by me in ISRO at individual level are as follows:

1. Meticulous planning of activities required before implementation.
2. Preventive steps are to be taken so that no problems arise later.
3. Such steps are to be taken well in time; not at a late stage, where implementing solutions become extremely difficult.
4. Mid-course corrections are also sometimes required.
5. Coordinated team effort is very important for success. All the team members require to work for the same purpose and goal.
6. Pro-active work culture leads to success. It does not mean there should not be any difference of opinion. Differences in opinions help in discussing merits and demerits of different approaches/options. With proper discussions, the final decision on the approach to be taken will automatically emerge and the leader of the team can easily take the decision. (This approach was well emphasized by Prof. S. Dhawan, who was a former Chairman, ISRO.)
7. Proper informal and formal communication between team members is a must.

Why not First?

Why we should be always second, third in the world in anything we do? Why not first? This is the question asked by Dr. Abdul Kalam also. Such an effort was taken by building the world's first twin satellites (IRS-1C/1D) in terms of their capability in providing best resolution pictures from space during 1996 in the civilian world. And nationally and internationally, their successes were applauded and appreciated. The mind must get tuned to strive for the first place.

Role Models in My Life

It is always good to look for a role model in one's life. I remember even now a couple of teachers who taught me well. In fact, I feel that even if you come across a couple of persons who have created an impression during some part of your life that is enough to change your life. I did not fall back on a single role model, but gathered the good traits from various persons, whom I have come across in my life. I have learnt from my superiors, colleagues, and subordinates. To mention a few, my mentors were Prof. U.R. Rao and Dr. Kasturi Rangan, who were chairmen, ISRO during successive periods. Similarly, I learnt many other traits from my colleagues. Those traits include dynamism, sincerity, and seriousness in what we do. I found many of my friends in ISRO possessed those great qualities.

What Is Required from Every Individual?

Responsibility: To have a responsible society, every person must be responsible. This can be achieved only by educating and training all citizens during their school/college days by parents and teachers and using different media and make them aware of individual's responsibility in a society.

Commitment to work: This ensures individual success and also collective success.

Commitment to excel: This is a must in all professions. We get satisfaction when we complete a task with this type of commitment.

To help others: Not everyone is equally capable. Others may require some help to achieve or complete the task to satisfaction. Wherever we can offer help, why not? We also derive good satisfaction by helping others.

Sincerity: We must be sincere and serious about anything we undertake or do. Automatically, success will follow. Additionally, others respect only such persons who are sincere to the purpose in what they are doing.

Contribution to society: Through our work, we contribute to the society in some way or the other. Youth is the time when our contribution can be maximum. There is also a question in my mind regarding the usually prevailing 'retirement' concept at the age of 60. There is no such thing as complete retirement during one's life. I feel, if one lives in this world, one must make contributions to the country/society (local or larger society) consistent with one's age and capability. The concept must be—we have no right to consume unless we produce or do service. Money should not be criterion to work after retirement, but it should be based on providing some service to the society. One must work all through the life to one's capability or liking. Mahatma Gandhi used to spin the wheel to help Khadi industry in addition to his voluntarily taken responsibility of being a moral teacher to the nation backed by his practice. Following the footsteps of our father of the nation, in a humble way, I started working as a teacher in an educational institution after my formal 'retirement' to share my experiences and knowledge with the students. I found it was liked by the students also as they were of the view that they get first-hand knowledge from a person who has worked on what he teaches.

How the Government or NGO's or Educational Institutions Can Help

The future of any country lies in its right population demography and capability of the population. To ensure this, a proper program is a must to get quality education and quality training for all citizens in the country which pave the way to result in quality products/service. Every educational institution must be accredited. Automatically, this results in economic growth and better standard of living. Each person is capable/talented in something which must be identified and improved.

Investment in improving the quality of human resources goes a long way to make the country better. Major responsibility for this lies with the Government (with the support of NGOs wherever required), as it has the power and financial resources to embark on such programs.

Opportunities Galore

In this present world, opportunities are more. With the knowledge explosion, there are so many new avenues opened up for progress. So many types of goods and services are required in this world. What is required is only that a person must be good at least in his/her chosen field. He/she can log on to that and try to excel. Excellence thrives, while mediocrity can only just survive. Let us work for a prosperous India of the future.

My Experiences in Creating Indigenous Technology Enterprise in India

P.S. Deodhar

Friends of Aplab and my near and dear ones often wonder why the Rs. 20 million Aplab Group has not grown to become a company worth a few thousand millions, despite its credibility for high-quality products, its quest for innovation, its steady exports to Western countries, its mastery over power conversion technology, and its business integrity. My answer to them is simple; principally, it is me and my upbringing. Second, it is our integrity and unfaltering resistance to compromise, while making deals with government buyers and inspectors. I must admit, before anyone else brings it up that I am responsible for ruining in the growth of my own venture, though unintentionally. For the growth of a company, the founder has to be ambitious about getting rich, and he needs to invest time and imagination towards that goal. Further, to achieve that goal, one also has to be ready to make compromises of all sorts.

Basically, I lack that kind of commercial ambition since it needs one to make 'deals'. My ego gets a boost from our technological successes. My mental make-up is that of a craftsman, and my passion has always been technology and its ability to provide a well-crafted solution. Satisfied customers have always been my honourable salesmen. Principally, it is they who helped Aplab to grow. My domain knowledge happens to be in electronics and engineering materials. I also had the benefit of coming from a 'Do-It-Yourself' (DIY) family. Aplab, therefore, restricted itself to high-technology areas like instrumentation and efficient power electronics.

P.S. Deodhar (✉)
Aplab Limited, Aplab House, A/5, Wagle Estate, Thane 400 604, India
e-mail: psdeodhar@aplab.com

P. Ghosh and B. Raj (eds.), *The Mind of an Engineer*,
DOI 10.1007/978-981-10-0119-2_31

Big Money Could Never Be My Goal

Chasing money never excited me nor was my ambition to be a businessman who would do any business that offers windfall profits. In fact, I prefer to keep away from those who would value me by my wealth or expensive lifestyle. However, there are those among the rich, whom I do hold in great esteem, for whom money has been a corollary of their great vision, extraordinary skill, and wisdom. They wear their wealth very elegantly and conduct themselves in a gracefully simple manner. One would never find them brushing shoulders with the current money-centric elite class.

What really excited me was developing products that my customers wanted and which other Indian companies could not deliver. My priority was always to put unrelenting efforts into manufacturing high-class products. My annual visits to Japan till the mid-1970s and then to Germany till the early 1980s helped me understand and appreciate the meaning of technological excellence. Those visits were educative, and they challenged me to attempt the same technological approach in India.

Aplab Grew Horizontally Rather Than Vertically

For the first two decades, our product development range covered a wide spectrum of industrial products, but no consumer product. The only exception was designing a 35.5 cm (14-in.) black-and-white television for the rural Indian market, which was a great success.

This happened during my tenure in Delhi as Rajiv Gandhi's advisor and as the chairman of the Electronics Commission. ET&T was a central government enterprise. In 1986, I designed a 35.5-cm B&W TV for it to market. It was an innovative concept where ET&T provided the technology, the total materials kit, and the ET&T brand name for free—the three aspects that prevent small-scale units from competing with big companies. I was proud that ET&T could capture almost 40 % of the local market using my 35.5-cm B&W TV design. I did the same to knock down the prices of personal computers to below Rs. 10,000, forcing Wipro and HCL to halve their selling price. Aplab had a wide product range and five product divisions, thanks to my greed to accept design challenges. Aplab therefore grew horizontally rather than vertically. A big chunk of profits went into new product development. During all these years of technological excitement, I gathered together many like-minded colleagues with a deep insight into the physics of power conversion and signal was on contemporary packaging design using appropriate materials. This had added to our abilities and our sense of excitement about what we enjoy doing. When I retired from Aplab in 2009, we had over 800 standard products for five market segments. Each vertical essentially had the potential to grow many times over, in size. While Aplab's market share was rather limited, the competition in each segment shows the scope we had to grow in size.

My Sense of Right and Wrong Often Stood in the Path of Growth

Aplab has had good and capable marketing managers, but they were handicapped. I often stood in their path of growth because it either involved compromising on specifications or required greasing palms. These colleagues were clever in market development but in India one also needs to be 'flexible'. This was especially necessary since 80 % of our business was with the government or government-controlled organisations. For our government to function, it needs the oil of bribes. My men had an uphill task—there could be absolutely no unethical compromises on product quality or commercial integrity in order to offer a cheaper price. I must, however, admit that Aplab too had to provide monetary lubrication to get paid our legal dues from such institutions. In the early 1960s and 1970s, I was lucky and 'the system' then was far cleaner, especially in the south India. Being a technical entrepreneur who offered well-engineered indigenous solutions helped me in the south. I enjoyed great benefits of being a pioneer. Till the early 1970s, we grew rapidly, doubling our size every year. IIT Bombay, BARC, TIFR, etc., in Mumbai, and Bharat Electronics, Indian Telephone Industries, NAL, and many others in the south were great admirers of Aplab. We were developing products for these companies, helping them to substitute imports with products delivering much more than their desired specifications. I recollect that, in 1967, Lt Gen AC Aiyappa, chairman of BEL, threw us a challenge to develop an AC voltage regulator to meet the K114E environmental endurance test. The product had to work at −40° C temperature in a moving military vehicle. Within four weeks, I went back to BEL with a sample that passed all the stringent tests. This helped BEL to replace a similar product from Siemens. Big orders followed from the Indian military for decades. The stories of developing solid-state ringers and tones for the Indian Telephone Industries for telephone exchanges or signal generators and microwave frequency counters for the military and DoT were very similar—we developed, got approval, and supplied. Aplab had almost no competition from within India, and imports were restricted. In 1975, we supported DoT's satellite earth stations with several high-power UPS systems at each location. The story with DoT's Telecom Research Centre (TRC) was no different. All the DC power modules for India's first Digital Exchange of TRC were developed and supplied by Aplab.

Aplab Learnt Right Engineering Practices by Selling in Germany

In 1972, I felt that Aplab was ready to offer its products to German customers. So we displayed our products at the Hannover Fair in Germany in 1972. We were probably the first Indian company to do so. Europeans passing by threw curious glances at me. It probably was as foolish as attempting to sell coal to Newcastle.

But we did find some venturesome German buyers. The first supply to a customer in Munich turned out to be a disaster. The buyer said that our products worked well but could not be sold in the German market. But he did help us to redo the packaging and compliance testing by showing us what his customers wanted.

My Eight Years Stint with the Government Was Painful and Frustrating

In 1982, I was sucked away from Aplab into New Delhi by Rajiv Gandhi, a close friend since 1974. His brother, Sanjay, had died in an air crash in 1980, and gradually, I began spending more time in Delhi with Rajiv at Akbar Road. In 1984, I left the management of Aplab in the able hands of my staff, well-groomed in the Aplab culture. I was away for over eight years but I did hinder Aplab's growth by advising them not to participate in any government business in which I was even remotely involved, one more instance of how my ethics were stunting the company's development. Aplab got hurt but I had no option. I had decided to work for the government, but had pledged to myself that I would not indulge in any unethical actions. I even refused a government salary by accepting a token that I had to. It gave me a nice feeling. I was free to call a spade a spade. I do not think Rajiv Gandhi was ever told about his mistakes, as bluntly as I did. But then, there is another face of Delhi that is now a common knowledge, its ugly face of corruption. Let me share my own encounter with the corruption. In 1985, I advised the government to float a tender for the bulk purchase of colour television tubes to save foreign exchange. India was importing these at US$76 a piece. My idea of tendering for five hundred thousand television tubes created fierce international competition. The lowest bid was US$64 and ET&T, of which I was the chairman, negotiated it down to US$63. We saved US$6.5 million for the country in one stroke, and we went on doing the same till 1989. Companies in Japan, S Korea, and France were the suppliers. What followed came as a surprise. One among those suppliers came to meet me and surreptitiously told me that he had set aside a dollar per TV tube and wanted to know how and where he should give it to me as my share. He was offering bribe after the supply was made. I called ET&T's the then financial director Mr Patro and told him that he should see how ET&T could accept the $50,000 he was offering. The supplier called to apologise and told me that this had been the prevalent practice. After this incident, I got labelled as 'clean'. People were warned to keep their hands off this guy. Indeed, it is obvious that such a reputation is no good for any ambitious businessman. Fortunately, my ingrained values stopped me from stooping that low.

The eight years in Delhi were a sad and painful experience for me. I found almost no one among the politicians or top bureaucrats, who was interested in nation building, something that demanded sustained well-planned efforts. We had remarkably brilliant bureaucrats and very sharp, cunning, and street-smart politicians. I found that the former were more concerned with their careers than the nation. Those among them heading the ministries barely had a year before they would be retired so the only future thoughts they had were about themselves and not the country. Most of the politicians had no vision for India. Their vision was limited to themselves, their region, their electorate, and their caste. The liberalisation of the 1990s had added further to our woes as a nation. Manufacturing had almost died. Commercial globalisation brought about an overwhelming change since it altered the social behaviour of people. Our society too had become totally money-centric. Nobility, sacrifice, selfless service, and wisdom were no more the measures of a respectable personality—instead, it was the wealth.

Our Government Does not Respect Technology

In the 60 years of my engineering career, I have come across innumerable bright technical entrepreneurs who, in spite of their handicap of limited means and resources, have developed very innovative products and processes with unique features and performance.

They have established credible technology businesses, but I have seen them struggling to grow. The government has not shown any interest in identifying them and giving them special attention even though any developing nation, in its own interests, must do so. In my early years, I was lucky to get such recognition from BEL, ISRO, and other institutions. I did assist them with my solutions and indigenously developed products. But there is no formal mechanism to utilise such innovators in our country, like in the US, Israel, and even China.

Our government is either not interested or is not sensitive towards them, and in the process, our country loses a chance to advance technologically. We have a Science Congress for our scientists to announce their work, but no Technology Congress where our technical entrepreneurs can present their work. I feel sad that those who govern this country do not know how to make use of technical entrepreneurs to drive national technological progress. Let me illustrate this point by recounting our experiences in two areas—smart card-based public payphones and automated teller machines. Both products were needed in huge numbers by India. What happened in both the cases still haunts me and makes me wonder what role corruption might have played in denying that business to us. Aplab was the first to offer India-made world-class ATMs and smart card phones. In another country, we would have been recognised as pioneers. We would have got a priority in government purchase programmes. But, sadly, both DoT and the nationalised banks used a commercial trick to disqualify Aplab, preventing it from even quoting a price against their public tenders for these products. We were thrown out for not being an 'experienced' supplier.

A 'Smart Card Industry Pioneer'

We at Aplab were quick in understanding the importance of smart cards in late 1989 after I saw, on the streets in Paris, smart card-operated public payphones. I knew that Semiconductor Complex Ltd (SCL) in Chandigarh had the technology to make EEPROM chips which I could use for making smart cards.

Aplab paid a big fee to SCL and got such a programmable chip made and developed the technology to embed this chip on to a plastic credit card. Soon, we developed a public payphone with a microcontroller. A unique communication protocol was developed to debit the card, depending on the rate applicable to the destination called. It was then put on trial by DoT in Mumbai for over six months. This established that our smart card public telephone worked fine. MTNL then allowed us to install hundreds of them all over Delhi, including several in the Parliament House. Thanks to my technology-savvy friends in MTNL, this happened without a bribe. MTNL did not buy our phones, but we put them up on the basis of a revenue-sharing formula. The International Conference on smart cards held in India honoured me as a 'smart card industry pioneer'. After successfully running them for 18 months, MTNL announced its plans to float a tender for 10,000 card-operated telephones. Everyone at Aplab was excited since Aplab had the pioneer's edge with a field-proven product. It amounted to Rs. 2 billion worth of business. I was called to meet the minister. I introduced myself to then minister (of telecom) Sukhram. He treated me well, and I told him our smart card story. DoT, however, knew that Aplab could not be their milking cow. So when the tender notice was released, DoT put in one qualifying condition about the vendor needing to have at least 50,000 smart card payphones functioning anywhere in the world.

With one stroke, Aplab was thrown out. The Government of India thus gave Aplab its first 'reward' for its pioneering work. Strangely enough, Aplab's smart card payphones got approved in Europe by Belgium Telephones, but they wanted international credit from India, and this I did not know how to organise.

Our Next Pioneering Product, ATM Machines, Was also Shelved

Next, I became excited by bank automation solutions. In 1996, I saw an automated teller machine in HSBC, which made me wonder why Aplab should not make ATMs. By then, Aplab had a good grip over computing techniques, and our IT team was very excited about having a go at ATMs. The heart of the ATM machine is the cash handling mechanism. De La Rue was the prime source of bank note dispensers to global ATM manufacturers such as NCR and Diebold, but they lost both these customers since they started making their own dispensers. So De La Rue readily agreed to support us with their long-proven cash dispensers.

The rest of the job involved designing a customer-friendly interface with a CRT screen, secure handling of the ATM card, and writing secure networking software. In a year, Aplab had the ATM machines ready. Again, it was a pioneering effort on the part of our design group. We sold over hundred machines over the next 12 months. We also had two unique design features. First, our machines were interactive with the user and guided the users in their own language. The second feature was the ability to take a picture of the person withdrawing cash from the machine. Both were not provided by our competitors, NCR and Diebold. With over 200 ATMs in the field, we were looking forward to banks buying our product.

But the Indian bankers were no different than DoT or MTNL. Tenders released by our nationalised banks included a condition that the tenderer must have at least 2000 ATMs installed and functional. In a clean sweep, Aplab was debarred by Indian banks from the ATM business. What a tragedy for our country! I met the finance and commerce ministers at the centre, who I thought would understand the merit of supporting an Indian company's pioneering efforts. After all, it was the Congress that was ruling, and I had served it in the 1980s. But, I just got a smile and a cup of tea from them besides of course a promise to 'look into it'.

No Regrets at All

There are many more instances for me to quote, but I gave you the two instances that could have made Aplab at least a Rs. 20 billion enterprise, if not bigger. Well, this is the cost one pays for being upright and not falling in line with the business practices in India. I must, however, conclude by saying that I consider the loss was India's and not mine. I feel very proud that we, in India, could do what others in the world were doing and that too in a small Rs. 1.5 billion mid-sized public limited company. I have no regrets today as I have my good reputation intact, which allows me to live with pride and contentment.

My Journey as a Materials Engineer

Samir V. Kamat

At the outset, let me confess that when I received an invitation from the INAE President for contributing an article for the INAE newsletter, I readily accepted without giving it much thought. I did not anticipate that writing about one's own life and work and making it interesting enough to hold the attention of the readers would be such a challenging task. But after struggling for months to finish the article (and often regretting on why I accepted the invitation), I can say that I am glad that I accepted the invitation because it gave me an opportunity to reflect on my life and career.

My tryst with engineering started very early in my life. My father was a civil engineer working with Gammon India Ltd., which is one of the premier construction companies in the country. He was responsible for building dams, bridges, and power plants all over India. These projects were usually in remote areas of the country, and thus, the only source of entertainment for a young kid like me was wandering around the construction sites. I thus had the good fortune of seeing massive civil structures evolve from scratch and let me tell you that is quite awe-inspiring for a child. I distinctly remember that at that time my dream was to follow in my father's footsteps and become a civil engineer and construct even bigger structures!! You can say that I was no different than any other young, impressionable child who thinks that the world begins and ends with what their parents do!

Most of these construction sites did not have good schooling facilities, and so my parents had no choice but to send me to a residential school. They chose an excellent school (Sanjeevan Vidyalaya) in Panchgani, a lovely hill station near the better known town of Mahabaleshwar in Maharashtra. The education there was well rounded with equal emphasis on academics, sports, and culture, and I still retain pleasant memories of my time spent there. Eventually, my parents (my mother) decided to set up a home in Bombay (or Mumbai as it is known today), and I completed my final years of schooling in King George, Mumbai, a very well-known

S.V. Kamat (✉)
Functional Material Division, Defence Metallurgical Research Laboratory (DMRL),
PO Kanchanbagh, Hyderabad 500 058, India
e-mail: kamat@dmrl.drdo.in

P. Ghosh and B. Raj (eds.), *The Mind of an Engineer*,
DOI 10.1007/978-981-10-0119-2_32

school. Gradually, visits to my father's construction sites became holiday-only (summer and Diwali) affairs, and as I became older, I began to realize the comforts and benefits of living in a city and how tough the life of an onsite civil engineer really is. I distinctly remember that by the time I was in 10th standard, my motto in life had changed to study "anything but civil engineering". However, in that era, the career options for any reasonably bright student were limited to "engineering" or "medicine". I had no aptitude for medicine so the choice was limited to engineering (other than civil of course). I would like to thank all my teachers in both the schools because it is my belief that the values that they instilled in me in my formative years played a major role in shaping my personality and helped me become more adaptive and independent.

To take the journey forward, I was extremely lucky to get through JEE as I did not specifically prepare for it and was content on focusing on my 12th standard syllabi. Times were simpler then, and an IIT admission was not a mind-numbing long endeavour as it is today. I remember visiting IIT Bombay campus on a rainy day in June in 1981 for counselling and was required to fill a form specifying twenty choices. At the time, my JEE rank I believed was good enough to get either mechanical or chemical engineering in one of the five IITs. As a result, I did not pay much attention to what my choices were beyond the 10th option and filled in random options on a lark. As luck or destiny would have it (IIT Kharagpur, metallurgical engineering), my 18th choice was where my next four years would be spent. The fact that I had to study metallurgical engineering did cause some initial panic, as I had no clue on what Metallurgical Engineers did for a living! My knowledge of engineering branches at that time began and ended with Mechanical, Electrical, Civil, and Chemical. And this was a time before "Internet" and "Google" so there was no easy way to obtain information on what Metallurgical Engineers did for a living. A friend claimed that his brother's friend's cousin had studied metallurgical engineering and was now working in a mine in Dhanbad. I was now cursing my luck and regretting my decision to not choose civil engineering as an option since working on dams and bridges and building massive structures above ground looked way more attractive than seeing the under belly of a mine! To make matters worse, the same friend (a self-acclaimed know-all) told me that among all IITs, IIT Kharagpur had the worst ragging.

With grave apprehensions, I landed in IIT Kharagpur in July of 1981 as a fairly naive 17 year old. I will spare you the gory details of the ragging I had to undergo!! (really speaking it was not bad at all). It so happened that 1981 was the first batch where the IITs were changing over from a 5 year to a 4 year B.Tech. programme. As a result, the first-year curriculum for the new programme was crammed to cover the syllabus of what was taught over two years for the five-year batch. The first year passed like a whirr, and before I knew it, I was in the 2nd year when the first departmental courses start. My first introduction to metallurgical engineering was a course on metallurgical thermodynamics. In cricketing parlance (especially for old timers), it was like starting your career facing Michael Holding and Andy Roberts on a fast and bouncy Sabina Park pitch. Luckily, I survived and gradually started to realize that metallurgical engineering involved much more than looking for metallic ores in mines!! The 4 years in IIT really flew by and were most definitely the best years of my life. Although I did fairly well academically (being the silver medallist in my batch), I would be not be stating the entire truth if I said that by the end of my final year, I was a true blue-blooded metallurgical engineer. In my own evaluation, I had a decent foundation but not much of a superstructure. This had less to do with the quality of the faculty at IIT Kharagpur metallurgical engineering department, which was in fact excellent, and more to do with my lack of maturity at that age to appreciate the finer nuances of the subject.

I was again at a crossroad; I had three choices: take up a job, study management at an IIM, or go abroad to do Ph.D. I chose the last option since most of my friends were headed abroad and I can say today (almost 30 years later) that I am glad I did!! I had a fellowship from The Ohio State University (OSU) at Columbus and my interest in and appreciation of materials or metallurgical engineering really commenced as a result of my experiences at OSU. For this, I would like to thank all my professors and fellow graduate students at OSU, but the bulk of the credit would go to my Ph.D. adviser, John Hirth. I have not met a more knowledgeable and nicer person before or since. His humility (despite his many achievements and stature) as well as his willingness to discuss problems with his students at their level is a character trait that I admired the most. I have no hesitation in saying that my career has taken the shape it has mostly because of my interactions with Prof. Hirth during my three years at OSU. However, one of my everlasting regrets has been that I failed to reach even 10 % of his funda levels!! Coming back to my Ph.D. thesis, those of you who are familiar with Professor's Hirth's classic book on "Dislocations in Solids" would have guessed by now that my thesis would be something to do with dislocations. You are both right and wrong. My main work was on studying the deformation and fracture behaviour of particulate reinforced metal matrix composites which were then just emerging as materials of great promise (unfortunately they never delivered on their promise). However, I also looked at how image and coherency stresses influence interface dislocations in multilayer heterostructures, which was a very important issue to be resolved for improved performance of microelectronic devices, as an appendix of my thesis. The interesting thing was that during my thesis defence, 90 % of the questions were from the appendix part!!

Dr. G. Sundararajan, from DMRL, happened to visit OSU (incidentally also his Alma Mater) and gave a lecture on the work going on at DMRL, just as I was completing my Ph.D. and contemplating on what I should do next : take up a faculty position in USA or India or work in a research laboratory. I was very impressed with the nature of research as well as facilities at DMRL and decided to apply to DMRL. Prof. P. Rama Rao, who was then the Director of DMRL, offered me an ad hoc scientist position and upon completing my Ph.D. and a short stint as a postdoctoral fellow at Washington State University at Pullman, WA, I joined the Defence Metallurgical Research Laboratory (DMRL) where I have now spent the better part of 25 years interrupted by a brief two year stint at the Group for Forecasting of Systems and Technologies (G-FAST) at DRDO HQ. I would always be grateful to Prof. Rama Rao as well as all the subsequent DMRL directors for the freedom and encouragement they gave me. This helped me integrate very easily in the laboratory and helped my career flourish. Along the way, I worked on several structural materials developed in DMRL for various defence applications, and these include particulate reinforced metal matrix composites, high-strength aluminium alloys, Al–Li alloys, titanium alloys, titanium aluminides, tungsten heavy alloys, ultrahigh strength steels, nickel-based superalloys, and silica/silica composites. In the initial years of my career, my research focus was on studying the correlation between processing, microstructure, and mechanical properties in these structural materials; however, as my career progressed my research encompassed other areas such as mixed mode fracture, mechanical behaviour at small length scales, and stress corrosion cracking. In recent years, I had the opportunity to lead and direct R&D on rare earth permanent magnets (REPM) as well as various advanced magnetic and ferroelectric materials in both bulk and thin-film form for various functional applications as well. No other institution in the world could have provided anyone the opportunity to work on such a diverse set of materials and areas. Additionally, the R&D work in DMRL is always application focused. Research culminating in an application is always much more satisfying, especially for me.

I will briefly highlight our team's work on REPMs as it encompasses the entire gamut from basic R&D, technology development, and laboratory-scale production. Among the front ranking high-performance REPMs, $SmCo_5$, Sm_2Co_{17} type, and NdFeB, alloys are technologically the most important. High residual magnetic induction and high energy product values are characteristics of these REPMs. They exhibit excellent combination of second quadrant (of B-H plot) characteristics and offer additional advantage of device miniaturization by virtue of their high energy product, because the volume of a magnet required for a given application is inversely proportional to its energy product. With these REPMs, it is possible to design for reduction in the size and weight up to 10 times for many electro-technical devices without compromising on their performance. Due to this advantage, they are used in a wide range of applications ranging from consumer products to very specialized areas of telecommunications, automotive, microelectronics, avionics, defence, space, and wind power. While NdFeB magnets are preferred for ambient temperature applications on account of their higher energy product, SmCo magnets (1–5 and 2–17 types) are used for applications where the magnets can experience elevated temperatures due to their higher Curie temperature and lower temperature coefficients of remanence and coercivity.

The rare earth permanent magnet technology is currently possessed by very few countries in the world because of various factors such as complexity of the technology, availability of raw materials (rare earth metals), and patent related issues. In fact, it should be noted that China not only has a monopoly (97 % of production) over the rare earth metal resources such as Sm, Nd, Dy, and Gd, which are the key raw materials for producing these REPMs but it also dominates the production and supply (50 %) of REPMs. Within the country, DMRL is the only organization that has worked assiduously to understand the basic science as well as establish the technology for producing these REPMs. The home-grown technology has been exploited to develop and supply speciality magnetic components for various DRDO and ISRO missions. Till recently, the laboratory facilities were sufficient to meet the demands of the strategic sectors. However, in recent times, the demand for these REPMs from the strategic sector has gone up tremendously. We are currently exploring the possibility of transferring of our REPM technology to Indian Rare Earths Limited to set up a pilot-scale plant to begin with. It will give me great pleasure, if such a plant is set up, as it will be the ultimate fruit of our R&D efforts.

Finally, I would like to end by saying that although my journey to becoming a materials engineer was not something I aspired and planned for, I do hope that my narrative will motivate at least a few youngsters to seriously consider materials engineering as a career option, by design.

My Experiences at IIT Kanpur

Sanjay Mittal

Student Days

I grew up in Dehradun and headed for IIT Kanpur in July 1984 to take admission in the B.Tech. programme in aeronautical engineering. Since it was the first time I was leaving home to stay alone, my parents wanted to inspect my new habitat and accompanied me all the way to Kanpur. I was myself quite nervous about starting at a new place and was very grateful for their kind gesture. At the same time, I was also very excited about the opportunity to study at one of the best institutes in the country. We took a train to Lucknow and then a UP Roadways bus to Kanpur. The conductor announced that "Ghantaghar, the bus station" has arrived and that we should get off the bus. It had also starting drizzling. When we approached the footboard of the bus, we realized that the bus is parked in a big pool of water and the only option was to step straight into a rickshaw. Indeed, Kanpur city had offered a glimpse of what to expect! We had substantial luggage and could barely fit into three rickshaws which took us to a bus to go to IITK.

To this day, the entry into the campus, through the hallowed gates of IITK, is fresh in my mind. Not just the physical landscape, the serenity of the campus captivated me. Both my parents and I were very happy with my new home. The classes began. It was very different than what I was used to at school. The pace of teaching was fairly rapid, and the material was pitched at a level that required coming prepared with some background work. Above all, unlike in my school, I realized that all my classmates are very bright and talented. Some of them were amazingly hardworking. A large number of them were interested in music, sports, and hobby clubs. Some had extraordinary leadership skills. Spending time with them rubbed off on me, and I owe a great deal to my friends who got me interested

S. Mittal (✉)
Department of Aerospace Engineering, Indian Institute of Technology,
Kanpur 208 016, Uttar Pradesh, India
e-mail: smittal@iitk.ac.in

P. Ghosh and B. Raj (eds.), *The Mind of an Engineer*,
DOI 10.1007/978-981-10-0119-2_33

in morning jogs, listening to and playing music, aeromodeling, etc. The campus was extremely vibrant. In addition to the Institute Annual Cultural Festival and Interhostel Festival, there were high quality events all-round the year. Even more impressive were our teachers. I had excellent teachers. We had plenty of opportunities to get to know them outside the classroom. It was very gratifying. Some of them delivered public lectures in the Institute, hosted by student-run societies. It was absolutely thrilling to hear a teacher delivering a lecture on a subject very different than he/she taught in the class. Sometimes, these lectures exposed us to the original research work of these individuals and their graduate students. As a youngster, it took me a while to realize that most faculty members spend enormous time on research/knowledge creation. What we see of them in the regular classes is just a part of the gamut of activities they pursue. Compared to the acquaintances that I was familiar with before coming to the campus, the simplicity and grace of my professors was at a different level. This already inspired me to pursue higher studies and explore, leading a similar life. The "four-year transform", as I often refer to my stay at Kanpur, exposed me to most of the components of wholesome education.

As a Faculty Member

After finishing my graduate school at University of Minnesota, Minneapolis, and about a year and half of postdoctoral fellowship, I was very fortunate to have the opportunity to come back to IITK as an assistant professor in June 1994. Help came in from all quarters of the Institute as I settled in. Everyone was extremely warm, affectionate, and offered help, many a times without even me having to ask for it. Within a matter of a few months, I was well planted and started on my research and teaching.

Teaching

Although I had been a teaching assistant during my graduate school days, teaching at IITK has been a different experience. The first class that I taught as an instructor was an elective for graduate as well as undergraduate students: finite element methods for fluid dynamics. This was in the winter semester and scheduled at 8:00 AM. A senior professor in our department attended all the lectures that I taught. Not just that he was present in each and every lecture, but that he occupied the centre seat in the very first row. Needless to say, it kept me on my toes. I ended up preparing my lectures not just for my students, but I had to ensure that I earn the respect of my senior colleague (and my teacher in my undergraduate days). In hindsight, this was the best training that I could have received. I also discovered, very quickly, that there are many a students in the class who are far more bright and quicker than me. The realization was very humbling. I figured that my approach as a teacher, for such students, has to be different. It is not so much about the content that I need to get across; I just have to motivate and inspire them. I realized that if I am able to do that, they will not just learn the subject, but also help the rest of the class see things better than I can do. The class relates better to them, than to me. Over the years, I have been able to work on this and evolve a model. In any course that I teach, I include a module of self-learning. Halfway through the semester, the students, in groups of 2–4, select a topic that is relevant to the course. They submit a short abstract along with the relevant references. After my approval of the abstract (I do a bit of reading myself and fine-tune their objectives and/or suggest additional/alternate references), they start preparing a term paper. In the second last week of the class, they make a 10-min presentation to the entire class. The number of slides is restricted to 10. I announce right at the beginning of the semester that there will be at least one question from each presentation in the end-semester examination. So, there is motivation for the students to carefully grasp each presentation. Most presentations are followed by a lively discussion. All the presentations and the reports are uploaded on a URL and are available to the entire class. Every semester, my students are able to teach me and the rest of the class something new. One of the other things that I have learnt during my stay at IIT Kanpur is that if I wish to learn a new subject, the best way is to offer an elective course on it. I have immensely benefited from this on a few occasions.

Research

An academic institute offers ample opportunities to ask questions and attempts to answer them. Students are the biggest resource. They bring in fresh ideas and very often a new way of looking at things that one has already been working on. Once I was able to nucleate a research group, I found that it is self-sustaining with very little effort from me. I just need to be fair and honest to everyone and ensure that the members of the group maintain a healthy relationship with each other. For this, sometimes I need to devote reasonably large time and effort to attend to personal problems that a student might be going through. The students learn from each other and, in fact, teach me many a things. Group seminars are very exciting where one or two students present their research. The presentation is profusely punctuated with very interesting discussions. My job is simply to ensure that the discussions remain healthy. The group members are always looking up to the mentor. It is very important to imbibe the right ethics, mostly by one's own conduct.

In my graduate school, I worked in computational mechanics. I continued working in that after joining IIT Kanpur. However, I also started doing laboratory experiments, a few years ago. Now, about 25 % of the members of our research group do experiments and they closely interact with others who do computations. This has added a new dimension to the learning and excitement to the group. Almost as a rule, our group starts on a somewhat new area of research every five years. The diversity and novelty of research problems, along with interactions with bright, creative students and the opportunity to collaborate with colleagues, who offer a complementary knowledge base, makes the journey very exciting and gratifying.

I believe that it is also important to consolidate our research findings. We, in our research group, mostly do it via archived publications. Compilation of the work as a report/article is usually the most educational activity as it offers an opportunity to look at all the results, comprehensively and critically. More often than not, we discover while writing that additional test cases are needed to arrive at a certain conclusion and for the results to be self-consistent. It requires loads of hard work to come up with a well-written manuscript even when all the data have been computed/acquired. It is an absolute delight when one is ready with it. Peer review of a manuscript offers critical feedback that one might have overlooked during the study/writing. Many a times, it also brings with it the much-needed humility.

The freedom in an academic institute enables one to attempt answering questions that arise from day-to-day life. I am an active badminton player and used to wonder as to why the play with a duck-feather shuttlecock is different than a synthetic shuttlecock. With the help of a few very motivated students, we carried out a computational fluid dynamics (CFD) study for these objects and put forth a hypothesis. The attempt to understand this practical life phenomenon was not only very educational, but also a lot of fun. Not to mention, the excitement at the badminton courts and the lively discussions with fellow players have been the added bonus. At present, we are investigating the aerodynamics of a cricket ball, both computationally and via wind tunnel experiments. One of the questions that we wish to address is—can a new cricket ball reverse swing? Our results show that it can, albeit at a reasonably large speed and for certain orientations of the seam with the flow. In another research along with a graduate student and a collaborator, a professor of civil engineering at IITK (who is also an extremely good friend), we investigated the flow of traffic in Indian conditions. We proposed a behavioural model for Indian drivers and modelled the traffic via equations for fluid dynamics, assuming that the dense traffic is a continuum. Unlike in the west, the general traffic in our country does not follow lane discipline. We have been able to explain quite a few phenomena arising out of this. My colleague did most of the work with respect to traffic modelling, and my contributions helped with the fluid model and its numerical treatment. Working together has its own joys.

Committee Work

I used to despise meetings and committee work when I joined IITK. It felt that these activities were driving me away from my endeavour to excel at teaching and research. Over a period of time, I realized that these meetings offer an opportunity to get a glimpse of the great minds of my colleagues whom I would not have otherwise interacted with because our teaching and research areas are so different.

Some of them are much more experienced and have given some of the issues, a lot more thought than I did. The opportunity to get a view of their thought process and ideas has been very educational and inspiring. Many a times, I became more aware of certain issues and learned how more experienced people handle them. It also developed a sense of appreciation for many individuals who work behind the scenes to make our system work. Most committees in our Institute have student membership. The views of the students are sometimes quite different than the members of staff. Knowing their views and being sensitive to them has helped me in becoming a better teacher (I think, although my students may feel otherwise!) and mentor. It has also helped me immensely as a parent in raising our own children and being able to appreciate their point of view. Another thing that I have experienced over the years is that creativity transcends areas of application. Creative individuals are creative in all scenarios and not just in the specific activity such as their research or hobby. They also have an excellent sense of humour. Meeting and spending time with creative and bright people is extremely refreshing. It is synonymous with positivity. I have had the opportunity to work in various committees. Two of them that stand out are (a) Organizing Combined Pre-Medical Test (CPMT) for the UP Government, where I was the vice-chair and (b) Chairperson of the Academic Review Committee at our Institute. While the former trained me in handling logistics in large systems, legalities, and public relations with the world outside the boundaries of IIT Kanpur, the latter gave me an opportunity to learn what engineering education is all about and how does one design programmes and curricula in these modern times where students aspire for more flexibility and interdisciplinarity.

Life Beyond Teaching, Research, and Committees

It is very important to be able to devote time to activities beyond the need of the profession and to be able to relax. In that sense, IIT Kanpur is heaven. The quality of life in the campus is extremely good with excellent social life, almost zero pollution (including noise), very secure, beautiful landscape, well-kept and well-maintained campus, and very well-maintained sports facilities. We have an active badminton club. We have been playing six days a week in the mornings. It keeps us physically and mentally healthy. I have been a regular to the Inter-IIT Sports Meet for the staff. That one week of fun, in one of the IITs, charges me up for the next year. There are ample cultural activities in the campus. Since I enjoy listening to and playing music, I have interactions with the Music Club. Many a times, they invite some of us as judges for their musical events. It is a lot of fun. One can take a walk or a bike ride in the lovely campus. It is most enjoyable. For those who want to pursue yoga/meditation, etc., there are regular camps/classes. Of course, I must also mention that the most fun that I have is at home and the credit for that goes entirely to my lovely wife and our two adorable children. At the end, it is the friends and family to whom I owe my happiness.

Journey I Enjoyed Amidst Challenges

S.C. Chetal

I was born on 10 January 1949 in a middle-class family. My mother, in particular, had a great influence on my education and compassionate feeling towards the needy and fellow human beings and inspired me to be dedicated to my job. I studied in a reputed government school in New Delhi, where the school teachers had shown extra affection and care for potential students who could get distinction in mathematics, physics, and chemistry. I passed the school with distinction in these subjects.

The Central Board of Secondary Education at that time had a system unlike present time where the 11th board examination was based on the entire teaching from 9th to 11th class. There was no concept of coaching for IIT examination at that time. God has been extra kind to me in remembering with ease chemical equations and mathematical formulae and confidence in examinations.

I got selected in IIT thanks to my performance particularly in chemistry and mathematics and was offered mechanical engineering at IIT Kanpur/electrical engineering at IIT Delhi. Not able to imagine living away from my family, in particular from my mother, and being advised that mechanical engineering is the best branch, I joined Delhi College of Engineering in 1965 in mechanical engineering.

To take a job or opt for higher education after passing engineering in 1970 was a tough decision. I got selected for MTech at IIT Kanpur but did not join. An important lesson was learnt of engineering when one of my friends who appeared along with me for IIT Kanpur was advised that one cannot afford to forget fundamentals. I joined IIT Delhi for MTech in thermal engineering. In the meantime, I was selected in Bhabha Atomic Research Centre (BARC) training school for a career in Department of Atomic Energy (DAE). After very prolonged discussions, I opted for the BARC training school, a decision I cherish as it has been very challenging and professionally rewarding.

The life in training school revolved around examinations on vastly different subjects on almost every Saturday. Though at that time, I felt why a mechanical

S.C. Chetal (✉)
F 38, Mansarover Garden, New Delhi 110 015, India
e-mail: chetalsc@gmail.com

P. Ghosh and B. Raj (eds.), *The Mind of an Engineer*,
DOI 10.1007/978-981-10-0119-2_34

engineer should be taught so much physics and control systems, with time I realised the importance as my responsibilities increased in the organisation. It is important for all of us to do our best to learn other disciplines and respect the same. I did my best to perform in all subjects. I completed the postgraduate training in nuclear engineering in 1971 with high ranking. Despite many other placement options, I was particularly lured by the challenges and importance of fast reactors, and joined the then Reactor Research Centre (RRC) at Kalpakkam, almost a village at that time and very far from my hometown Delhi. From August 1971 (beginning of my job) to January 2013 (superannuation), I was at Kalpakkam and held different positions at Indira Gandhi Centre for Atomic Research (earlier RRC).

After spending my first two years in engineering experiments in support of fast reactors, I joined the design team. I was engaged in the design of heat transport system of Fast Breeder Test Reactor (FBTR) in particular for sodium-heated steam generators, the component known to decide plant availability. My senior engineers had enormous confidence in me and it made me work hard. I am extremely satisfied and feel proud that the steam generators and other equipment designed as well as system modifications carried out by me for FBTR have been functioning to the desired level since the beginning of FBTR operations.

For FBTR, when the bids were opened for the rupture discs, safety device to relieve pressure resulting from large sodium water reaction, the offers were far beyond the estimated value. The then Director and Principal Design Engineer decided to opt for indigenous development and entrusted the job to me. The rupture disc design primarily involves sheet metal forming to spherical shape without any die, theory of elastic buckling, and material hard enough to cut the stainless steel disc on reversal at set pressure. With the support of my draughtsman and unskilled labour, the development was successfully accomplished after a number of trials. In this journey of development, enormous moral support came from my seniors. All the rupture disc assemblies manufactured in-house were installed in FBTR. With time I realised and advised my young engineers that Indians could accomplish many challenging engineering tasks. The principal ingredients are confidence of seniors and technology denial by other developed nations.

During the course of my interactions with my Principal Design Engineer, I noticed him performing back-of-envelope calculations. I got inspired by this approach. With gift of memory of numbers and formulas in strength of materials, fluid mechanics, and heat transfer, I started checking the design of my young engineers as I grew up in the organisation. I am of the strong opinion that every engineer should develop taste for approximate estimates through analytical approach through the use of formulae even in the present era of every activity solving by ‘software’.

It was realised at the early stage of my design activity that one needs to acquire expertise, and accordingly, I spent on regular basis considerable time at home to read design codes, background journal papers, and effect of imperfections on structural integrity of mechanical components. The expertise acquired has been of great help in taking decisions on non-conformances in the manufacture of reactor components based on ‘fitness for purpose’. My interest in design codes increased with the passage of time, and I could spare time even when I was Director. It led to a number of modification proposals for both American and French pressure vessel codes.

Based on confidence in design and construction of FBTR, the Department of Atomic Energy took a tough decision to opt for indigenous design and construction 500-MWe Prototype Fast Breeder Reactor (PFBR). I was associated with this project since the beginning and have enormous satisfaction to have contributed to this project in a number of ways. This includes conceptual design, detailed design, material selection from fuel cladding to condenser tubing, R&D, regulatory clearances from regulatory authorities, and dealing with deviations during manufacturing. There is feeling of pride that I had played a crucial role in encouraging and inspiring the Indian industries to successfully manufacture the reactor components, first of a kind, to very challenging specifications.

A great challenge and opportunity came in my career when I was asked to lead a composite team to select the materials and to define R&D for materials development for 500-MWe PFBR at the young age of just 26 years. I enjoyed learning the subject of materials, in particular mechanical metallurgy. I worked very hard to get grip over the subject of materials. The complimentary knowledge of design, materials, and relationship of materials with design has been of great help to me in carrying out a number of technical activities outside the Department. This includes the design and manufacturing of retorts for zirconium sponge plant and titanium sponge plant, and materials selection and its successful development for advanced ultra-super-critical thermal power plant.

Based on my interactions with engineers who have been exceptionally successful, I will like to sum up by stating that every engineer should aim to follow ten ‘E’ commandments, as applicable, viz. (i) Enjoy your professional and home life, (ii) be Effective in every activity you undertake, (iii) look for Excellence, (iv) do not be Emotional in both grand success and utter failures, (v) be Experimentalist, (vi) put Extra Effort to get success, (vii) upgrade your Education continuously, (viii) be Ethical, (ix) Express your views clearly, and (x) design an Economical product.

Reminisces from the Past

Sushmita Mitra

I was born into a highly educated family in Kolkata on a cold December afternoon, the only child of Dr. Maya Mitra and Dr. Girindra Nath Mitra. The doctor had wrongly pronounced me dead, even before I could see the light of the day. But, luckily for me, I survived.

My father was a scientist with the Indian Council of Agricultural Research (ICAR). My mother was a professor of botany at Bethune College in Kolkata—a leading women's college from where the first women graduates of the British Empire came out. Whatever I am today, be it in daily life or in the professional world, I owe it completely to my mother's untiring sincere efforts, relentless sacrifice, and careful grooming. She had been my "supermom," adeptly managing both home and work with equal dexterity.

From her childhood, my mother rebelled against the concept of preference for a male child that was prevalent in those days. She revolted against getting married at an early age and be burdened with loads of children and desperately wanted to study. In spite of being forced to come to India as a refugee from Burma, during the Japanese bombing in 1942, she built her life from scratch. She had the ambition to finish college, and eventually obtained a Ph.D. in 1960—also receiving the Agarkar gold medal of the Calcutta University in 1962, for her pioneering contribution to research in Botany. Both my parents also had the rare privilege of independently publishing in the reputed journal *Nature*! This was amazing and awe-inspiring, particularly for a woman of those times. Setting aside such a promising research career, she decided to marry and raise a family.

After all women are always expected to have first priority toward their family obligations, in this patriarchal world, and second comes their professional interests. It is pretty difficult for a woman to concentrate with devotion to her studies, once she starts family. While many men continue to be cared of by housewives, after their mothers, professional women cannot afford this luxury! Therefore one has to arrive at an offset, between work and home. It is how one can optimally achieve this

S. Mitra (✉)
Machine Intelligence Unit, Indian Statistical Institute, Kolkata 700 108, India
e-mail: sushmita@isical.ac.in

P. Ghosh and B. Raj (eds.), *The Mind of an Engineer*,
DOI 10.1007/978-981-10-0119-2_35

balance, and upon how much it matches one's objectives in life that leads to satisfaction or happiness.

I was the daughter of a working mother. In the late 1960s to the early 1970s, when I was still a kid, the concept of working women was not that prevalent. Most of my friends returned from school to homemaker moms. But that never made me unhappy, as she always made up by spending quality time with me. With her encouragement and support, I grew up in a healthy educational environment. My father's career at ICAR implied that he spent much of his working life outside Kolkata, in Shillong, Cuttack, and Hyderabad. Under this circumstance, my mother took care of my education as well as all other needs. I was proud that *ma* was a professor in a college, that she took me along with her in the numerous botanical excursions which she organized for her students, that I could visit so many places to meet and see so many interesting people and things, that she brought me plenty of exciting gifts, that she could look after my studies, and most importantly that she could still shower unlimited affection during the times that she spent with me.

Our surroundings were still quite green, with the nearby *Tallah* park being full of different kinds of trees, bushes, and flowers, along with a beautiful lake. It was so enchanting that it made me imagine Rip Van Winkle lay sleeping under some bush in some hidden corner. But I never managed to locate him, no matter how hard I strived. There was a transmission tower with a red star-like light twinkling at its top. *Ma* said it was from the radio station of Calcutta. I imagined people sitting there and speaking whatever we heard on the radio. Television was yet to arrive in India, and we had wired antenna at home to receive the transmissions into our valve-based radio. Soon darkness used to descend, with throngs of fireflies flickering in all directions. The park transformed into a fairy land. Where, oh where, have those fireflies gone? Have they disappeared with my childhood? They have taken my innocent dreams with them. Urbanization has taken its toll. Today, the greenery is depleted. No more do children have time for such small pleasures of life.

Every Thursday, our school was closed, and there was nobody at home under whose care I could be entrusted. So that was the day I carried my work to Bethune College with my mother. Although I never officially studied in either Bethune School or the Bethune College, yet I practically grew up over there. I still recall with pleasure how I enjoyed my lazy afternoons amidst the lush natural greenery within the campus, exploring new plants, flowers, and butterflies, while she attended to her responsibilities in her office. To a city-bred kid, confined to the concrete jungle, these small pleasures were like whiffs of fresh air. I trundled across the vast expanses of the college playground on my first bicycle. Incidentally, the garden was under my mother's care, and she knew each and every tree round the campus. There were also the laboratories of the Botany, Zoology, and Chemistry departments, which held out their wonders before my amazed eyes. Needless to admit, I was an obedient child who never gave her any reason to be dissatisfied with my behavior.

Baba got deputed to the International Rice Research Institute (IRRI) in *Manila*, The Philippines, for a year. *Ma* arranged for her six-month leave in college, met the principal of my school regarding my forthcoming absence, and then both of us travelled to join my father. Sister Elizabeth (the Principal) had marked out my lessons for *ma* so that my studies would not get disrupted during our stay. In those times such trips were rare for Indian kids like me.

Our cottage was surrounded by gardens, at places unkempt. The landlady had ten kids. They stayed in an adjacent house. There were butterflies of different shapes, shades, and hues, with multifarious beautiful designs on their wings, flitting about. My friend Bobby and I built huts thatched by dry palm leaves collected from the garden. In the evening, Bobby said "Let us go and watch TV." I was a full eight-year-olds from an urban English-medium school, but I had never heard the word "TV" in India. I assumed "watch TV" could be some game. So I followed my friend to a Filipino house where we soon squatted in front of a mini-cinema emanating from a box. The show involved "Popeye the Sailorman." This was my introduction to television which I later learned to identify.

Time flew by and I entered high school. I always had good grades and caused no worries anywhere. *Ma* located announcements regarding the National Talent Search Examination, conducted by the National Council for Educational Research and Training (NCERT). In those days, online Internet service was yet to arrive, and one had to solely depend on newspapers for any information followed by the subsequent physical collection of relevant forms and brochures. Eventually, I was awarded the NTS scholarship. I passed ICSE and ISC from Auxilium Convent School and Calcutta Girls' High School, respectively, topping my class. I had great friends, with some of whom I am still in touch. And, surprisingly, most of them are successful in life today—in their independent professions. Perhaps, this is because the parents who sent their daughters to English-medium schools those days were illuminated enough not to discriminate between boys and girls in terms of their career.

Thereafter, I joined Presidency College to study Physics Honours. The magnificent staircase, along which the luminaries of Bengal had once ascended, held me awestruck. Here, I had the privilege of studying under stalwarts like Prof. Amal Raychaudhuri and Prof. Shyamal Sengupta. My father suddenly passed away at the age of 53 years, due to cerebral hemorrhage, and the entire responsibility of the family shifted on to my mother. ICAR offered me a clerical job as I was still an undergraduate. So I declined the offer to continue with my studies. *Baba* could not live to see me even become a graduate.

As far as my education was concerned, *ma* had always been happy and deeply satisfied. It was she who provided me expert guidance and help in organizing my flow in the right direction. She had searched out the upcoming field of computers as a possible place to explore. I secured admission to B.Tech. in computer science of University of Calcutta. Incidentally, I had also cleared the admission test for the integrated five-year M.E. course of the Indian Institute of Science. However since it was just a couple of years that *baba* had died, I was unable to leave my mother and go. In Raja Bazar Science College, I met the renowned Prof. Arun Choudhury. It was he who made us understand the nuances of electrical circuits, with the voltage drops and current flows at complicated junctions. In class, we were always wary of being made to go to the blackboard and calculate the potential difference between specified points.

After ranking first in B.Tech., I cleared GATE to gain admission to M.Tech. in IIT Kharagpur. I went and took admission, also visited the hostel, and eventually searched out an excuse not to go—as I felt homesick. So back I went for M.Tech. in the University of Calcutta, where again I ranked first. I was awarded the *University Gold Medal* for my performance.

Around the time of my project dissertation, I came across an advertisement on our notice board regarding a project assistantship at Indian Statistical Institute under Dr. Sankar K. Pal. Many people discouraged me from applying, saying that the topic was pattern recognition and hence not related to computer science. It was Dr. Samar Sensarma, a classmate of Sankar*da*, who insisted that I join the project. So I came to ISI and met Sankar*da*, who asked me several questions related to my family and financial background. He was trying to ensure that if I joined I should not discontinue before completing my Ph.D. Of course, this was never an issue with me since both my parents were already Ph.Ds. That was 1988 and the first time that I set foot on the hallowed corridors of ISI. Thereafter, I left my GATE fellowship to join the project.

I started working for my Ph.D. in neuro-fuzzy pattern recognition, with a CSIR Senior Research Fellowship under the supervision of Prof. Sankar K. Pal in 1989. After my mother, he has been the next great source of inspiration in my career. He has mentored me in my professional life and been a guardian throughout. He has always allowed me complete freedom in going about my research in a fully independent manner.

Over my entire academic career this far, in spite of gaining admission in elite educational institutions outside Kolkata, I had never been able to leave home. It was mainly because of my homesickness related to the recent sudden demise of *baba*, the extremely small size of our family, and my very special relationship with *ma*. When I was eventually selected for a DAAD fellowship in 1992, for working with Prof. H.-J. Zimmermann, even my teachers (including Sankar*da*) were skeptical as to whether I would at all be able to spend the scheduled period in Germany. At this stage, my mother intervened, warning me that if I dared to return before completing my work, then I would be sent out straight away. So off I went, the first time away from home and all by myself to a distant foreign land. I was at the RWTH Aachen University in Germany (the largest technical University in Europe). The initial things that struck me were the empty expanses of roads, sparse existence of surrounding humans, extreme punctuality of people as well as transportation, the beautiful snowfall, and the warm hospitality in German homes. And sure it was an overall wonderful learning experience, both academically and otherwise—during the period of 1992–1994.

I completed my Ph.D. in computer science from the ISI in 1995. I have been working there since 1991, publishing extensively in reputed international journals, including IEEE, and rising up the academic ladder to the level of a full professor. My research, on neuro-fuzzy computing and its generic hybridization with other soft computing paradigms, has been internationally acclaimed. I received the IEEE

Neural Networks Council Outstanding Paper Award in 1994 for my pioneering work in this direction. One of the examiners of my Ph.D. thesis suggested its publication as an authored book. I have written several books—Neuro-Fuzzy Pattern Recognition: Methods in Soft Computing; Data Mining: Multimedia, Soft Computing, and Bioinformatics, both published by John Wiley; and Introduction to Machine Learning and Bioinformatics, published by Taylor and Francis—beside a host of other edited books. My research also resulted in several fellowships—including those of the IEEE, INAE, and NASI. I am associated with the editorial activity of several international journals and have chaired many international conferences.

Often *ma* used to inspire me to explore possibilities in a direction involving a fusion of her subject botany with the informatics of mine. And I am amazed and glad today, standing in the twenty-first century, to realize that this is just what I am doing in the integration called bioinformatics. I only wish she were beside me now to share the joys in my professional domain.

Ma retired from Bethune College in March 1995, after thirty-seven long years of sincere, relentless service toward education. A lovely daughter was born to me in October 1995. I had my Ph.D. viva just one month after that. I had married someone, who I believed had "loved" me for around 18 years. But he ran away with the dancer wife of somebody else, leaving my 7-month-old daughter for me to bring up alone—despite the fact that I had made a significant contribution as his ladder to progress both academically and financially.

The logistic support provided by my mother helped relieve a lot of the burden of child rearing and allowed me to concentrate further on my research. It provided important emotional support to my daughter as well. Whatever *ma* had been forced to forego in her own academic career after marriage, in her effort to concentrate on her family and on me her child—she wished to see those fulfilled in my professional life. And hopefully I did not fail her.

Now that I was established in life, I often needed to go here and there on academic assignments. *Ma* wished to accompany me in most cases, not only because she loved to travel but also since she wanted to be by my side wherever I went round the world. Therefore during most of my national as well as international academic visits, I strived to bring my close-knit family along. As a result, we managed to travel together to a number of exotic destinations around the world, and in the process had the opportunity of accumulating and adapting to amazingly interesting experiences. We got to know so many wonderful people from different countries and cultures, having different customs, cuisines, and observing different ways of life. I have travelled to most parts of the world, often on lecture tours as well as on visiting professorship and collaborative research assignments.

My mother passed away almost suddenly in the summer of 2006. A part of me perished with her.

Today, I have several students, many of whom are established as faculty of reputed institutes. My students form my extended family—a trait I have imbibed from my mother. My daughter has grown up and is studying B.Tech. in Kolkata. Often I go around the world and am presently serving as an IEEE Distinguished

Lecturer. My research interests include natural computing, data mining, bioinformatics, medical imaging, and pattern recognition. The institute provides me full freedom in doing whatever I like with respect to my research.

I have not faced gender discrimination in any phase of my education and/or professional career. Wherever I went, I have always been a vocal proponent for women's cause. My mother has been a powerful influence in my life in this direction. She had *dared* to think differently. It was because of women like her, who walked before us, that we are able to achieve what we want with relative ease. Today, I strive to imbibe in my daughter a similar conviction.

Life encompasses three major lotteries. You cannot choose your parents. You think you optimally select your spouse—but often this turns out to be an illusion. You can neither choose your children. Those who are successful in life try to make the best use of these attributes. Those who are not-so-successful complain. As the shadow of my life grows longer, I feel happy with whatever I have received—for, when you want nothing you begin to have everything.

Engineering Challenges—A Personal Account

Vasant Manohar

I had a natural inclination for engineering, and when I passed Inter Science from Elphinstone College Mumbai, I joined VJTI and obtained my B.E. (Mech. & Elec.) (Hons) degree with first rank from Bombay University. At that time, VJTI had very good faculty and many of them were picked up for the new IITs being started. To give an example, Dr. P.K. Kelkar, head of the electrical department was appointed director of IIT Kanpur and later director of IIT Mumbai and awarded the Padma Bhushan.

I went to Imperial College in London for a master's degree in electrical engineering with the Nathubhai Mangaldas Scholarship of Bombay University. I was also awarded the Metropolitan Vickers Scholarship by Imperial College. This made my life in London quite comfortable. After completing my studies with a D.I.C from Imperial College and an M.Sc.(Eng) degree from the University of London, I joined English Electric Co. in UK in their Machine Development Department.

Here, I had my first lesson in professional responsibility. I was asked to study a report on machine oscillations prepared by the Research Dept. of English Electric as I had worked on a similar problem at Imperial College. It was clear that the research findings were totally wrong and I said so in my draft conclusions. Since I was new, I requested the department head to have my comments issued jointly with another engineer as its conclusions would have serious repercussions. He told me: "If you have any doubt please withdraw your comments, if not, you should not hesitate to take personal responsibility for your conclusions." The comments were issued as a departmental report in my name, and there was quite a commotion in the Research Department whose findings had never been questioned before. However, in the end, they accepted their mistake and officially withdrew their report.

On return to India in 1958, I joined the Tata Power Co. One of the first assignments was the modernization of the 6 × 8 MW Bhivpuri hydro station that was commissioned in 1922. A systematic study was undertaken by a team, starting with the storage dam, the complete water conductor system, the turbine generators, electric switchgear, etc. Based on this, a step-by-step upgrading program was drawn

V. Manohar (✉)
Goenka House 16, Walkeshwar Road, Mumbai 400 006, India
e-mail: vnmanohar@yahoo.com

P. Ghosh and B. Raj (eds.), *The Mind of an Engineer*,
DOI 10.1007/978-981-10-0119-2_36

up. This required an overall view as well as great attention to detail to specify new equipment to fit in with the existing equipment with minimum disturbance to station operation. The dam was strengthened using the Coyne method; the water conducting tunnel was cleaned of debris and the lining repaired; and the steel penstock pipes were cleaned internally and externally and painted with corrosion preventive coal tar paint. New pelton wheel turbines were installed on the old shafts and new windings fitted in the generators raising the station capacity to 6 × 12 MW. A completely new control room and outdoor switchyard were also added. The refurbished station gave reliable service at its upgraded capacity for over 30 years. This was an excellent example of life extension of the whole power station at minimal cost.

At this time, Tata Power Co. decided to add a 150 MW thermal unit (largest in India at the time) to its Trombay station to meet increasing power demand. Ebasco Services of New York were appointed consultants, and a local team, many of them fresh graduates, was assembled to work closely with them. I headed the electrical design group in this team. After the design work was completed, I, as well as many others in the team, were deputed to the project site at Trombay to manage and supervise the construction activities. This combination of design and construction experience was highly beneficial to the young team.

With great foresight in 1962, Tatas decided to start an independent firm of consultants in partnership with Ebasco named "Tata-Ebasco Consulting Engineering Services," and I and other members of the core team were transferred from Tata Power Co. to Tata-Ebasco who took on the engineering of the 150 MW Trombay unit and also other projects such as the 2 × 140 MW Dhuvaran station of Gujarat State Electricity Board, projects for other Electricity Boards, many captive power projects and later nuclear power projects for the Department of Atomic Energy. It soon established itself as the premier consulting firm in India. In 1968, Tatas took over the share of Tata-Ebasco and the firm became "Tata Consulting Engineers."

TCE also diversified overseas with work in Nepal, Laos, Sri Lanka, Algeria, UAE, Malaysia, Iraq, Iran, Nigeria, Tanzania, and others. In Iran, TCE had a long-term assignment for training of engineers of Tavanir, the Iranian power utility. I spent 6 months in Tehran, to start off the electrical training part of the assignment. It was a novel experience to work in a different country with a different language, culture, and work ethics. We had a large team to help Tavanir in design of substations, transmission lines, and the operation and maintenance of thermal power stations. Later, I was in charge of the Tavanir assignment and had to visit Tehran frequently during the Shah's regime, during the revolution and also after the revolution. The TCE group was drawn from TCE, Tata Power, and some State Electricity Boards, each with different work cultures and it had to be molded into a team. As we worked as a team in close cooperation with the Tavanir staff as well as their management, we were asked to continue our work through the regime change in Iran.

TCE specialized in taking up one-off projects under its Special Projects Division that was also under my charge. This required design from first principles and threw up many challenges. One of the first projects in this area was the equatorial mount radio telescope for the Tata Institute of Fundamental Research (TIFR). The concept required a 530-M-long antennae array consisting of 24 parabolic structures each 30 M wide connected by fine stainless steel wires to make up the antennae surface. It was to be placed at a sloping angle of 11°, equal to the latitude of the place and rotated at a speed corresponding to the earth's rotation so as to track the celestial object being studied. A suitable location was found near Ooty in Tamil Nadu.

At a meeting with TIFR, instead of asking consultants the usual questions about prior experience in telescope design, Dr. Homi Bhabha asked our structural design team: "I want this design to be done in India. Can you do it?" After an internal review, we accepted the challenge and the design was done in 1963 in close coordination with TIFR keeping all scientific requirements in mind (Fig. 1).

The radio telescope was also fabricated in India, and has operated flawlessly. It is still in operation providing significant research data in radio astronomy.

I was part of an internal study group to consider options for meeting the transmission requirements of Maharashtra as the grid was proving inadequate to meet future demands. The study concluded that a major interconnection between western and eastern parts at the next higher voltage level of 400 kV was essential and it would allow bulk power transfer from coal mine-head power stations in Vidharba to load centers around Mumbai and reduce coal transport by rail. The results were presented at the CIGRE conference in Paris in 1974. MSEB decided to go ahead with such an interconnection, and TCE was awarded the pioneering task of the design of the 700-km 400-kV transmission system to interconnect Koradi near Nagpur to Kalwa near Mumbai. This was the first time that the 400 kV voltage was introduced in the Indian grid. I was the project manager for this, and I went to

Fig. 1 Radio telescope at Ooty

Bangalore and visited the High Voltage Engineering Department of the Indian Institute of Science and the Central Power Research Institute (CPRI) in Bangalore to see whether any test data on 400 kV equipment or lines was available that would help in the design work. However, no such help was available, and we had to fall back on published data from France and USA as the basis of our work. This illustrates our common problem that investments made in research do not provide results that can be used for practical application.

With TCE's considerable experience of 220 kV systems and using published work for reference, we were able to complete the design with confidence. During the design, we developed a very neat layout for the 3 bus 400 kV outdoor switch yard using pantograph isolating switches. This design was copied by the Central Board of Irrigation and Power (CBIP) and issued as a 400 kV switchyard reference standard.

TCE continued to be selected for unique project assignments; one such was the Giant Meter Wavelength Radio Telescope (GMRT) built at Narayangaon near Pune for TIFR. This is totally different from the Ooty telescope and has an alt-azimuth mount and comprises 30 fully steerable 45-m-diameter dishes. Of these, 14 are located in a central array within an area of 1 km^2 and remaining 16 in a Y-shaped layout over a large area with an interferometric baseline of 25 km. Each steerable dish is made up of 16 parabolic trusses. All dishes have to move in unison to track the object under observation, and the radio signals obtained are correlated using specially developed software (Fig. 2).

After much brainstorming, a unique design was evolved comprising "Stretched wire Mesh Attached to Rope Trusses (SMART)". This resulted in a very lightweight and low cost design.

Fig. 2 Giant meter wavelength radio telescope (GMRT)

Fig. 3 500 MW thermal power unit at Trombay

Another challenge, of a different kind, was the first 500 MW thermal power unit in India for Tata Power Co. This was a quantum jump from the 200 MW size that was the highest till then. TCE were appointed consultants and Ebasco Services of New York the review consultants for this World Bank funded project. As a policy, the Tata Management decided to use the services of the review consultants to the minimum to build TCE capability and confidence to undertake such megaprojects.

TCE had to work in close coordination with the Government of India, the Central Electricity Authority and BHEL, as this unit would be the forerunner of a number of 500 MW units in the Indian power sector. BHEL entered into a collaboration agreement with Siemens of Germany for the manufacture of the 500 MW turbo generators; the steam generator was also ordered with BHEL who had an ongoing collaboration with Combustion Engineering of USA (Fig. 3).

It also provided an opportunity for in-depth study of the latest technology available internationally, and we selected modern static electronic control systems, automatic turbine start up, computerized data acquisition systems, SF6 circuit breakers, etc. The challenge of undertaking a "first of its kind" project in the country motivated the whole team to deliver top quality work that was praised by the review consultants in USA. The unit was commissioned in 1984.

As I moved above the Chief Engineer level, there was less involvement in individual projects and more on overall management of the organization and overseeing all 4 offices of TCE and other specialties such as industrial and chemical, water supply and sewerage, environmental, and construction. I also opened new TCE offices at Delhi, Hyderabad, and Pune.

Another aspect was manpower planning, recruitment and training and making sure the new staff got assimilated quickly as part of project teams and imbibed the TCE culture. To manage over a 1000 engineers who are all first-class graduates from good universities and have their own specialization requires an open and participative form of management. Further, it requires strong coordination to ensure that quality and technical standards are maintained throughout the organization. This was achieved by documenting TCE experience in the form of design guides to be used as the basis for all work. The design guides had to be updated periodically to incorporate latest design trends. We adopted the ISO 9001 Quality Management System that gives a detailed written down procedure to ensure quality. It mandates reference to latest guides and standards and documented checking and verification of all work output, compulsory annual review at CEO level and verification by the Certifying Agency. It required a lot of effort to put in place such a system and get everyone familiar with it, but it has been a great help in ensuring uniformity and quality of output.

Consulting work is very interesting as each project is different, and each subsequent project involves improved technology. Even when projects are similar, the site conditions have to be taken into account and also the requirements of each client who has his own preferences and priorities.

The rapid growth of TCE was also matched by the development of indigenous manufacturing capacity over the same period. From an initial group of 30 engineers, TCE grew to over a 1000 by the time I retired in 1998. I am happy to say that the growth story continues, and it today has over 4000 engineers and has a large part of its work from overseas operations and continues to maintain its status as the premier engineering consultancy organization in India.

After retirement, I joined Volkart Foundation as a Trustee. This exposed me to the world of non-governmental organizations (NGOs) that operate in India. There is a very large number of NGOs working for social causes, particularly in the field of education, health care, help to the disabled, women, and child welfare, village upliftment and the like. While some are well funded by major donors, most are small NGOs run by dedicated individuals and their friends and colleagues who work tirelessly for the betterment of those in need often spending their personal savings for the cause. Such selfless work is not known to the public as the media does not report their activities as it only looks for sensational stories.

I have also kept up my interest in technology. Recently, I made a study of the power distribution system in India that is bedeviled by huge losses and load shedding in both urban and rural areas. It is an intractable problem because of political interference and no workable solution has been found by the government or the distribution companies over decades. I have proposed a practical method of overcoming this taking all constraints into account that is simple and straightforward to implement and hope that it will be adopted by the distribution companies. The work has been published as a paper entitled "Virtual Power Generation in the Rural Sector—Agricultural Demand Side Management" in the June 2015 issue of the International Journal of Combined Research and Development.

From Academia, with Love

Ashutosh Sharma

> *Behold the turtle, he makes progress only when he sticks his neck out.*
> —Bruce Levin

An important theme that does copious rounds in academia, industry, numerous committees, and meetings is why academia–industry interactions, including that vital issue of technology transfer, are either at lower than the desired intensity level, or happening very slowly or are even nonexistent! The problem is clear and immediate, its desirability is contested by none, and yet its solutions are often mired in the long sermons to both the industry and the academia buried in the reports never twice visited! For what these may be worth, here are my 2-bit, 1-D thoughts on the issue, which is admittedly a complex matter with its roots, shoots, and branches that cover a wide scale from education to our mindsets to the way we do business. Of this complex and vexing issue, I have chosen to tightly focus only on one aspect of academia–industry bridge-building here; that of technology incubators and technology startups in the academic spaces.

How do successful universities around the globe do it? For one, it would be very unrealistic to expect a professor to raise resources for research, generate new ideas, micro-manage research, guide students, write papers, keep accounts (!), teach effectively, go to the conferences and meetings, worry about her publications and H-index, review papers, do administration and then also do significant technology development and transfer and possibly see its commercial upscaling at the end of the day. Many of these imply chasing many companies and people, who have a limited interest often only in a fully developed technology or product off the shelf, and to further convince the industry about the economy of scale and the market for it! A professor with a ready-made technology available on the go is rare and this path to the tech transfer simply does not and will not happen in sufficiently large measures. Further, it is equally unrealistic for an industrial scientist or engineer to

A. Sharma (✉)
Department of Science and Technology, Technology Bhavan,
New Mehrauli Road, New Delhi 110 016, India
e-mail: ashutosh@iitk.ac.in

P. Ghosh and B. Raj (eds.), *The Mind of an Engineer*,
DOI 10.1007/978-981-10-0119-2_37

be enthused simply by the sparkle of arcane ideas, without a clear appreciation of how it might fit in the commercial scheme of things. This coupled often with a lack of awareness about the latest technology directions, and the low priority often accorded to real R&D by many of our industries, may hamper a meaningful academia–industry exchange. A further impediment is the eternal conflict between the activity of new knowledge generation and the need for a time-trusted robust technology with a clear commercial potential, but requiring no new scientifically challenging inputs.

But then again, perhaps the oft emphasized tension between the basic and the applied research and technology translation is more imagined than real, fueled at least partially by our cultural and educational conditioning of hermetic knowledge silos. Much of what goes under the garb of basic research today relies rather heavily on incremental analysis rather than synthesis. Being toolbox-centric, it often ignores the problem solving, which necessarily requires interdisciplinary and multidisciplinary approaches that connect distant dots rarely connected before. Of course, one needs new and profound knowledge generation for the longer term good of the society. However, whether incremental 'basic research' is intellectually more fulfilling or desirable as compared to problem-centric 'applied research' remains a moot point which will not conclude either way in the foreseeable future. What is clear is that in basic research, emphasis has to be on the profound, on the novel, and on quality, with ample risk taking without an undue emphasis on the numbers. In the applied research leading to innovation, the risk has to be minimized by carefully budgeting for the unknowns based on a creative synthesis of the known. In any case, it is clear that it is not an individual, but the academia, society and the nation as a whole that have to strike a healthy balance between knowledge generation and its use, between invention and innovation, between converting financial resources to knowledge and back conversion of knowledge to resources for the societal good. While debates about the basic versus applied, private versus public good, invention versus innovation and so on rage on forever, one simply cannot abandon the middle path. We also cannot wait for the large-scale changes in our educational, cultural, structural, and policy matrices before we start moving forward on an effective academia–industry connect.

An effective industry–academia dialog leading to concrete gains can happen most naturally and efficiently with the rise of a streamlined system of student-driven startups, which take industry relevant ideas from academic laboratories, get funding (including from industries), translate them to the well-identified applications, and develop near-to-the-market prototypes. Technology transfer is too important an activity to be left to the professors alone, but requires dedicated innovation ecosystems interfacing the universities and industry with the sole mission of translating the common interests and needs.

What our universities lack foremost and in sufficient quantities is a robust and deep-rooted ecosystem of technology incubators and startups that form the most natural interface between the industry and academia. The strength of a composite material is often only as good as the adhesive strength of its buried interfaces, bolstering of which requires an additional layer serving as the twilight zones of compatibility between the two otherwise incompatible materials. Startups can and do serve the role of this 'compatibility layer' by having a shared nomenclature and concerns between the academia and industry, by harmonizing the longer term vision of the academic research with the immediacy of the industrial goals and by tempering sometimes unrealistic expectations on both sides.

The reasons for the lack of startup culture are many: starting from the school (read coaching) and university education that emphasize a fragmented knowledge base catering to routine examinations rather than innovation, risk-taking, entrepreneurship, effective communication and even common sense! Then, there is always the hankering after the so-called job security and the lack of professional confidence, not helped in the least by the prevalent model of education, that prevents most students with a middle-class background to explore other options. From my own experience, I have found students most eager to plunge into starting a startup when a reasonable baseline earnings or salary could be guaranteed for the first couple of years to stabilize the venture. While there are now several government-supported schemes that go halfway to that end, participation of target industries at an early stage would be a very welcome move. The benefit to the industry will be leveraging of enormous intellectual and infrastructural resources and manpower at the disposal of our universities at a lower cost than in the in-house industrial R&D (when it exists) plus access to state of the art in a technology area. Early stage Venture Capital in India for technology-driven startups is rather scarce. The government and industry would thus have to partner much more. A partial model is Stanford's investing in student startups like a VC (http://techcrunch.com/2013/09/04/stanford-university-is-going-to-invest-in-student-startups-like-a-vc-firm/). Since nothing succeeds like success, we need a critical mass of successful startups as the role models for the next generation. Perhaps, some of the support by the research funding agencies such as the DST, DBT, CSIR ought to be earmarked for the projects that also incubate a student startup or have clear plans for one.

The time for the wakeup call is now already past. It can continue to be business as usual, without the new shimmers of optimism, excitement, and opportunities for our young engineers, and at a great cost to the technology readiness of our nation. On the other hand, many a splendid opportunities wait in the middle-of-the-way synthesis of the extreme aspirations symbolized by the academic and industrial R&Ds.

> There are only two mistakes one makes along the road to truth; not going all the way, and not starting at all.—Buddh

Postscript: Since my move in January 2015 to the Department of Science and Technology, Government of India, I have had reasons and opportunities to meditate again on some of the issues sketched in this article and on possible actionable items. Here are some possible measures, plans, and programs that address the needs and opportunities in this direction. Items 1–3 below have been initiated (August 2015) and the rest are being worked on. For the sake of time and brevity, the descriptions below are in the Upanishadic rather than the Mahabharatic narrative style!

1. Enhancing the Quality of Basic Research or Moving Away from the Incremental to High-Risk, High-Gain scenario: A new program that specifically calls for research proposals on significant problems requiring new knowledge generation or new interpretations/synthesis/insights, wherein the high-risk elements are specifically recognized and possible approaches to address them are reflected on. It may fail, which is acceptable (!), but if it does succeed, it should produce non-incremental, conceptual advances.
2. Funding of industry-relevant science and technology R&D in academic entities: The major challenge is how to recognize if something is of real interest to industry? The surest way to determine the industrial relevance is when an industry provides financial support to a project in an academic space. A new governmental program that matches the industry support to synergize the flexibility of industrial grant with the long-term infrastructure needs can leverage the best of both. Of course, the non-triviality of the scientific/technology aspects of the activity needs to be vetted (it should not be a metaphorical equivalent of 'kick-starting a pump!') and the industry money has to be there *before* the governmental support is released!
3. Participation of industry in the significant national missions (climate, solar, water…) including the newly launched National Supercomputing Mission where industry, industrial, and corporate R&D including education industry can play pivotal roles in sharing of resources, manpower development, joint academia–industry projects, and commercialization.
4. Creation of around 200 new Technology Business Incubators housing around 8000 startups, largely in academic spaces together with the strengthening /creation of the whole innovation value chain from sourcing of ideas, motivating and training students, hand-holding, business plans, early capital, and even exit strategies.
5. Facilitating an early identification and connection of a startup with a relevant industrial partner to allow greater focus on technology needs and opportunities and building of confidence. Groupings of industries such as CII can play an important role in the process.
6. Fellowships for students that incubate a company renewable for a period of three years based on the yearly progress review. This will remove one major hurdle for a potential incubatee, viz., the minimal financial security required in resolving the dilemma of job versus startup?
7. While there are several fellowships (e.g., J.C. Bose) to recognize excellence in basic research, there are none in the academic space for recognizing and encouraging demonstrated and significant technology development, transfer, commercialization, and startup activities. A limited number of (Kalam/Bhabha?) fellowships to the faculty/academic scientists can address this important issue.
8. Some enabling policy issues: (A) Should technology startups qualify for R&D tax benefits applicable to industry? (B) Should technology startups qualify for

the Corporate Social Responsibility contributions? (C) Could there be dedicated SEZs for R&D companies allowing leveraging of resources? (D) Separate exit strategies for startups?

In academia, I discovered that issues and insights, commonplace among the scholars, are viewed as highly controversial and even as 'heresy' in the churches.—John Shelby Spong

The House of Education

M.S. Ananth

"Many of us have visited ancient monuments, which inspire solemnity and awe. The labour of generations of architects and artisans has been forgotten, and the scaffolding erected during the construction has long since been removed. You see only the perfection of the completed whole. However, if you enter such an edifice that is still partly under construction, you hear the sound of hammers, the reek of tobacco and the mundane conversations of the workmen. You realise that these great structures are but the result of giving to ordinary human effort a direction and a purpose" [1].

In some respects, the "House of Education" (HoE), to borrow a metaphor from the renowned physicist Robert Oppenheimer's description of the House of Science [2], is similar to an ancient monument that inspires solemnity and awe. It is a vast house. Some wings are complete. The scaffolding has been removed, and you see the perfection of the completed whole. Some parts are still receiving their finishing touches. From time to time, someone rearranges a piece of furniture in these parts to make the whole more harmonious. And then, there is the scaffolding being built for adding new rooms, in fact new wings to the House.

The HoE is also very different from any monument. It is in many ways confusing to the novice and yet indescribably beautiful. It is never complete and is not built to a preconceived plan, but with a wonderful randomness suggestive of unending growth and improvisation; there are no locks; there are no shut doors; everywhere there are the signs of welcome. It is an open house, open to all comers—novices, visitors, mere passers-by or workers [2].

You enter the HoE with some teachers to guide you through the fully completed rooms explaining to you how they were built. Some rooms are simple in design others are very complex. When you have seen and understood the design of a fair number of these rooms, you "graduate" from school. Most will then move on to undergraduate education and begin to see rooms that are less finished. You will still have a teacher guiding you through these rooms. When you graduate from college,

M.S. Ananth (✉)
Department of Chemical Engineering, Indian Institute of Technology Bombay, Powai, Mumbai 400 076, India
e-mail: ananth@iitm.ac.in

P. Ghosh and B. Raj (eds.), *The Mind of an Engineer*,
DOI 10.1007/978-981-10-0119-2_38

some of you, like your teachers, will take on the job of guiding others through the finished and less finished portions. Others will continue into research, help build new rooms or even help set up the scaffolding for a brand new wing.

The Idea of a University

The research university is a collection of people engaged in "the pursuit of learning" in a special manner as a cooperative enterprise. It is a body of scholars, each devoted to a particular branch of learning. It is not a machine for achieving a particular purpose or producing a particular result; it is a manner of human activity. Thus, scientific activity is not the pursuit of a premeditated end; nobody knows or can imagine where it will reach. What holds the participants together is not a known purpose to be achieved, but the knowledge scientists have of how to conduct a scientific investigation [4].

The university today is a major part of the HoE. It is built on the ideas of the renaissance thinkers who made three basic assumptions: the lawfulness of the material world, the intrinsic unity of knowledge and the potential for indefinite human progress. They believed that the underlying unity in the diversity around us can be unravelled only by scientific inquiry across both the natural sciences and the humanities. We still take these most readily into our hearts … (not the) fragmentation of knowledge [5]. It was von Humboldt in the nineteenth century who advocated the present model of the research university. While research brings passion to teaching, the latter rejuvenates the researcher. The two must go hand in hand in the university.

Teaching and Learning

Shri Aurobindo [6] sets down a few interesting principles that should form the basis of all education. I would like to discuss these principles, restated in current idiom, in the light of more recent literature on education and the famous split-brain experiments of the 1970s that have provided valuable insight into the learning process.

"The first principle is that nothing can be taught; the teacher is a helper and a guide; his business is to suggest not impose; he does not impart knowledge but shows the student how to acquire it for himself …". Maria Montessori [7] resonates with this principle: "… education is not something which the teacher does, but that it is a natural process which develops spontaneously in the human being. It is not acquired by listening to words, but in virtue of experiences in which the child acts on his environment." Montessori's method depends basically on creating environments that are conducive to experiential learning. Technology is a great enabler in this regard, and we can now create virtual environments in which learning is effectively experiential. Faculty will soon be expected to also design and create such environments while continuing to teach in the chalk-and-talk environment of the conventional classrooms.

The second principle is that the mind has to be consulted in its own growth … not forced to abandon its own dharma. Educational psychologists talk about the personal response and commitment of the student, without which knowledge remains an un-integrated, inert piece in the mind of the learner. Personal response involves developing opinions and convictions rooted in one's system of beliefs or dharma. Aurobindo's second principle urges us therefore to consult the student's mind in its reorganisation of personal beliefs.

The third principle is "to work from the near to the far, from that which is to that which shall be … to begin with the basis of man's nature (in the environment) to which he is accustomed". It behoves us to use examples in teaching from the environment that the student is conversant with rather than from a borrowed unfamiliar one. Our poor documentation often makes us resort to examples from the unfamiliar west.

A fourth principle is that "… suppleness and comprehensiveness of the mind is increased not by the number and variety of subjects for study but by diverse approaches to the same subject". This applies equally to the benefit of solving the same problem in diverse ways rather than solving diverse problems. Finally, it is commonly accepted that repetition is a necessary and integral part of education because, at any given time, "only a part of the congregation is listening". It is also recognised that practice leads to perfection, an advice more useful for students preparing for an examination that for those who wish to enrich their minds.

The Paradigm Shifts in Education

A well-rounded education should impart to the student knowledge, know-how and character [3]. The former enables one to understand what one learns in relation to what one already knows. It can be organised into intellectually tight compartments that are conveniently taught as courses in a conventional curriculum. Know-how on the other hand is the ability to put knowledge to work. It requires the purposeful organisation of knowledge from different areas of learning and is taught through elective courses on technology, design, project work, industrial training and other opportunities for individual initiative and creativity. Character traits are easy to recognise, but character-building processes are difficult to define and implement. Educational institutions try to provide the atmosphere, in which the students, during their education, will develop character traits such as honesty, truthfulness, integrity, initiative, competitiveness, self-esteem, leadership and the ability to work both alone and as part of a team.

The paradigm of education has always been that "knowledge is power". The change in the paradigm with time has been about who should have access to this power. India evolved a unique system of predominantly oral education called gurukulam, meaning literally, "education in the home of the teacher". By the fourth century BC, education had shifted from the home of the teacher to the monasteries (predominantly Buddhist). The paradigm of education until this time was that knowledge should be in the hands of those with character. It is printing that made mass education possible and the industrial revolution that made it a prerequisite for gainful employment. Knowledge now had to be in the hands of those serving the rapidly growing industrialised economy. The ICT revolution and the World Wide Web have completely democratized learning. In India, we are moving all the way from gurukula to sishyakula or "education in the house of the student (with gurus distributed over the WWW)".

Traditionally, the major activities in the HoE have been teaching and learning and character building. In the global knowledge economy of today, universities are expected, in addition, to help increase the gross enrolment ratio (GER) and to encourage innovation and entrepreneurship. We discuss here briefly the universities' response to this expectation in the Indian context.

Increasing the GER

To compete successfully in the globalised knowledge economy of today, it is generally accepted that the GER in higher education should be at least 30 %. Some figures in this context are as follows: India 15, China 30, US 80 % and so on. Given the size of our population, going from a GER of 15 to 30 % by traditional means will require, inter alia, the creation of one brick-and-mortar institution every week. Furthermore, there is a severe shortage of faculty that cannot be overcome in a short time. The creation of a virtual university is India's only option. A prerequisite for the virtual university is the creation of open courseware and massive online education.

The National Programme on Technology Enhanced Learning (NPTEL) was initiated by IIT Madras, supported by the other IITs and funded by the MHRD in 2003. Today, NPTEL is the world's largest collection of open courseware with the over 700 modular courses each in web/video formats. For the interested reader, the website: http://www.nptel.iitm.ac.in has more details.

Learning and Creativity

Shri Aurobindo also argues that "… for thought to be creative it should be the natural and logical synthesis of all points of view … the totality of one's thoughts becoming a dynamic and constructive force depends on the level and universality of the central idea of mental synthesis." This idea of synthesis is related to the modern idea of the complementary role of intuition and logic in creativity.

Roger Sperry, a 1981 Nobel Laureate in Physiology and Medicine and his co-workers, Gazzaniga and Levy have given us some insights into the process of learning. The two sides of the brain think in fundamentally different ways: the left thinks in words and uses step-by-step logical sequences. The right brain on the other hand uses visual images and intuition to draw conclusions. The creative process itself has four discernible stages [8]: the preparation stage of information gathering by the left brain; the incubation stage, during which the right brain tries to see the "whole picture"; the illumination stage, in which the right brain's insight and intuition generate possible solutions; and the verification stage, in which the left brain logically tests the solutions and then organises and elaborates the correct solution into a finished form.

This synergistic relationship between the left and right halves of the brain is the real basis of creativity and learning. The freedom from logic and structure is what makes the visual thought process of the right brain so effective in generating ideas. Since most of the ideas so generated fail, when tested logically, the left brain is equally important in the creative process. A striking example of the synergy between the right and left brains is visible in the joint work of Ramanujan and Hardy. The freedom from logic and structure is what made the visual thought process of the Ramanujan so effective in generating the "central ideas of mental synthesis". Since the ideas so generated had to be proved logically, Hardy's struggle "to wrest rigorous proofs" was equally important in the creative process [9]. Our tradition suggests that humility and faith in God are prerequisites for an insightful visual thought process—an idea that we shy away from because we cannot explain it to our "scientifically minded" colleagues.

Education in the university emphasises logic. Albert Einstein remarked: "The intuitive mind is a sacred gift and the rational mind is a faithful servant. We have created a society that honours the servant and has forgotten the gift". Students have intuition and we should nurture it. They should realise that while the right brain is the true source of creative ideas, it is the confirmation of one's intuition by one's own logic that leads to self-confidence. Educational institutions should bring together "unlike" minds, unlike in disciplinary training, in cultural origins and in attitude and right-dominated as well as left-dominated brains.

Managing Creativity

In today's global economy, innovation and entrepreneurship are central to a nation's successful participation. Research universities are the main source of creativity, and in the globalised world of today, they need to manage their creativity well.

Discovery and invention have been the driving forces for university research. Innovation on the other hand includes a crucial economic component. It is often about extracting value from a creative understanding of what is already known.

It has everything to do with commercial success and drives the economy [10]. It is successfully managed by the "idea factory" approach—bringing unlike minds together, creating the right atmosphere and giving them freedom but carefully structuring interactions [11]. Minds can be unlike in terms of cultural background, disciplinary background or attitude. While science is universal, the scientist has a cultural background and hence many prejudices. Bringing together minds from different cultural backgrounds can help overcome at least some of them as exemplified by the success of the US graduate schools in generating innovative ideas. The remarkable success of the Bell Labs in the post-war decades of the twentieth century is an example of innovations triggered by bringing together multidisciplinary groups of researchers to work on problems of societal importance. The university research parks bring together minds with different attitudes: faculty with knowledge of fundamentals; students with their spirit to conquer the world; and S&T personnel from the industry with their ability to convert ideas into marketable products.

The IIT Madras Research Park

The University Research Park (URP) is basically a property-based venture located near a university campus with a difference. It creates a local concentration of skills and technology and promotes innovation, competitiveness and entrepreneurship. It helps convert research ideas into innovative technologies, houses R&D of companies, creates and nurtures start-ups and drives technology-led regional economic development. Wessner [12] has a wealth of information about university research parks.

Louis Pasteur is reported to have said, "… discovery is the result of chance meeting a prepared mind …". A significant fraction of IPRs in the 1990s in Silicon Valley have names of IIT alumni associated with them. It appears that IITs have been preparing minds and chance has been meeting them in Silicon Valley. This is one of the reasons why IITM sought the permission and partial financial support from the Ministry of Human Resources Development (MHRD) in 2002 to set up the IIT Madras Research Park (IITMRP) as a Section 25 (non-profit) company. Simultaneously, IITM also approached the state government for the grant of 11.5 acres land and associated infrastructure in the vicinity of its campus. The funds were raised from five sources: a grant from MHRD, a loan from a bank, 20-year rentals, internal accruals and alumni donations. The aim was to create 1.2 lakh m^2 to house R&D of companies (85 %) and incubates (15 %) and create opportunities for access to venture capital, legal advice etc. The IITMRP (first phase of 40,000 m^2) was launched in 2010. In its very first year, over 50 patents were filed. The website—http://respark.iitm.ac.in—has more information on the IITMRP.

The Role of Industry in Education

The university seeks unity in the wild diversity around us, while the industry thrives on differences. After all the industry cannot sell its product competitively unless it is superior in some respect to similar products that other industries produce. The industry has an important but a limited role in education. This lies in sharing some of their empirical knowledge of specific systems, giving "reality-check" lectures on design in the appropriate courses and in participating in the definition and execution of projects. By addressing the student chapters of professional associations, they can create an awareness of innovative and useful multidisciplinary applications of what is learnt in the university. In research, the interaction is of mutual benefit and the most effective interactions are through the University-based Research Parks described earlier. In all its interactions with the industry, the university will do well to remember that the values of the university are different from those in the marketplace. Monetary incentives, as the late Charles Vest remarked, "are the last gasp of an (educational) institution in trouble" [13].

The Value of Values

Of the three components of education—knowledge, know-how and character—character has now become the most important attribute for the survival of civilisation as a whole. One way to build character is through education in values. This has taken place in India since ancient times by means of informal structures of learning that nurture the youth into specific cultural attitudes about right and wrong [14]. These attitudes were inspired by the great scriptures and epics, which have shaped the fabric of the whole culture. In practice, this education was mediated through the extended family, the public discourses about the perennial struggles of good and evil and the practical decisions of the village panchayats and in recent decades by the law. The first three appear to be breaking down, while the fourth is too technical to be a substitute for true education in values.

It is perhaps more logical to teach college students the "value of values" than to teach them values per se [15]. Non-violence is the only universal value. Students should learn that the value of non-violence is far greater than the value of money or of power or of the fulfilment of one's desires. They are welcome to acquire these non-violently. Lord Krishna tells us in the Bhagavad Gita, "If you walk the path of dharma, I will walk with you, otherwise you walk alone". Walking alone is not a sustainable proposition for human beings.

The Need for National Pride

I would like to conclude with a plea for revival of national pride in our youth. India went through a long period of colonial governance that left us with little pride, less self-esteem and an identity crisis. Traditionally, Indian culture has been associated with non-violence, an appreciation of the arts, pursuit of knowledge and wealth in that order. Non-violence is more than the absence of physical violence. It is the absence of malice and hatred. Our nationalism has been described as the only such movement that was free of hatred of fellow human beings. Since independence, we have consciously faced five important debates and we in India should be proud that, in each case, we chose wisely and resisted the temptation to take the easy way out: democracy versus totalitarianism, secularism versus fundamentalism, defence versus development, globalisation versus self-reliance and centralisation versus federalisation. Even when it comes to detailed achievements, we often fail to recognise our great achievements in the face of non-ideal competition: the white, green and brown revolutions leading to self-sufficiency in food, milk and oilseeds, remarkable achievements in higher education, defence, atomic energy and space, in manufacturing and in software—reason enough for us to be proud of modern India.

References

1. Lewis N, Randall M. Foreword to "Thermodynamics". New York: McGraw Hill; 1961.
2. Oppenheimer R. Science and the common understanding. The Reith Lectures; 1953.
3. Tribus M. Essays on education, "Quality in Engineering and Business Education".
4. Oakeshott M. The voice of liberal learning; 1989.
5. Wilson EO. Consilience among the great branches of learning. Daedalus. 1998;127:131.
6. Aurobindo and the Mother. On education. Pondicherry: Sri Aurobindo Ashram; 1997.
7. Montessori M. The absorbent mind. Chennai: The Theosophical Publishing House; 1949.
8. Blakeslee TR. The right brain. Berkeley Books; 1986.
9. Kanigel R. The man who knew infinity. Rupa and Co.; 1961.
10. Schrage M. Quoted by Colby S. in http://colbys.blogspot.com/2004//invention-or-innovation.html.
11. Stewart TA. The wealth of knowledge. New York: Currency Doubleway; 2001.
12. Wessner C. Report of symposium on "Understanding Research, Science and Technology Parks: Global Best Practices". Washington DC: National Academies Press; 2001.
13. Vest CM. Pursuing the endless frontier. Cambridge: The MIT Press; 2005.
14. Stackhouse ML. Teaching values in South India: an Experiment in Education; 1988.
15. Swami Dayananda Saraswati, talk in IIT Madras, Sept 2007.

Need of Promotion of Innovation in Indian Engineering Institutes

V.G. Gaikar

There has been a consistent debate and usually anguish on the rankings of the indian engineering institutes every year whenever the list is published by one organization and other, irrespective of the criteria used in such surveys. In local discussions too, in both academic and professional circles, the quality of the graduates coming out of hundreds of government and private universities in different disciplines, not just engineering, has been another point of concern. The engineers are produced by hordes in India, thanks to the liberal policies of the monitoring bodies like AICTE, in giving permissions to start new engineering colleges or new degree courses. Engineering colleges do not have to fulfil the criteria of having an industry associated with it for starting a degree course, unlike medical colleges which should have a tieup with a hospital for practical training of the doctors.

The clamour for an engineering degree is understandable in India where the degree is a major step in climbing a ladder of social status. A huge number of aspiring youngsters appear every year in entrance examinations for the most coveted seats in IITs, NITs and the state universities. Because of a large number of engineering institutes offering seats in plethora of disciplines, admission to engineering courses is almost guaranteed, whether the candidate has an aptitude or not. If every one of 1.3 million engineers that the country produces each year is a well-qualified and well-trained engineer, then these human resources should have put India long ago on the path of becoming the largest economy with the highest per capita income. But alas, many of these graduates fail even to find satisfactory job. It must be also understood that only a fraction of this human resource goes for creating knowledge as Ph.Ds. The manufacturing industry still despairs over the poor employability of fresh engineers because it finds these graduates are not trained for the jobs. The institutes may not be able to cater to the needs of a specific company and give only general broad based training in a particular discipline.

V.G. Gaikar (✉)
Department of Chemical Engineering, Institute of Chemical Technology,
Nathalal Parekh Marg, Matunga, Mumbai 400 019, India
e-mail: vg.gaikar@ictmumbai.edu.in

P. Ghosh and B. Raj (eds.), *The Mind of an Engineer*,
DOI 10.1007/978-981-10-0119-2_39

It becomes the responsibility of the industry then to give specific job oriented training to the new recruited engineers.

The question of poor employability of the engineering graduates may arise from a number of factors, starting from poor preparedness for the engineering profession to quality of training imparted in engineering schools where teachers themselves have no exposure to industrial practices and, finally, the evaluation system which reinforces learning by rote. It is common in most private engineering colleges to contractually appoint the fresh graduates from the same school, if they are unemployed.

The poor understanding of the students at the entry level depends a lot on the way science is taught in pre-university years, particularly in XI and XII classes. The focus in these two years, which may even predate into the last two years of the secondary school education, is on passing examination with as high a score as possible, but not necessarily with better understanding. There is little scope in the course structure for brining out the innate ability of an individual to learn/analyse science or apply equations of abstract mathematics in daily life. The race for the top few percentiles, to get into the 'best' engineering institutes, suppresses the desire to learn anything new on his/her own. In these four formative years of education which are also the most impressible years of teenagers, the creativity is suppressed. It is learnt, sadly, that it is not necessary to be curious.

The Institute of Chemical Technology (ICT), Mumbai, conducted comprehensive diagnostic tests on all its new entrants, at both UG and PG levels, under the Technical Education Quality Improvement Programme (TEQIP) which is supported by Ministry of Human Resources and Development and the Government of Maharashtra. The tests were conducted to evaluate numerical ability, language skills, creativity, visualization, self-awareness and some of standard personality tests. The analysis of about 600 students threw up many surprises. Although a sizable number of this population could handle manipulation of numbers, only 10 % could score moderately well in creativity and spacial awareness. Only one or two out of the 100 could give accurate description of 3D world around them that they daily travel through. The worst score was in the language skills, with majority showing very poor reading and comprehension skills. This group of study comprised of a very small number, but knowing that these entrants came with very high scores in the entrance examination, one can extrapolate the findings to other institutes which are poorly endowed with faculty or infrastructure. As academicians as well as practicing engineers, we need to wonder what causes such a decline in the quality of the students at the entry level. It cannot be simply the lack of ability of individuals. Inadequate education imparted to them in schools and junior colleges could be just one factor amongst many others. And most importantly, whether we, as engineering teachers, do any thing different to provide real value adding education *before* these students graduate as engineers or we continue to reinforce their rote learning?

The current set of examinations for the entry, itself, is not really challenging enough to identify the true potential of these engineering aspirants or to make them to think out of box of set rules. And once they are in, they still have to be remoulded to take up the challenges in their profession. Unfortunately, the university examination system is also slowly following the same pattern as the higher secondary examinations. The emphasis again is on passing the examination and not on learning professional and other skills for life. Most engineering students remember the equations as recipe as there is little scope given to them to relate these equations with real-life experiences.

ICT had done another analysis of the performance of graduates in the final year with their scores in physics, chemistry and mathematics (PCM) at their XII examination, because these subjects were considered for admission to engineering courses. It was difficult to spot any relation. On the other hand, the performance of these graduates showed a strong linkage with their scores in languages and non-technical subjects in XIIth examination! It is important to know that there are no coaching classes in languages! If the students do well in languages, it means on their own they find their ways of 'learning'. Is coaching in specific subjects then killing the curiosity and 'learning' process of these youngsters? Interestingly, there is profound conformist attitude, if the parents are highly educated as against the candidates having background of no formal professional education in family! Do we as more educated parents unknowingly also expect the kids to follow the well trodden path to success as we understood it?

The Internet and telecommunication might have made 24 × 7 connectivity possible, but reading habits of the students are at negligible levels. The writing skills are reduced to one-liners and comprehension does not go beyond one paragraph. In academic fraternity, the despair about the quality of students is as intense as that in industry about quality of the fresh graduates. If at the basic level, the situation is so alarming, what shall be the conditions of the postgraduate education and research in engineering institutes? Merely increasing the number of graduates and post graduates and even Ph.Ds. is not enough. It may even be counterproductive if the quality is not matched to the requirements.

Research is now a regular activity at major universities and academic institutes in the country. MHRD is promoting research in a very big way. It is understood that all IITs put together need to churn out 10,000 Ph.Ds. every year! It is said that there is direct link of number of Ph.Ds. in engineering with per capita income of the country. All TEQIP-supported institutes, in particular, and NITs also have increased their research activities significantly in the second phase of the TEQIP. Most of this research, however, ends into research publications, conferences, seminars and theses. And many more are presented in so-called international conferences at obscure places.

Converting this research output into transferrable technology is hindered by complete lack of cohesive infrastructure today in the country. Research and process development is not equal to technology, unless all loose ends of the process and product are tied up. It is unlikely that such infrastructure shall be built adequately in near future in the country, barring a few examples of technology parks, such as one in IIT, Madras, and a few centres of excellences in specific areas. In this regard, IITs are often and better funded because the funding is provided directly by the central government. The state-level universities, however, suffer hugely as there is either no confidence in their abilities or there is little knowledge of their performance at the MHRD level. Some of the engineering institutes, like ICT (former UDCT,

University of Mumbai), have made their name in equal measures as IITs in specific areas. It is high time that MHRD takes a more objective view for supporting better performing state-level institutes by providing generous special funding, of course, with necessary accountability on deliverables.

I must add my own experience with one of the MNCs who have been working on 'open innovation' concept. As a first step, the company scrutinized the research spectrum of the country for its own needs and identified individuals and institutes with necessary expertise. After due diligence, these researchers are invited to give a talk and collaboration is suggested to them. The projects are then supported with clear understanding on sharing of intellectual rights and financial terms. Such an approach, as a policy, is missing in the country and we keep supporting routine research work where success is sure for expected results. But rarely, we fund high-risk innovative projects even when they are on sound scientific footing. We indeed need a complete change in the way the R&D projects are supported. If necessary, we should have search committee rather project approval committee for funding high risk, but excellent research with individuals with proven track record to work on projects of immediate and long-term importance to the country, with no administrative hassles attached. If necessary, such individuals can be invited to form a core group to deliver the innovative products and processes. The different theme-based task forces on government bodies should also deliver by working together by identifying individuals and institutes for innovation where the research output should be converted into technologies, useful for society and industry.

The Fraunhofer Institutes in Germany are classic example of how basic research can be converted into value. There are several theme-based Fraunhofer Institutes throughout Germany which coordinate with Max Plank Institutes involved in fundamental science. The work done in Max Plank Institutes finds its way into Fraunhofer Institutes and then into industry. It is not surprising to see a physics professor building his/her own NMR unit in his laboratory which is picked up by an industry and packaged and supplied to laboratories in India over a period. Much of the analytical equipments in India are imported from such countries where the emphasis is on building them. It could be worthwhile to analyse the percentage of funding provided by DST, DBT, CSIR and other agencies that has gone into import of ready-made equipments for research in India. It may be necessary to import these units to avoid time delays, but it also reduces our own capacity. And interestingly, it is difficult to suggest in the project proposal, that the PI wants to build own unit for the project. If the unit is built in-house, the cost could be brought down by a factor of 10 and imagine the confidence and capacity building of such individuals.

MHRD can create centres of excellences, in specific areas of engineering with the mandate of developing technologies based on research conducted in different organizations. The major mandate of the centres should be to develop technologies for the industries to improve local economy. Shutting down industries when they have no access to better technologies is not the solution because it also leads to loss of livelihood for local people. The academic world also has to take a notice of the problems faced by such industries, in terms of the technology employed and provide affordable solutions to overcome the problems. At the same time, industries have to work with local institutes, supporting the efforts of the academia to arrive at a workable and economical solution for their own problem. Also it should be understood that there is nothing like free solution. The industry must be ready to pay for the efforts of the academic institutes. Partnership of industries in industrial and academic consortia is almost unheard of in India, because industries look for their own interests and intellectual property rights. They are wary of loss of their IPRs. But if there are common causes, such as water pollution, air pollution, low-cost energy generation, manpower training, then formation of an industrial consortium for specific product or process, can bring down the development cost for each company. Also involving the academic and research institutes in the group gives exposure to faculty and students alike to industrial perspectives.

Innovation shall be a key parameter today in order to survive in the global competition. The academia and research fraternity need to work together to develop competitive technologies and products that can support Indian industries as well as society in general. It is unlikely that one institute will have expertise in all areas and cooperation among different institutes and even industries, to share expertise and resources, may become imperative.

The MHRD had invited the author, a couple of years ago, to brainstorm on the activities of TEQIP to improve industry–institute of interaction, along with IIT, Kanpur; IIM, Calcutta; and Indian school of Business, Hyderabad, amongst other institutes. ICT is known to have nurtured organic links with chemical and allied industries for long and has been recognized by AICTE and CII in 2013 and 2014 as the best institute for its symbiotic interaction with the industry in chemical engineering and technology. A concept to develop a research and technology park in collaboration with partner institutes and industry in Western India was presented by the author in these meetings. The need of establishment of innovation and technology park was acutely felt by all participants to convert the research done in different laboratories.

The innovation networking of TEQIP institutes in the State of Maharashtra has been a result of the meetings at MHRD and the brainstorming sessions held at different institutes in Maharashtra. The project was whole heartedly supported by the State of Maharashtra and the National Project Implementation Unit (NPIU), TEQIP. It was envisaged that the innovation networking would use current expertise and infrastructure available at the partner institutes to develop/create prototypes for technology development and transfer. It was also hoped that the spirit of innovation would be spread to other states to enthuse young engineers and technologists for entrepreneurship. ICT has become now a hub for networking with other institutes under TEQIP to promote this spirit of innovation.

ICT took the initiative to form a virtual network of institutes in the state which promised to bring together different engineering disciplines to build products and prototypes. Although, initially planned for Western India, the scope was reduced to the State of Maharashtra because of funding pattern of TEQIP where the state was also expected to put in its own share in the project. ICT's close relationship to the chemical and allied industries has led, in the past, to relevant research programmes with a high level of innovations, large consultancy programmes, a dynamic curriculum development process and a high level of involvement from the industry. In chemical engineering, its contribution is well recognized by its peers and its highly motivated and qualified faculty and talented students have an outstanding history of academic achievements. ICT thus presented a role model to promote the networking activities. Since ICT has expertise only in chemical engineering, it invited institutes having expertise in other disciplines to partner with it in the proposal which was formally put up to the TEQIP/MHRD.

After a delay of few months, the final funding was approved by NPIU through TEQIP, for the innovation networking in Maharashtra in March 2014, but funds were received only in June 2014. Out of original 9 partners, finally 5 institutes, Institute of Chemical Technology (ICT, Mumbai), Veermata Jeejabai Technological Institute (VJTI, Mumbai), Dr. Babasaheb Ambedkar Technological University (DBATU, Lonere), Sardar Patel College of Engineering (SPCE, Mumbai) and Shri Guru Govind Singh Institute of Engineering and Technology (SGGSIET, Nanded), signed MoU on 1 April 2014 to launch formally the innovation networking with 14 projects. The project is coordinated by the author with his colleagues and Ph.D. students in ICT.

The selected projects were such that each partner institute had to bring in its own expertise for the project. It was not easy to convince the partners to participate in the programme. While discussing with different institutes, the participation of the individuals became more important than collective vision of the group from any institute. This has to be learnt as a lesson from this programme. Most faculty members that we talked to preferred to work in their own confined boundaries, fiercely protecting their interests. The hurdle also seems to be because of the language barriers, not of English, but the language of each discipline. Each engineering branch follows its own terminology and teaching each other was a task itself. However, the final team emerged from these parleys, and over a set of lectures on the need of innovation at various institutes. The final five institutes accepted the spirit of innovation in its true form. We realized the need of at least one champion from each place to pursue the idea of innovation in collaborative manner. The participation was thus kept entirely voluntary.

Our second hurdle was getting project staff to work on these short-term projects. We faced the same problem as the industries, i.e. lack of talented and motivated assistants to work on these ideas. At this time, we took a conscious decision to involve the UG students from all partner institutes. This was a risk as we were working with inexperienced bunch of youngsters, but here was the biggest surprise for all of us. The undergraduate students, who participated, were the most enthusiastic lot that just jumped on the opportunities offered to them, that is when we stared rethinking about the ability of the engineering graduates. Until now, they had a very few opportunities, if at all, to work on something on their own. These projects gave them a chance to apply to real systems, whatever they had learned so far. This experience also gave us a hope that the situation does not seem to be as bad as we thought earlier. We need to challenge these youngsters. They actually thrive on the challenges. If they get an opportunity to do things differently, we can change the mindset of the entire population of UGs towards their learning. The projects were quite a hit amongst these teams, who started learning from each other. We conducted design workshops at different institutes at frequent intervals where different team members could ask questions or had to answer questions of other members. They had to go back to drawing board each time they realized that their idea has some problem or other. Working on calculations and designs again and again to meet the requirement of the projects became common. Engineering solutions had to be found for every problem encountered.

There were deadlines set up by me as project coordinator on each project, to complete the basic designs, to prepare a bill of materials, complete with entire set of specifications, online submissions of reports and discussions for designs and input by each team member. Anyone from the team could bring in an idea and argue over it. Each one had the freedom to look for resources, build more networking and bring skills if the team was lacking in it. Many of these students learnt new techniques, and new softwares on the job, upgrading themselves continuously as they struggled to learn new areas and other disciplines. If they had not studied the subject earlier, they learnt it from their seniors and brought themselves at par with them. For us, the experience has been satisfying, knowing that we have started a new movement among these students. At DBATU, Lonere, they considered this training equivalent to industrial training. Right from the beginning, we did not put restrictions on the students who wished to be a part of the project. Everybody has been welcome and we had probably a number of UG students working on these projects, each member doing his/her bit. A few Ph.D. students from my own group pitched in with their experience and provided mentorship to the undergraduates. Communications by the e-mails and WhatsApp, among the team members, exhorting each other to do more to beat the deadline has been common. We could use these communication tools very well.

I wonder if this approach can be adopted formally for education and to promote learning in our engineering education system, where the project building becomes an integral part of the curriculum. Case study-based teaching can build confidence of the graduates as it provides a holistic approach to engineering education. It also means a different teaching methodology which can challenge the students to do their best. The teacher will not just pass on the information to students in the lectures, but facilitate the learning from each other while working on a project. It provides an avenue to interact with other disciplines and accountability as a team. The basic initial design stage on paper was cleared in just five to six weeks where many designs were drawn, modified and thrown away but at a point, we had to freeze the designs and the projects moved into building prototypes as per the designs. There was no 100 % guarantee of success at the stage, but we knew that we would have to improve on the job.

We faced another hurdle. We had agreed to not purchase any readymade equipment under this scheme. The components of the system, however, could be purchased. But buying the components, as small as resistors costing a few rupees every time team needed them, was difficult, considering stringent rules of purchase of any WB-supported project. Finally, we decided to combine requirements of all projects, as mechanical, electrical, chemical and special materials. Our Ph.D.

students' teams fanned out throughout the city finding the vendors who could give the materials, collecting price lists and registered them with the TEQIP-networking office as we had to follow the e-store system of ICT to place the orders. The entire purchase was being coordinated at ICT for all partners who were located at remote places. The panel of the vendors was made and registered with the Purchase cell for the procurement of components. The process took more than a couple of months to collate all the information, selecting the vendors with value-added services without additional cost. This had streamlined the purchase of all small items and fabrication procedures. Here is another lesson that we learnt, and so the student teams. Accountability of the project funds with specific requirements was identified at the beginning of the project execution. Each partner institute could send the requirement and the system could supply it within the reasonable time frame, from approval to PO generation to purchase and payment. The accountants and auditors are unfortunately, completely different lot. And their review comments led to delays that ate up a lot of time and, therefore, the enthusiasm of the youngsters. The young students were most impatient with the administrative procedures and it adversely affected the project's progress. For bigger items, we had to follow the centrally located procurement management system which almost killed some of the projects. It was only because of persuasion of ICT, that much procurement was completed and I must admit that a couple of my Ph.D. students wholeheartedly helped me in the procurement and fabrication processes. The entire exercise turned into human resources development, project management and inventory management projects, apart from the product engineering. One of the project assistants will be using this experience for his project management internship.

We also realized how difficult it was to get standard materials for building the products. Unlike in USA or Europe, where one can simply walk into a DIY store to buy every item of the shelf, we had great difficulty in indentifying the suppliers. Many of the vendors did not want any thing to do with the University fearing the paper work that they had to do. Some of the suppliers flatly refused to provide a material unless we order them in thousands! Most items are still imported and local vendors were not even aware of their availability. We had been working on a tight schedule and also on tight budget. In four months from the starting of project, the projects moved into fabrication stage and with in three months, the first project prototype was ready. Today, we have 10 projects in the final stage of testing, two patents are already filed and more are to be filed soon as extended dead line of 15 May 2015 is approaching.

The success of these projects was not with the project investigators but with the bunch of youngsters and Ph.D. students who sacrificed their vacations to work on the ideas. They have taken ownership of the projects as they have been promised the partnership in the project outcomes. They fiercely guarded their designs and kept working furiously to meet the deadlines set for them. They learnt corporate working, team work, keeping the cost factor low, machine designs for easy maintenance, protecting own IPRs and respecting those of others. We see now among them healthy respect for other disciplines and appreciating the efforts that each one had put in. It does not mean that we had no problems. There were many.

The UGs could work only when they had no lectures and examinations. Some of the teams changed completely. In one project, the entire team but one member abandoned the project without any reason. But a lone member, a girl student, kept on working. I had to talk with the faculty investigator, to keep the project on for the sake of the spirit of one innovator. She did her part to her maximum ability, but now we will complete the other parts with another team. What was disappointing was the majority of the students showed indifference towards the innovation activities. They are just happy to pass examination. Not all faculty members were enthusiastic for putting additional efforts in the activity. The institutes, as whole, took cursory interests in providing necessary infrastructure or support system. It is only the persistent efforts of the few students and faculty members that the innovative products have come out of the entire project. Unless the benefits, tangible as well as intangible, of such activities are visible, it will be difficult to get participation of all stake holders in such activities. But those who participated have benefitted a lot. There was a change in attitude of the UG students who participated in the projects towards their profession. They are able to relate their theory classes with practical applications and thus are 'learning' better. Some of them were happy to do something different which they could not have done in the normal course. A bunch of electrical engineering students from Sardar Patel College of Engineering stands out because of their efforts to learn new things and build a controllable system for microwave reactor. The coordination amongst students had visibly increased and instead of competing, they complemented each other.

This experience probably throws up an answer to the question raised earlier, what we, as teachers, can do about the declining quality of graduates. The answer is probably involving them earlier in their profession, particularly to solve real problems as they progress through their classes. It may not be possible to teach every thing in the class, but learning on job would provide a better education. And each of them may be credited for their contribution in a continuous assessment. The idea is to make the learning an enjoyable experience and at the same time appreciating their efforts in learning. It is believed that such an approach would prepare graduates to meet the demands of the profession when they join the industry and industries' concern of employability of the graduates will be taken care of.

There would also be the possibility of the team members, with their stake in the development, opting for entrepreneurship for manufacturing and marketing their own products. The management of the engineering institutes may have to take a role in promoting the innovation laboratories with all support extended for supply of materials for building such prototypes with clear understanding that the innovators will have a stake in the final IP rights. The participation of technical and support staff, in machining and fabrication, can be also welcomed with appreciation in suitable form. The most important part is identification of right type of product and process that can have value addition to students' education. If these products and processes are built at affordable cost, it would help local economy and society. A great care is needed to identify the innovation in each idea that the students may bring to the table. We need to engineer the engineering education to meet the demands of the country by appropriately modifying the curriculum to incorporate the innovation activity in the syllabus.

If this spirit of cooperation and collaboration catches up with other institutes, we will be willing to share our experience. I have been conducting ideation and innovation workshops at the partner institutes for a while to see whether additional collaborative projects can be built. My own Ph.D. students conduct hand-on tutorials at these colleges for help them in learning the tools necessary for designs, including molecular modelling. We get an overwhelming response immediately, but sustaining the interests over long time has been an arduous task as there are several external factors that can kill the entire innovation spirit. Holidays, examinations, regular assignment submissions in colleges, non-availability of standard materials, lack of teachers, unwillingness of faculty members/management to participate in the innovation activities and biggest of all, the procurement and accounting procedures all can kill the project completely. It appears that we know what is needed, but the process of doing is tiring if you need to chase multiple centres for one job. At this point, the lack of a supporting ecosystem is felt acutely by all of us.

Ultimately, we believe that this will lead to innovations that are affordable in local communities, businesses and education. There still remains the question of role of the government, particularly MHRD, in the process. It may not possible for a country with limited financial resources, to establish full-fledged research parks costing crores of rupees over a short period of time, as Germany has done for its Fraunhofer Institutes in specific areas. But coordination amongst different institutes might help to work in the form of virtual networking. A set of networking innovation parks at different locations in the country, with dedicated set of people who can facilitate the growth of innovation, still remains the acute need to promote the innovation using the largest resource we have, i.e. the engineering students in their third and final years. The success of this small pilot project under TEQIP can probably rekindle the proposal that ICT put forward to MHRD for establishment of innovation parks in four corners of the country.

This experience has taught me many more things which I would not have learnt in my own laboratory. And given a choice, I would be happy to work again with different institutes in the country to build their own local networks.

From Physics to Engineering Physics

K.L. Chopra

I was born in 1933 in Chahal Kalan—a small village near Gujranwala in Pakistan. I used to walk about 4 km to my high school and read in the light of an oil lamp. My favorite sports were kabadi, wrestling, and playing free-for—hockey on a dirt road. I learned to swim in the village pond.

Our science teacher taught us general science in the 9th grade. He was very particular in demonstrating some simple science experiments to us as a part of our practical class after the school hours. I was the topper in the class and I missed one such session. The teacher was furious and told me that experiments are the soul of science and that I will not become a scientist unless I learn to do experiments. This harsh advice has impacted my future career as an experimental physicist so much so that I believe what I see or I can verify experimentally. After being promoted to the 10th class, the tragedy of partition of India took place and our family, like millions of others, had to hide and save our lives from fanatic mobs by moving on foot with little belongings to nearby military-supervised camps. After months of moving from one refugee camp to another, our family reached Delhi where I got admission in DAV school in tents for refugees near the Golemarket. I topped the 10th grade examination. With a few career options open even to a topper in those days, Physics was the most preferred choice which is what I did to graduate with B.Sc. (Hons) and M.Sc. (1954) in Physics from Delhi University.

We were taught a lot of Physics in Delhi University by some of the best-known professors of Physics at that time in the country. But, there was little excitement or inspiration since there was no attempt to relate the Physics taught to us to anything around us which we could appreciate. For example, we were taught in great details the theory of dynamics of a spinning top but were never told its connection to a practical application of a gyroscope. Photons have no mass but have energy and momentum. When I asked how momentum arises and how we know it, I was told to find it out myself in the library. I found out that a very simple experiment of observing the rotation of a very light vane in vacuum on absorption of light on its blackened side demonstrated nicely the transfer of momentum by the photons. We

K.L. Chopra (✉)
M-70, Kirti Nagar, New Delhi 110 015, India
e-mail: choprakl@yahoo.com

P. Ghosh and B. Raj (eds.), *The Mind of an Engineer*,
DOI 10.1007/978-981-10-0119-2_40

were taught crystal structures of materials but we were not told as to why we need to have this information. Or, why Si and Ge having a diamond structure are important electronic semiconductors. Even more dull were the laboratory experiments which we were expected to perform but which had little to do with what Physics was being taught to us. Experiments, such as measuring the viscosity of water, or the observation of Newton's Rings did not stir any imagination or excitement.

Inspirational Experience

After receiving M.Sc. in 1954, I was selected for a World University Fellowship to pursue a Ph.D. program in Physics at the University of British Columbia, Canada. After a journey of one month by ship from Mumbai, I reached the University of British Columbia in Vancouver, Canada, in September 1954. My would-be supervisor, Prof. James Brown, told me that the quantum behavior of a zero viscosity superfluid liquid helium which flows out of any container against gravity via a micron size thin film would be an interesting Ph.D. project. However, I told my supervisor that before starting my research, I would like to learn to work in the workshop with my own hands. I was taken to the departmental workshop as an apprentice. I learned to use machine tools, different lathes, soldering, welding, glass blowing, silvering of glass Dewars, some electronic instrumentations, etc. As was the practice in those days, I built my own vacuum system out of glassware, with a mercury diffusion pump. I prepared silvered glass Dewars for liquid helium and liquid nitrogen. Liquid He temperatures well below 2 K were measured by using a thin India ink dot written on a piece of paper as a temperature sensor. A simple Wheatstone bridge circuit was used to control low temperatures obtained by pumping (to reduce vapor pressure) as well as by adiabatic demagnetization. An ultrasonic transducer was used to move liquid helium in narrow channels. Micron size channels for liquid helium flow were fabricated by first filling a thin Cu–Ni tube with bare copper wires and then by drawing down the tube through successively decreasing size of dies. Beautiful hexagonal shape channels so formed became the subject of my very first research paper. Working with a very competent and helpful technician to liquefy helium, and to recover evaporated helium gas, to purify it, and to test its purity before recycling it were experiences of sheer delight.

Observing the superfluid liquid helium (a macroscale scale quantum fluid) getting out of its container against gravity (a quantum phenomenon) was akin to a spiritual experience for me. Sometimes, such experiments lasted up to 48 h. My supervisor stayed with me all the time and brought packed food for both of us. This experience of conducting exciting experiments in collaboration with a supervisor as a coworker and mentor left a deep impact on me. Then and there, I made up my mind that, given an opportunity, I would like to set up a laboratory where students and faculty would work together like colleagues, learn to work with their own hands, learn from experiences of others, and above all appreciate the need to translate the knowledge for something useful. Little did I know at that time that my dream would be fulfilled eventually in the form of an Institution which I will build and which came to be known globally as "Thin-Film Laboratory" (TFL) at IIT Delhi.

After my Ph.D., I had a stint as a postdoctoral Research Fellow of the Defence Research Board of Canada at the Royal Military College, Kingston, Canada. I set up facilities to verify the recently announced BCS theory of superconductivity by measuring ultrasonic absorption in superconductors. This was followed by a stint as a Max Planck Fellow in Fritz Haber Institute, Berlin, West Germany, where I built a liquid He temperature X-ray diffractometer from scratch to determine any asymmetry in electron density distribution in hexagonal metals.

Thin Films: A New Frontier

An extraordinary development took place in my career during 1962. Some reported observations of unusual electronic and optical properties of vacuum evaporated ultrathin films of several materials showed great promise of emerging as a new frontier of science and technology. I got an opportunity of working on the subject with Philco-Ford Scientific Labs, Philadelphia, and Ledgemont Laboratory of Kennecott Corp, Lexington. As a senior scientist, I was given a free hand to conduct research in any direction which could be of interest to the industry concerned. Some US scientists claimed that ultrathin (nanometric size) oxide films sandwiched between two thin metal films exhibit nonlinear current–voltage behavior which meant one could make very cheap small size active electronic devices. The properties of the films and devices so obtained were found by all investigators to be not very reproducible which led some persons to conclude (perhaps jokingly!) that thin films are the fifth state of matter (besides gas, liquid, solid, and plasma). We must remember that at that point of time, the processes of deposition of thin films by evaporation and sputtering were rather primitive and there were no nanoscale analytical techniques to determine the chemical, metallurgical, or physical structure of thin-film materials. Nevertheless, the promising future of this new research area led private industries to invest heavily in R&D which made it possible to achieve considerable improvements in the control of thin-film deposition processes. The arrival of a commercial electron microscope at that time also helped a lot in structural studies of thin films. As a result of the research efforts of many of us, it became clear that properties of thin films of any material (and similarly nanomaterials created by any ab initio process in today's context) depended on how the atomic species are created, how these are transported, and finally how the atomic species get together in their random walk process to create matter ab initio. It was established that nanometric thin films are created by a critical size nucleation process, followed by growth processes which are primarily determined by the minimization of surface, volume, and strain energies of the growing nuclei. Since these basic birth stages of nucleation and growth processes are affected by a large number of deposition parameters, reproducibility of the properties of thin films (as also any nanomatter) depended on how reproducible and controlled these parameters are. Further, these studies revealed how to tailor useful and exotic properties of thin-film materials by varying these parameters. This knowledge led to the growth of a large number of thin-film technology industries globally which made it possible to use thin films in numerous applications ranging from micro- and nanoelectronics, magnetics, photonics, integrated optics, solar cells, hard and tribological coatings for surface engineering, optically selective coatings, quantum standards, etc. Today, thin films of appropriate materials are an important and integral part of any sophisticated opto-electronic device. Thus, Thin-Film Technology is considered as the mother of what is popularly known today as "Nanotechnology."

I published my research findings extensively and was granted 4 US patents. As early as 1962, I published a paper, the first of its kind, on a very low-efficiency Au/NbO_X/Au Thin-Film Photocell (a solar cell as it is now called) in IEEE Journal. I discovered current-controlled negative resistance switching in sandwiched thin oxide films (a topic which is being revived for commercial applications). Our pioneering work on the changes in the structural, electrical, and optical changes of GeTe chalcogenide thin films at elevated temperatures led private companies (such as Moser Baer) to develop writable CDs. I developed and patented a duoplasmatron ion beam source for vacuum sputter deposition of hard optical films/coatings. Two decades later, this patent came to the rescue of a US-based multinational which was dragged to court with over one billion dollar legal suit by a rival company for violating their patent for sputter-depositing durable laser mirrors used in laser-guided weapons for defence. By citing my patent as an evidence of an existing knowledge in the field, the company won the court battle. The news attracted headlines in the New York newspapers.

The exciting ecosystem of research in industry-oriented research laboratories in USA inspired me to challenge myself to author the very first book in the field. At the age of 32, I spent the next three years in my spare time to write a Treatise "Thin-Film Phenomena." Published in 1969, the book was a grand success. Known as the Bible of the field and translated into Russian, the classic book is still read widely all over the world. Learning about my work in the new area of thin films, the Director, IIT Delhi, inquired whether I would be interested in joining IITD. The Institute offered me a very attractive position of a Senior Professor of Solid State Physics and also offered to pay all my relocation expenses by air for the family. Such a generous offer by a premier institution in India was unthinkable in those days. I could hardly refuse this golden opportunity to fulfill my dream of setting up a unique research laboratory in India. We arrived in Delhi on August 23, 1970. My wife and I met the Director of IITD on the 24th morning. Warmly welcoming me to IIT Delhi, the Director handed over personally the keys of the bungalow at 10, West Avenue, IIT Campus bungalow which became our future residence for 17 years. This warm welcome and reception led me to resolve to burn the bridge to my attractive US position (from which I had taken leave) and to justify the trust posed in me by IITD.

The Birth of Thin-Film Laboratory (TFL)

I was shown around the Physics department and, in particular, two large empty rooms where I was supposed to set up my research laboratory. A Ph.D. student, who had decided to join me even before I joined IIT, was keenly waiting to talk with me. I told him that if he was prepared to accept the rough and tough role of a pioneering scientist, he was welcome as my student. Thus, both of us started together to clean the floors of what would become the globally famous Thin-Film Laboratory (TFL). I had mentioned in my recently published book that growing polymer films epitaxially and doping them appropriately would be a fruitful area for R&D in developing new materials. We devised a very simple solution growth technique with very little equipment to deposit thin films of well-known polymers such as PVC, PVB. By incorporating different metals in the polymer films, we created what came to be known as semiconducting "metallopolymers." We received our very first grant of Rs. 30,000 as a DST sponsored project to synthesize conducting polymer films which boosted our spirits and our credibility. We were well ahead of time in creating semiconducting polymers but were not good chemists. The discovery of conjugated polymers some 8 years later led to the Nobel Prize discovery of conducting polymers by Japanese and two American chemists. No, we were not disappointed. Our simple Solution Growth Technique became a popular technique for synthesizing and studying polymer thin films by numerous workers globally and our original papers were cited widely.

The number of students keen to join TFL for Ph.D. and M.Tech. projects started increasing rapidly. We needed to build all—purpose vacuum systems for thin-film research. A workshop was created in TFL out of whatever old tools were available within the institute All students, male or female, had to learn to use the workshop for building equipment with whatever was available anywhere in junk, waste, or scrap form. The students built several vacuum systems for evaporation, and for sputtering. Since large metal vacuum chambers were not available anywhere in the market, our students bent stainless sheet into a cylindrical chamber by sheer muscle power and welded it. Improved diffusion pumps, crude pressure gauges, electron beam gun, sputtering modules, quartz crystal and optical monitor, ellipsometer, DLTS, EBIC, spray pyrolysis system rapid quenching setup, among others, for thin-film deposition by different techniques were created by *jugaad* (an Indian version of innovation).

After his visit to TFL during 1976, Dr. Arcot Ramachndran, the then Secretary of the DST, asked me what DST can do for TFL. I asked for a Nanoanalytical Facility. Dr. Ramachanndran lost no time in sanctioning Rs. 15 lakhs. I persuaded a US company which, at that time, was developing a Scanning Auger Microprobe-cum-Electron Spectroscopic Chemical Analysis, to give us the instrument for experimentation and evaluation at a nominal coast of Rs. 15 lakhs. The SAM–ESCA was the first such instrument anywhere in India we maintained and used this instrument extensively for almost two decades. We received some more analytical equipment such as an electron microscope and spectrophotometers under the Indo-UK aid agreement. With further development of facilities, TFL became a foundry to develop a large number of home-made physical, chemical, and electrochemical thin-film deposition techniques and a range of micro- and nanoanalytical tools. At one time, TFL was considered globally as one of the few best thin-film R&D facilities under one roof.

Our research contributions were appreciated globally. TFL attracted lots of visitors—foreign dignitaries, academics from all over the world, and several Nobel Laureates. The TFL was open and functioned 24 × 7. The Nobel Laureate Dr. Abdul Salam visited TFL in one late evening. A theoretical physicist himself, he was so impressed to see our work that he asked me if we could set up a similar facility in the International Centre for Theoretical Physics, Trieste, Italy, of which he was the Director at that time. Two prominent Japanese scientists came all the way only to verify our claim of having created transparent conducting ZnO films for the first time. German scientists visited at night to look at our work on thin-film CdS/Cu_2S solar cells. The Vice-President of IBM Research Centre at Yorktown Heights, USA, spent six months of his Sabbatical leave in TFL. The list of prominent visitors is endless.

Invitations to me for lectures came from all over the world. Requests for engaging my students for postdoctoral fellowships came on telephone from many scientists abroad. The CEO of ULVAC—a leading Japanese Vacuum and Thin-Film Company—requested me to send three of my students as PDFs on a 5-year contract. During my visits abroad, I spent some time to consult at the Research Centres of IBM, Westinghouse, and ARCO in USA. The pinnacle of our global recognition came when TFL was asked to hold the 7th International conference on Thin Films in 1987. With 700 participants, including three recent Nobelists, from all over the world, the Conference held in Vigyan Bhavan is still remembered by the world community for the high quality of presentations, superb arrangements, and hospitality—all taken care of by the TFLians.

Significant R&D Contributions of TFL

Some of the original and prominent scientific contributions from TFL during my time are listed as follows:

- Developed Semiconducting Metallo-Polymers
- Developed graded refractive index multilayered coating—the first "photonic crystal" (as it would be called today) of its kind-for nearly perfect reflection mirrors
- Established rigorously electron transport processes in thin metal and amorphous semiconducting films.
- Developed ZnO-based bulk and thin-film varistors. The technology was transferred to WS Insulators, Chennai.

- Developed scientifically a chemical bath deposition process for CdS films—which some call it a Chopra process—used globally by all thin-film solar cell industries today
- Developed, for the first time, transparent and conducting thin films of ZnO which in its bulk form is a well-known insulator. Such ZO–TCO films are now used extensively by thin-film solar cells and other optoelectronic industries without, unfortunately, giving credit to us
- Discovered structural and optoelectronic changes in various Ge-chalcogenide films which has led to the manufacturing of CDs
- Discovered giant photograph contraction effect in Ge-chalcogenide films and demonstrated its lithographic and reprographic applications
- Detailed study of the Physics of thin-film CdS/Cu_2S solar cells
- Developed nanotube-structured aluminum oxide template for synthesizing nanostructured optically selective coatings for solar–thermal applications. Creation of such templates of various materials is now an established industry
- Developed hard metal carbide and nitride coatings for surface engineering and machine tools

Since patenting of innovations was not a priority in educational institutions at that time, we missed an opportunity to patent several significant innovations which were adopted in due course by global industries. However, TFL attracted industrial consultancy assignments and thus interacted strongly with the industry in India. Various consultancy jobs, and technology development and transfer assignments were executed by our students under the supervision of the faculty. The first consultancy assignment in IITD was offered to TFL by the then well-known razor blade company manufacturing 7-O'clock shaving blades. We were required to improve the smoothness, sharpness, and life of blades by coating nanometric Cr films by a sputter-deposition process. The project was successfully executed and our students tested the performance of the blades by regular shaves in the morning inside TFL. Thus, began an era of Coated Blades.

Some of the interesting consultancy and translational research projects undertaken by us were as follows:

- Sputter deposition of Cr on 7 O'clock razor blades (Malhotra Razor Blade Co)
- Study of nanostructure of imported silica powder for rubber tyres (Good Year, Delhi)
- Thin-film CdS photo-cells (manufactured by Patel Enterprises, Bangalore)
- Thin-film color coatings on ophthalmic glasses (Laxmi Opticians, New Delhi)
- Manufacturing of electron microscope grids (Montek Industries, Chandigarh)
- ZnO varistors (manufactured by W S Insulators, Chennai)
- Moire gratings for lathe machines (HMT, Bangalore)
- Thin-film strain gauge for Roorkee University
- Thin-film IR detector for DRDO
- High-power electron gun (manufactured by VICO, Delhi)

- Magnetron sputtering module (manufactured by VICO, Delhi);
- Optical monitor for thin-film deposition (Hind High Vacuum, Bangalore)
- Optically selective copper black coatings for solar–thermal applications (Jyoti Ltd. and BHEL)

A Unique TFL Ecosystem for Engineering Physics

The conduct of original and high-quality research by Ph.D. students, training and mentoring M.Tech. students, mentoring by postdoctorals, and translational role of all for providing consultancy service and for technology transfer required a new model under Indian academic conditions. It is called for a seamless environment for moving from physics to engineering physics. A unique ecosystem was evolved which made it possible for TFL to grow into an Engineering Physics Institution for learning, mentoring, working collaboratively, hand holding, family-like group activities, innovations, and entrepreneurship. Some of the salient features of this system are as follows:

- Keeping TFL neat and clean was the responsibility of all—students, staff, and faculty—under my personal supervision. The cleaning activity was carried out every month on Saturdays and was followed by snacks and tea for all in a convivial get-together.
- TFL was open all days, including national holidays, on a 24 × 7 basis
- Total responsibility for running and maintaining major instruments in good working conditions rested with the faculty and students assigned for each facility.
- Before starting any research work, every student was expected to spend a few months to learn to handle workshop tools, learn to know who was who and who was doing what and why in the TFL. The success of this homogenization process of the student was evaluated by me before being allowed to proceed to start his/her work independently.
- Irrespective of who the official supervisor(s) of the student was, he/she had access to all the TFL facilities. At the same time, every student was expected to help any other student who needed help.
- Coming punctually to TFL in the morning was mandatory for students. Random discussions regarding their work were held with individual students during the day. Though working in TFL in the evenings after dinner was discretionary, most students worked at nights. I visited TFL frequently after dinner largely to chat with the students working at night.
- Publication of any research paper by a student or PDF was allowed only after rigorous discussions. Students were expected to write drafts of the paper The paper was submitted for publication only after approval following thorough discussions.

- On completion of Ph.D., every postdoctoral fellow (PDF) was obliged to spend at least 1–2 years in TFL to mentor our M.Tech., and Ph.D. students. Whereas Ph.D. students concentrated on research for a Ph.D. thesis, M.Tech. students were expected to demonstrate a working device or instrument, or a proof-of-concept of a new device. The PDFs learned how to mentor and work with others and also they were exposed to the process of translating research ideas to a useful product or process. Our research projects provided fellowships to PDFs and they were assured of other PDF assignments abroad. Exceptional achievements by anyone were rewarded with a special prize.
- The TFL group, which became as large as 40 persons, including several faculty members, had a weekly packed—lunch meeting for 2–3 h. Everybody brought a packed sandwich lunch and we sat together to eat it. By rotation, a group of students had the responsibility of preparing and serving hot—I mean piping hot cups of coffee to all of us. Using the consultancy money earned by TFL, excellent facilities for making tea/coffee were set up. At the weekly lunch meeting, students and faculty had an equal opportunity and freedom to talk about any problem or subject related to his/her project, availability of any instrument, any new global developments in the area, any controversial views on academic matters, etc. These meetings were invariably full of heated debates and excitement.
- Seminars in TFL by the students and faculty were held regularly and ritually on Saturday afternoons. These seminars often led to intensive and spirited discussions. Students were encouraged to participate in such discussions
- Very frequently, a visiting distinguished academic/researcher visitor from India or abroad joined our group lunch. His/her sandwich lunch was sent by my wife along with mine from my house. Discussions by students with such visitors were encouraged and were often quite animated. These face-to-face discussions with prominent scientists in the field generated a lot of confidence among our students
- TFL was the first destination of a visit by distinguished political and academic guests of IITD. While taking the visitor around, I expected the students to explain as to what they were doing and why. This nurtured student's confidence and sharpened their communication skills.
- Any event calling for shared joy and happiness provided an opportunity for a group dinner party. Such events included the publication of a first research paper, award of a degree, offer of a PDF from abroad, any form of professional recognition, marriage. The party was held at my residence with cooked food brought from outside by the students. With students, faculty and their wives sitting on the floor of my large drawing room, the dinner was followed by a long jam session wherein individuals recited poetry, sang, told jokes and stories, etc. I contributed a few Urdu couplets that I was fond of writing in my younger days. My wife contributed jokes. The group experienced a very warm and intimate family togetherness

- The students, faculty, and their families looked forward to periodic picnics arranged and managed by the students. The larger TFL family enjoyed picnic food as well as sports.
- The TFL code of conduct included punctuality in all activities, equality among all researchers, transparency, duly earned credit for authorship, strict discipline and disciplinary action against any form of scientific misconduct, plagiarism, or conflict of interest. Toughened by this kind of discipline, TFLians cite this attribute as one the most important contributing factors to what they are today.

Mentoring of Entrepreneurship

During my 17 years at TFL, I supervised/co-supervised 55 Ph.Ds. and some 60 M.Tech. projects. All my students have done well in a variety of technical professions as teachers, researchers, entrepreneurs, and Chief Executives Officers. We published over 300 research papers and a large number of review articles in international journals of repute. Several of our papers became classics by virtue of citations. One of our reviews on Transparent Conducting Oxide Films had the distinction of being the most cited paper in thin solid films for over two decades in a row. Our two monographs, namely Thin-Film Solar Cells and Thin-Film Device Applications, first in the field, were co-authored with my two students, Suhit Ranjan Das and Inder Jeet Kaur, respectively.

There was a global demand from reputed academic institutions to employ my students as PDFs. The feedback from those who invited my students for PDF was that my students were enterprising and indeed were up to any challenge, even in new areas of research. Over a dozen of my Ph.D. students have chosen to become entrepreneurs and they have set up their own successful industries in their related areas, both in India and in abroad. Milman Thin-Film Systems created by Milind Acharya in Pune, India, Flexible Solar Modules (FLISOM) established by Ayodhya Tiwari in Switzerland, Coatings Mantra established by Sunil Kumar in Australia, information, science, and technologies (Instech) Consultants by Jagriti Singh in Australia, RF Array Systems Inc by Chandra Deshpande in India and USA are some notable examples of such companies. Several global thin-film solar cell companies are being steered by my former students.

The IIT Kharagpur Challenge

I served IITD as Head, Physics Department, Head, Centre for Energy Studies, Dean, Post Graduate Studies and founder Dean, Industrial Research and Development (IRD). As a Dean, I was keen to translate my enthusiasm for nurturing innovations on a larger platform of the institute. The Department of Science and Technology offered all financial support and land to IITD to set up a Science Technology Entrepreneur Park (STEP). Although I was very enthusiastic about the project, the then Director of IITD was not so keen on this venture which he considered as being beyond the purview of the institute. And, then in 1987, a totally unexpected event took place. I, a Physicist, was persuaded by the Ministry of Human Resources and Development to take up the challenge of heading and reviving the run down IIT, Kharagpur—the mother of other five IITs, as I call it. Apprehensive though I was, I accepted the challenge and started head-on with academic, management, and governance reforms and innovations The blueprint for these changes was essentially the same as that for creating TFL which included: leading from the front by action and by example, total transparency in all activities of the institute, treating faculty, students and staff as partners in progress and in reinventing of the institute, and strict enforcement of ethical values and codes of conduct for all stake holders.

Despite being busy with the IITKGP affairs, I continued my research albeit at a slow pace. Several research students and some faculty members helped in setting up a state-of-the-art Microscience Laboratory. During my tenure of 10 years at IIT, I co-supervised 4 Ph.Ds. and co-authored a book on "Vacuum Science and Technology" with Professors V.V. Rao and T.B. Ghosh. We explored new and emerging areas of science and technology of thin films. We developed a chemical technique for synthesizing nanopowder of multicomponent oxides on a large scale. The technology was transferred to ACC.

The creation of a STEP in IIT Kharagpur was topmost in my mind. The WB government gifted 100 acre land area near the IIT campus for the proposed STEP. Mr. Jyoti Basu, the Chief Minister of West Bengal and I laid the foundation stone of STEP which was sponsored by DST. This STEP, the only one among the IITs even today and one of the 12 STEPS in the country, is presently doing well and is self-sustaining. Several IITs have subsequently followed the example of IITKGP by setting up different versions of tech-parks such as an incubator, technology business incubator, and technology park.

Some of the novel and innovative ideas adopted by us in IITKGP for translational work through academia–industry interaction are:

- Members of the faculty, alumni, and students were encouraged to become entrepreneurs in STEP so as to explore the commercial viability of their research outcome in IITKGP. This was an extraordinary step that was taken for the first time in the country. Similar concepts have now been adopted by most IITs
- Faculty members were allowed to earn any amount from their successful ventures in the STEP as long as accounting, transparency, and IIT's share of turnover were assured through the STEP management.
- Joint ventures of IITKGP with selected industries in the area of expertise of IIT faculty were initiated
- The industry was persuaded to sponsor Chairs so that the Chair Professor would conduct research work in areas of mutual interest. Both Joint Ventures and

Industry Sponsored Chairs were unique concepts in Indian academia at that particular time. Both concepts have now been emulated by other IITs.

After retirement from IITKGP, I have continued to persuade many technical and academic institutions in the country to take a step toward any form of STEP as an integral part of training and nurturing technical manpower. I have helped establish Incubators in St. Xavier's College and West Bengal University of Technology in Kolkata.

Concluding Remarks

A transparent, liberal, flexible, seamless, multidisciplinary, interactive, and ethical ecosystem has enabled my numerous students and faculty colleagues working as a group to contribute to our journey toward translational and transformational research in physics and engineering physics. I had the privilege of working with many co-workers. I stand tall on their shoulders and owe them a debt of gratitude.

Physicist or Metallurgist

R. Krishnan

I obtained my B.Sc. (Honours) and M.Sc. in physics (X-rays and Crystallography) from the Madras University and subsequently while working at BARC, the Ph.D. from the Bombay University. Are you a physicist or a metallurgist was a question that was posed to me during my early career. In fact, after working at the Metallurgy Division of the Atomic Energy Establishment Trombay, for about 5/6 years, and having presented a few papers at the Annual Technical Meetings of the Indian Institute of Metals and publishing papers in its journal, the 'Transactions of IIM', I applied for membership of the IIM, which was promptly turned down because I did not have a basic metallurgy degree. Those were the days when 'Materials Science' was not yet prominent as a discipline. I knew that if the IIM was to recognise me as a metallurgist by profession, then I should get an overseas recognition first. Based on my Ph.D. on 'X-ray line broadening studies on cold worked metals and alloys', and my publications, the Institution of Metallurgists, UK elected me a Fellow, considered equivalent to a Ph.D. in metallurgy in UK and elsewhere. After this, IIM did not have any difficulty in accepting me as a member. I was also made an Editor of the Transactions of the Indian Institute of Metals. Later on, I was even elected as the President of IIM in 1993. I was conferred 'Lifetime Achievement Award' instituted by the Ministry of Steel and Mines in 2014.

The Transition

How did I make the transition from a physicist to a metallurgist? After my M.Sc. in X-rays and crystallography, Prof. G.N. Ramachandran who was the Head of the Department of Physics at the University of Madras wanted me to continue there for my Ph.D. and I had started my work. But then he had had a nervous breakdown and did not show up in the laboratory for a few months. At that time (1957), the Atomic Energy Establishment Trombay started a training programme and was looking for

R. Krishnan (✉)
F-3, Parkside Manor, 5th Main, Defence Colony, Bangalore 560 038, India
e-mail: rangachari.krishnan@gmail.com

P. Ghosh and B. Raj (eds.), *The Mind of an Engineer*,
DOI 10.1007/978-981-10-0119-2_41

graduates and undergraduates in different disciplines for work in the nuclear energy programme of the country. Because I had the privilege of travelling free in Indian Railways thanks to my father being in the Railway service, I chose to attend the interview at Bombay. I was selected and joined the first batch of the AEET Training School programme. On completion of this one-year orientation course, I was posted to work with Dr. P.K. Iyengar in neutron diffraction, in the Nuclear Physics Division. Just 4/5 days prior to my joining the Nuclear Physics Division, Dr. Ramanna, who was in charge of the Training School, told me to meet Dr. Brahm Prakash, Head of the Metallurgy Division. BP told me about the CIRUS reactor programme and that Dr. Bhabha wanted AEET to fabricate half of the metallic uranium fuel elements required for the reactor. One of the critical factors to be assessed in metallic uranium fuel is the texture or the crystallographic orientation. Dr. K. Tangri, who had done his doctoral work on texture studies in the USA, was heading the Physical Metallurgy Section. As I was the only person from the training school who had the requisite background, BP asked me to join in his division. With continuous smoke emanating from BP's cigarette and puffs from Tangri's pipe, I was a little dazed and I told them that I would think it over and meet them the next day. I met BP the next day and told him that I would join the Metallurgy Division. He said he would keep me happy and he did. This was the starting point of my conversion from a physicist to a metallurgist.

My BARC Days

I started my work right earnestly and collected several samples of fuel rods from the Fuel Fabrication Facility, (code named 'Fagots') to determine the texture and to confirm that quenching from the high-temperature beta phase had led to a random texture. The samples have to be metallographically polished and my senior M.K. Asundi helped me a lot in this regard. I also learnt my basic metallurgy from him. As the work load was increasing, BP told me that if I required, he would post a physicist from the 2nd batch of training school to work with me and thus V.S. Arunachalam joined me. I should mention here that the in-reactor life of the fuel rods made by us in Trombay was twice that of a Canadian fuel element, as tested by the Canadians.

I should also mention an interesting anecdote in this connection. Bhabha used to review the progress of uranium fuel fabrication, to which I was always invited. In one of the meetings, when some wrong interpretation was presented, I got up and started talking. Bhabha stopped me and asked BP who was sitting next to him: 'I say, did we ask him to say anything?' BP told him that I am working on textures in uranium and that I should be heard. Bhabha then allowed me to proceed with what I wanted to say. Later on, when the source of natural uranium for fuel making was changed, the fuel rods started contracting instead of the anticipated expansion. I examined the texture of the fuel rods again and suggested a small cold-drawing operation after beta quenching. Rods that underwent this treatment behaved better with no shrinkage/growth. This work earned me a promotion almost without an interview.

I was sent to CEN de Saclay in France for training for a year. I did not find the training at Saclay of any use and as I was left alone, I started working on stacking faults in uranium and predicted how faults can occur in alpha uranium, which was confirmed experimentally later on. After 6 months at Saclay, I switched to Laboratoire de Physique de Solid at Orsay to work with Prof. Andre Guinier, on X-ray small angle scattering of irradiated quartz. I found that under neutron irradiation amorphous regions started forming in small pockets, while the crystalline structure was retained. This surprised Prof. Guinier. This stay at Orsay enabled me to have a few fruitful discussions with Prof. Friedel and Prof. Castaing.

On my return to Trombay, I found that Dr. N.M. Parikh had joined the Ceramics Section after his doctorate from MIT, USA. He told me that I should look up Warren and Averbach's papers on X-ray line broadening and that may be a good topic for my doctoral work. I should thank him for this as I did take his suggestion seriously and registered for my Ph.D. in physics with the Bombay University with Dr. Raja Ramanna as my guide. I submitted my thesis just after two years in 1966, but had to wait for a year for my thesis defence. A major satisfaction from the X-ray line broadening studies was that our work on cold-worked uranium was accepted for publication in Acta Metallurgica. This is the first paper to be published in this prestigious journal from an Indian research group.

At that time, Dr. V.K. Moorthy was working on sintering studies on UO2, used as a fuel in nuclear power reactors. He found that UO2 powders derived from different routes showed differing sintering behaviour. I told him that we need to characterise the powders better and suggested looking at the crystalline structure of these powders. I did a lot of line broadening studies and estimated the particle size and the degree of crystallinity in the powders and related them to their sintering behaviour. Powders which were fine in the range of 20–40 nm sintered to a higher density at lower temperatures. I wish now that I should have used the term, 'nano particles' then.

Subsequently, I was asked to look after electron microscopy work. We started from scratch and learnt preparation of thin films of metals and alloys to get familiarity with the techniques of microscopy and electron diffraction. The group got strengthened when P. Mukhopadhyaya and Srikumar Banerjee joined the team. We studied phase transformations and structure property correlations on a wide variety of alloys based on zirconium, titanium and some steels. We published several articles in Acta Metallurgica and Metallurgical Transactions. The basic research carried out enabled us to solve the production problems faced by the Nuclear Fuel Complex when they did not obtain the strengths expected of Zircaloy tubes. Subsequent to my departure from BARC, I was happy to learn that my former colleagues had a major role in establishing the flow sheet for the production of Zr–Nb tubes. I should also mention here that Banerjee once questioned me about the usefulness of doing basic research and how does it help the country and the public. I told him how things get connected and he realised later that it was essential to do basic research to understand the behaviour of materials so that when problems are faced, they could be solved by the researchers. I am happy that subsequently he became the Chairman of the Atomic Energy Commission and Secretary, Department of Atomic Energy. He was my second colleague in reaching the top position in the Government, the first one being Arunachalam, who became the Scientific Advisor to Raksha Mantri and Secretary, Department of Defence Research and Development.

BARC was the first to acquire a scanning electron microscope in the country in 1974 and I was asked to look after that. I had initially trained G.E. Prasad, who became an expert in using this instrument and in interpreting the micrographs. He had carried out a lot of studies of different nature, including failure analysis investigations with Asundi. ESCA and Auger spectrometer was the next big instrument that I handled at BARC. We were planning to carry out a few surface analytical studies of nuclear materials. I was particularly interested in evaluating the effect of additions made to Zircaloy 2, to understand its surface characteristics.

My Transition to DRDO

Dr. V.S. Arunachalam, DG, DRDO, was looking for a Director for the Naval Chemical and Metallurgical Laboratory (position to fall vacant on 1 March 1985) located at the Naval Dockyard in Bombay. He first convinced me to join NCML and later talked to Ramanna to release me on deputation to DRDO. After his concurrence, he had to seek approval from Shri. P.V. Narasimha Rao, who was then the Defence Minister. The minister was away in Russia in the last week of February 1985 and on his return on 27 February, Arunachalam got my appointment as Director of NCML approved and flew to Bombay on 28 February morning, got Ramanna's approval and took me to NCML and introduced me to the staff there. I mention this in detail because only Arunachalam could have done it in such a short time; no one else!

NCML was a relatively small laboratory in DRDO with staff strength of about 300. While there were excellent facilities available in the laboratory, not much of basic research was being carried out, except in one or two groups. I found out later that research was not a priority in many DRDO laboratories. I introduced a research culture in NCML by asking two of my senior colleagues to register for their Ph.D. in IIT Bombay and IIT Kharagpur in metallurgy. Both of them got their degrees within three years. Another scientist obtained his Ph.D. from IIT Bombay for his work on PZT transducers and he was able to get some good publications in international journals. For the first time, I made the paints group to publish their work in an international journal. I initiated some new programmes on naval bronzes, brasses and molybdenum-bearing steel. I was happy that in a short period of 3 years, I had changed the complexion of the laboratory.

I was the chairman of the Materials Advisory Committee for the 'Advanced Technology Vessels' programme. I was responsible for the Indian Navy rejecting the age old AK 25 steel as the hull material which the Russians were recommending for our nuclear submarine; they then came out with a much better material called AB grade steels with very good mechanical properties, weldability and corrosion resistance. The Defence Metallurgical Research Laboratory later on indigenised these grades of steels. I consider the induction of AB steels as my major contribution to the nuclear submarine programme.

One day in March 1988, Arunachalam told me that I should come to DRDO Head Quarters to takeup the position of Chief Controller R&D. I resisted initially as this was a pure desk job, but I yielded. At one stage, I was assisting more than half of the technical laboratories in DRDO. I was also in charge of the human resources development. But, I got bored doing this desk job as CCR&D and requested Arunachalam to send me back to a laboratory. He gave me a couple of options and I chose to be posted as Director of the Gas Turbine Research Establishment at Bangalore. I was not a gas turbine specialist, but materials problems of the engine contributed to more than 50 % in the engine development. Kaveri design incorporated the use of a variety of nickel base super alloys and titanium alloys. Thus, I was sure of making my contributions in the development of the engine.

Unlike NCML, GTRE, a major systems laboratory had more than 1500 employees had built the after-burner for the Orpheus 703 engine, but that was not accepted by the Air force because of its overweight. It had built two technology demonstrator engines designated GTX 37-14U and GTX 37-14UB, the former being a straight jet, while the latter is a bypass version. This had given confidence to GTRE to bid for building the flat-rated gas turbine engine (Kaveri) for the Light Combat Aircraft. The design of the Kaveri engine was based on a mixture of Eastern and Western philosophies. Obviously, several problems did surface. Prior to my joining GTRE, reputed jet engine manufacturers such as Rolls Royce, Snecma and General Electric had shown interest in participating in our engine development programme, but did not materialise. After my joining GTRE, Mr. Peter Chipporus, a retired chief designer from GE came for discussions and based on his suggestion, the number of stages in the high-pressure compressor was increased from 5 to 6. Also, it was brought out that the engine design dimensions specified correspond to the operating conditions and one needs to take into account the thermal as well as the centrifugal expansion to arrive at the dimensions for manufacturing. Another problem pertained to dimensional distortion due to residual stresses arising out of welding/machining. I assisted in developing a stepwise annealing schedule to reduce the distortion to the minimum. Most of the super alloys required for the manufacture of various discs and blades were initially imported, but later MIDHANI developed all the alloys which were certified by CEMILAC, the agency for certification of materials for airworthiness. From the designers' point of view, it was necessary to generate adequate data of the materials in use and hence a separate facility, Aeronautical Materials Testing Laboratory, was established at Hyderabad near Midhani.

While most of the engine testing facilities was available, we had to go overseas to get some specialized testing done. Russia offered these tests at a lower cost, but procedural delays were considerable. GTRE had 5 engine test beds and it was used to test the prototype only once a day. I suggested that it has to be done more number of times in a day or otherwise we may not be able to meet the project schedule. And it was done. My moment of happiness was when the first prototype of Kaveri was test run in 1994.

I should also mention here that even though we had turned down the offer of collaboration from other engine manufacturers earlier on, in one of my visits to Moscow in 1994, our ambassador in Moscow Mr. Ronen Sen told me that after Glasnost, Russian experts are available at a paltry honorarium of US$1000 per month. He said that if I need a couple of them, he would arrange to send them to GTRE. I did not have the authority for this and the ambassador's suggestion did not fructify. If that happened, then may be Kaveri would have been flying in LCA now.

I had just 6 years in GTRE and personally felt that part of the problem that GTRE was facing in the project work was due to the lack of adequate research experience on the subsystems of the engine. Having worked in BARC, which has a high level of research culture, I initiated a few basic research programs. The then DG, DRDO A.P.J. Abdul Kalam, wanted me to continue beyond my retirement date (31 December 1995), but I had told him a year earlier that I would quit on reaching my superannuation date which I did.

Post-retirement

Post-retirement, I assisted a NRI in setting up a laboratory named 'Shiva Analyticals' for chemical and metallurgical analysis at Hoskote, near Bangalore. I was the CEO, drew up the plans, got the building built, procured the necessary equipment and recruited scientists to run the laboratory. As there was a cash flow problem, the metallurgical analysis part was getting delayed, and hence I quit the organization (1998), because Chemistry was not my forte.

I then assisted Agastya International Foundation (www.agastya.org), an NGO created to inspire a scientific temper in disadvantaged rural school children. My part was more in training the rural teachers. I still continue to be associated with Agastya but to a lesser extent.

In the meanwhile, V.S. Arunachalam who founded a not-for-profit NGO called 'Centre for Science, Technology and Policy' (www.cstep.in) asked me to join them as an advisor. I worked there for a couple of years during 2011–2013 and assisted them in writing one report on 'Rare Earths and Energy Critical Elements' for the Ministry of Mines, Government of India and another report on 'Engineering Economic Policy Assessment of Concentrated Solar Thermal Power Technologies for India' for the Ministry of New and Renewable Energy.

I should also mention my association with K.K. Sinha and V. Ramaswamy (both Fellows of INAE) in bringing out a document titled 'Successes and gaps in metallurgical R&D in India' for the Indian National Academy of Engineering. I enjoyed writing this very much. Presently, I am interested in 'High Entropy Alloys' and 'Low Energy Nuclear Reactions'.

Concluding Remarks

I would like to mention that for a successful career, one needs to be lucky to have good bosses. I was lucky in having understanding bosses such as M.K. Asundi, Brahm Prakash, C.V. Sundaram, P.K. Iyengar and K. Balaramamoorthy in BARC and V.S. Arunachalam in DRDO. It does not mean that I was always lucky but I survived.

Commercialization of Indigenously Developed Technologies

M.O. Garg

Introduction

The current rate at which scientific and technological changes and innovations are taking place worldwide is creating many challenges while providing enormous benefits to humankind. Science and technology are closely interlinked with our lives, and the studies and developments in both these areas are essential for human development. We all are also aware of the positive impact on our lives brought about by the advancements in these fields.

Technology commercialization is a process of transferring know-how generated in the research centers to the industry with a business model of increasing profits. Hence, technology commercialization has to be thought of in relation to appropriate technology transfer activities to increase the success rate of technology commercialization.

The successful commercialization of new technologies and processes from laboratories to industry in order to reap fruits of investments in R&D for the larger benefit of the society, however, remains an area of concern.

The Importance of Technology Commercialization

In a society based on knowledge, technology innovation is being appraised as the most important factor for national competitiveness and corporate competitiveness. As a result, governments of various countries around the world are becoming increasingly involved in technology innovation with a keen interest and active involvement in technology commercialization. Technology innovation can be seen

M.O. Garg (✉)
Indian Institute of Petroleum, PO IIP, Mohkampur, Dehradun 248 005, India
e-mail: director@iip.res.in

P. Ghosh and B. Raj (eds.), *The Mind of an Engineer*,
DOI 10.1007/978-981-10-0119-2_42

in two ways: technology development and technology commercialization including technology transfer (technology spread concept). Domestically and internationally, up until the 1980s, it was believed, in line with R&D expansion, that technology development policy would bring technology innovation. However, at present, the recognition that developed technology itself is not the result but that technology must create additional value through proliferation and commercialization has set in, and technology commercialization policy has taken the core position of industrial technology policies of each country.

The global position of any country depends on its technological development. It is essential to develop and implement indigenous technologies in the country to realize true and sustainable independence. India is an expanding economy and expected hub of manufacturing industries. Considering this, there is a tremendous potential to develop and commercialize indigenous technologies in India. Research institution (academic, corporate) and industries and supportive government policies can play an important role in commercialization of indigenous technologies.

Research Institutions in Downstream Oil Sector in India

In view of the authors' expertise in the downstream sector, this document would focus on issues related to technology development and commercialization in the downstream oil sector. However, the issues that have been discussed are of broad nature and are equally applicable to the upstream sector without any exception.

There are various institutions in India who are currently engaged in research and development in downstream oil and petrochemical sector. These are as follows:

- Corporate Research Centers
  - Indian Oil Corporation Ltd.
  - Bharat Petroleum Corporation Ltd.
  - Hindustan Petroleum Corporation Ltd.
  - Engineers India Ltd.
  - Gas Authority of India Ltd.
  - Chennai Petroleum Corporation Ltd.
  - Reliance Industries Ltd.
  - Etc.
- National Laboratories
  - CSIR-Indian Institute of Petroleum, Dehradun.
  - CSIR-National Chemical Laboratory, Pune.
  - CSIR-Indian Institute of Chemical Technology, Hyderabad.
  - CSIR-Central Institute of Mining and Fuel Research, Dhanbad.
  - CSIR-North-East Institute of Science and Technology, Jorhat.

- Academic Institutes
  - IIT Kanpur, IIT Mumbai, IIT Delhi, IIT Guwahati, IIT Kharagpur, IIT Chennai, and IIT Roorkee.
  - Indian Institute of Science, Bangalore.
  - Institute of Chemical Technology, Bombay.
  - The National Institutes of Technology.
  - BITS, Pilani (various campuses).
  - Other universities and institutes.

The academic institutes broadly focus on the basic research, some of which is related and is of interest to the hydrocarbon industry. Most of this research ends up in publications in peer-reviewed journals and may be of use in the industry, perhaps, not in the near future. Several of these institutes have collaborated with the industry or the corporate research centers to undertake joint research leading to the development of a technology. In this case, while the academic institutes provide valuable inputs in basic sciences, the industry/corporate research centers provide the vital input for its industrial application.

The corporate research centers have largely concentrated on industry-specific short-term research which is of immediate application. Several of these corporate laboratories also provide valuable support to the corporate operations in terms of evaluating catalyst, troubleshooting, value addition to certain streams, and high-end technical services. Several technologies have been developed by these laboratories and are available for commercialization.

The national laboratories such as Indian Institute of Petroleum, National Chemical Laboratory, and Indian Institute of Chemical Technology have invested heavily since the middle of the twentieth century in creating a world-class infrastructure which is designed to carry out both basic and applied research and technology development. In this respect, CSIR established the Indian Institute of Petroleum in 1960 with the primary aim to support the oil industry in terms of cutting-edge research and technology development. Over the last 54 years, this institute has developed and commercialized very large number of technologies, some of them against international competition.

Networking

There is an attempt to network among the above organizations with the purpose to leverage expertise across the organizations with the primary aim to develop a technology. There have been both successful and unsuccessful attempts in the past. The successful attempts such as collaboration between EIL and IIP, EIL and IIT Kanpur, EIL and IIT Mumbai, IOC R&D and IIT Delhi, etc. have led to the development of cutting-edge technologies. There is an urgent need to felicitate a

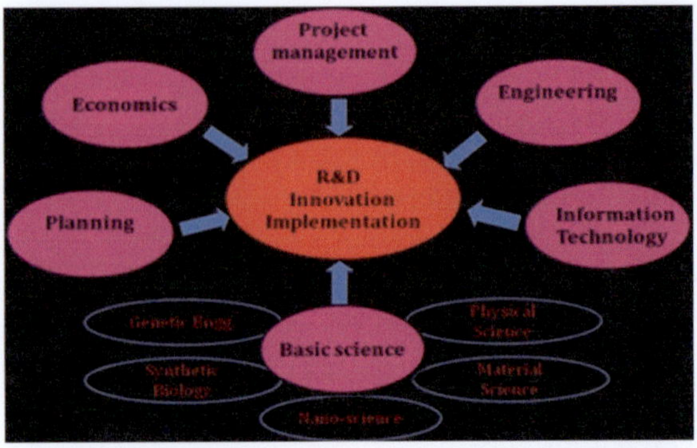

Fig. 1 Diverse skills needed—not available under one roof

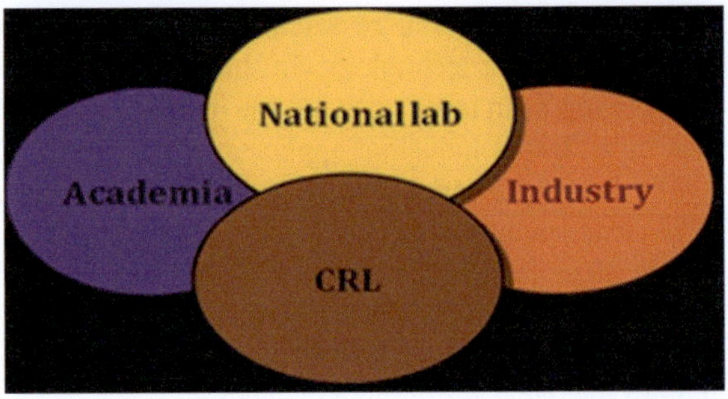

Fig. 2 Networking among academia, research centers, and industry

system by which this networking becomes more natural and productive. There is also an urgent need for collaboration between the corporate research laboratories and the national laboratory system; in fact, this collaboration can accelerate the development and commercialization of technologies; some of these technologies can find a place in the organization itself. Also, as depicted in Figs. 1 and 2, in view of the fact that in today's research, a large number of skills are required which are of diverse nature, and such collaborations are indeed a necessity to create a useful technology or a product. Above all, such collaborations are essential to create a critical mass required to deliver a competitive technology.

Funding of Research

At present, there are several agencies which fund research as well as have funds for technology development and also have schemes for demonstration of the technologies.

These funding agencies are listed below:

Government Agencies

(a) Department of Science and Technology (DST);
(b) Department of Biotechnology (DBT);
(c) Ministry of New and Renewable Energy (MNRE).

International Funding

DST has agreements with various countries around the world for collaborative research. Some of the major countries are as follows: USA, UK, Germany, Japan, Australia, Norway, etc.

There are calls for proposals at frequent intervals highlighting the areas of interest by the two countries. The academic institutes and the national laboratories are encouraged to apply. This is a huge opportunity for leveraging international expertise to develop world-class technologies.

Funding by OIDB

OIDB by itself and through its various agencies such as CHT, PCRA, and DGH funds research encouraging development and commercialization of technologies. CHT through Scientific Advisory Committee to the Ministry of Petroleum and Natural Gas has funded a very large number of projects over the last 35 years; a few of them have already been commercialized.

Corporate Research Centers

The corporate research laboratories such as Indian Oil Ltd., Bharat Petroleum Corporation Ltd., Hindustan Petroleum Corporation Ltd., and Engineers India Ltd. provide their own source of funding for their research. Several of these collaborate with academic institutes to leverage their expertise. In this case, the corporate laboratories provide funding to the academic institutes/national laboratories.

It has been seen that there has been no dearth of research funding. In fact, over the last 10 years, it has been seen that there is an acute shortage of projects and proposals which are eligible for funding. Overall, there has been an acute shortage of new ideas for funding.

Commercialization of Technologies

As mentioned above, it is extremely important for the true independence of a country that it commercializes its own research as well as commercializes the same worldwide. In fact, the ultimate purpose of any scientific discovery and invention is in its commercialization and that is the ultimate dream of any scientist. This requires a seamless integration of science, engineering, and economics.

The successful commercialization of a technology has its roots in the following:

- Innovativeness;
- Should survive strong competition;
- Should address volatile market;
- Should provide profits even in thin margins;
- Should meet dynamic product specifications;
- Should meet tight timescales; and
- Should be environment-friendly.

The planning at the inception of technology development should take care of the above aspects on a long-term basis.

Major Concern of Customers

There are several issues which act as roadblocks for commercialization of indigenous technology. The first and foremost is the requirement of our industry for a proven track record. It has been seen that whenever the industry puts out a tender or an enquiry, one of the essential requirement is that there should be two or three units which are operating with similar capacity. This has a highly negative impact on commercializing anything new. Also, the industry looks for a proven track record with similar feed and similar capacity. However, there is an indigenous technology which has been commercialized, but since it may not be processing similar feed or is not of same capacity, it does not qualify to compete.

The industry also has a perceived notion that the indigenous technology is not state of the art and is inferior to an imported technology. This is a very strong negative perception in the minds of the people in the industry. There is a general feeling that if it is indigenous, it has to be second rate.

The industry also has a concern whether the developers of indigenous technology will be able to provide a long-term support to the technology and are capable of hand-holding and troubleshooting whenever required. In fact, they are even concerned about whether the developers would continue to invest in the further development of the technology to keep it competitive throughout its life cycle.

Issues in Technology Development Which Need to Be Addressed

While some of the above concerns of the industry could be genuine, several of these, however, are only their perception. It is, however, important to note that the developers of the technologies should offer a complete technology solution. Such complete technology solution is indeed available to the industry from the foreign licensors. There is a need by the developers to address such concerns. In fact, in addition to the development of core technologies, the developers should also focus on developing enabling technologies such as material of construction, process engineering, catalyst regeneration, analytical support, effluent treatments, environmental impact, and catalyst disposal. This is just a short list of areas which need to be looked at by the developer and a solution offered as a complete technology package.

It is often seen that the developer announces the development of a technology just because he has developed a catalyst. This is indeed a beginning. The development of the reactor system, the flow sheet, and the complete basic design engineering package is what constitutes a technology.

Barriers/Constraints

There are genuine problems and constraints faced by the developers in India while developing and offering a complete technology solution. As compared to the developed nations such as Europe and USA, there is a complete lack of facilities in India for demonstrating the development know-how. Unlike the West, we lack heavily in pilot plant facilities. There is a need to create several general-purpose pilot-scale facilities which can be used to demonstrate technologies. There is a lack of cutting-edge engineering research which can enable high-end engineering designs, and overall, there is a lack of engineering support. Also, the country lacks in database for carrying out technocommercial feasibility of new research ideas. Above all, there is complete lack of facilities in the country for scale-up of indigenous catalyst and its manufacturing.

While the perception of the industry may be real, it is also important that the developers are not provided with the right kind of infrastructure and support to be able to overcome not only the concern of the industry but also the barriers and constraints faced by them in commercializing research and to run that most important last mile.

How to Make It Happen (Commercialization of Indigenous Technologies)

The research and development activities carried out by the corporate research laboratories, national laboratories, and academia, either individually or in collaboration, can be clearly divided into the following three categories:

1. Technology for production of certain products and chemicals.
2. Commercialization of products such as catalyst, adsorbents, and additives.
3. Specialized technical services such as energy optimization, product improvement, and training.

The following paragraphs provide the foundation on which government policies can be framed and enforced which will lead to increasing commercialization of indigenous developments and know-how. The discussion is with respect to the above three categories.

Technology for Production of Certain Products and Chemicals

There are a very large number of technologies, which have been developed, and several of these have also been commercialized. There is a need for separate policy and government intervention based on the following:

- **Technologies which are already licensed and commercialized**

National laboratories such as IIP, NCL, and corporate research laboratories of Indian Oil Corporation Ltd. and Bharat Petroleum Corporation Ltd. have already commercialized a few technologies; the first to be commercialized was the IIP/EIL sulfolane-based BTX extraction technology at BPCL, Mumbai, in the early 1980s. Since then, several technologies related to solvent extraction, visbreaking, delayed coking, FCC, hydrotreating, etc. have been commercialized. It is seen that in spite of the availability of a successful and proven indigenous technology, the industry keeps on repeatedly importing these technologies from abroad. Sometimes, these technologies are imported under the pretext that the indigenous technologies are not processing the same feed or are not of similar capacity. These are simple technical issues which can be addressed and overcome, since, as a part of overall development of a technology, these are naturally addressed. On other occasions, it is seen that the already commercialized technology is repeatedly imported because it is a component of a larger complex on which the industry seeks guarantee from the licensor. With the right policies and the intervention of the government, this can be easily avoided and resolved.

It is suggested that the government should not allow any import of technology if an indigenous technology is already operating successfully and has a proven track record. Needless to say, the developers of these technologies secure the same by taking IPRs and over a period of time try to build an impressive portfolio which should be taken notice of.

- **Technologies which are ready to be commercialized**

There are several technologies that are available which have been developed at the bench/mini pilot-scale level and can be easily scaled up with zero risk. Generally, these technologies are ignored because of the requirement of proven track record. Given the right kind of encouragement by the government, these technologies can be brought to the marketplace, and in fact, once proven, these have the potential to be marketed worldwide. It is suggested that in order to provide the incentive to the industry to accept and commercialize such technologies, the government should announce zero custom duty on imported equipments and zero excise duty on indigenous equipments that are procured to commercialize these technologies. This would provide strong economic incentive to the industry. In addition, OIDB can provide escrow or hedge funds to cover any perceived risk in the unlikely event of non-performance of the technology.

- **Technologies requiring scale-up**

There are, indeed, several technologies which have been proven at the laboratory scale and also at the mini pilot scale, but need a large-scale pilot plant for demonstration and to generate scale-up data. These technologies generally involve a catalyst, and it becomes necessary to run a demo unit for long duration in order to establish the effectiveness of the catalyst as well as any regeneration requirement, etc. It is suggested that funding agencies such as OIDB and CHT should identify such promising technologies and fund the creation of such pilot plants preferably within an operating unit or a refinery so that the feedstocks are easily available as well as the products can be usefully utilized. This would also reduce the cost of the pilot plant since the infrastructure and utilities would be available from within the refinery. Moreover, since the pilot unit will be operating inside the refinery, there is a strong possibility of the industry adopting the technology.

Commercialization of Indigenous Products

The research and development units of the oil sector as well as the national laboratories have developed a very large number of specialty products such as catalyst, adsorbents, additives, and performance chemicals. It is suggested that proper policies should be put in place to commercialize such products after due diligence and comprehensive performance comparison with imported products. Once these products are commercialized and are proven in the industry, the government should make a policy of not importing these products but insist on the developer to continuously innovate and develop such products in line with international standards.

It is seen on several occasions that however, the catalyst developed indigenously has been used very successfully by the industry, but the procurement department of the oil industry still tries to go for competitive bidding and procures an imported catalyst based on pricing. I believe that the pricing of an indigenous product can be negotiated and the developer would be more than happy to match the price of his product with the corresponding international product. It is important to mention that India should be developed as a manufacturing hub of such specialty products, such as catalysts and chemicals so that the same can be exported to both developing and developed economies of the world.

Specialized Technical Services

The oil industry, to remain competitive, seeks a number of specialized technical services from both India and abroad. In view of the high energy cost, large emphasis is on energy efficiency improvement, performance improvement, global benchmarking, etc. For this purpose, the industry seeks services from companies such as Shell Global, Solomon, and Japanese Consultants. Within the country, companies such as Engineers India Limited and IOC R&D, and national laboratories such as IIP and NCL have a large database and expertise which can be utilized. In case such study is conducted for the first time, it should be made mandatory to involve Indian organizations along with the foreign consultants to absorb the know-how and to develop it further, specifically dovetailing it to our requirements. A very successful example is the transfer of advanced control technology by EIL/CPCL from SETPOINT in the late 1980s. With the right encouragement, the Indian companies would be able to come up to the satisfaction of the oil industry and moreover the entire knowledge and the know-how, that otherwise is given away by the industry to the foreign consultant can be avoided. It is important to note that the foreign consultants while doing the projects learn tremendously from us and enrich their own data bank. On the top of it, they charge for their services. Once the Indian companies develop this expertise, the same can be also offered to several refineries at least in the developing economies.

The Indian Institute of Petroleum has made a remarkable contribution in terms of training oil industry personnel in various subject areas. Such activities should be encouraged, and training of people abroad or bringing the foreign trainer should be avoided to the extent possible.

How Does India Lose (If Indigenous Technologies Are not Encouraged)

It is important to discuss the fallout of not encouraging commercialization of indigenous technologies and repeated import of foreign technologies. Although the oil industry might debate the need to import technologies to remain competitive at a world level, it is important to realize that this would cripple slowly and finally, permanently the research base of the country. As mentioned before, the only dream of a scientist is to see his/her technology commercialized. In the absence of support from the industry, the scientist will take an alternative route to glory in terms of publishing their research, thereby proving how good they are individually. The research that they would publish will be picked up by companies abroad who will then convert them to a technology and the same would be exported back to India for which we will have to pay huge license fees. This also destroys the team spirit and a team effort, which is required to develop a technology. Publishing papers is more often an individual pursuit and does not, to a very large extent, contribute to the development of the nation and its independence with respect to a technology.

Development and commercialization of technologies also encourage the developers to protect their invention by taking patents in India and around the world. All developed economies and the companies in such economies, such as Shell, ExxonMobil, UOP, AXENS, IFP, Chevron, and Lummus, have a huge portfolio of patents worldwide. They also aggressively protect their intellectual property, many a times, even with litigation, in case they discover infringement of their IPR. IPR or patent is a true indication of the country's wealth in today's economy. Repeated import of technologies will systematically end this effort by the research organizations.

It has been seen in the past that in some sectors of the economy, namely defense, nuclear, and space, there has been technology denial and these sectors have strived to expand their own research base to overcome such a barrier and the success stories are several and known to everybody. Time is not very far when India may be denied cutting-edge technologies in the energy sector as well, particularly those which are green and can replace fossil fuels effectively and sustainably. This may happen perhaps three decades from now, but by that time, we would have already crippled our R&D infrastructure and reduce to a paper publication machine. It is about time that we do not allow our research institution particularly the one in the corporate research laboratories and the national laboratories to degenerate. This, therefore, needs a strong intervention by the government to support and encourage indigenous technologies.

Finally, I would like to add that strong support to indigenous technology by the industry will encourage the developers to bring new and innovative technologies at a faster pace to the marketplace without the fear of being rejected for one reason or the other. This, in turn, will create a strong innovation system in the country and develop and support the complete research infrastructure, which we so dearly lack. It may be mentioned that in the energy industry, it is extremely difficult to predict future technology innovations; a comprehensive innovation-based technology supply ecosystem is thus absolutely essential to survive around the year 2050 and beyond.

Conclusion

India is the fourth largest processor of crude oil with 22 refineries processing close to 235 MMTPA of crude oil. We export more than 70 MMTPA of petroleum products. Today, India is an acknowledged hub of petroleum refining, petrochemicals, and other specialty products. All of this is based on imported technologies. It is high time that India promotes its own technologies to manufacture petroleum products and petrochemicals and export these technologies and licenses them to companies abroad.

India also has a very large technical base in terms of highly skilled manpower as well as large number of academic institutions at the level of IIT, etc. as well as a strong system of national laboratories. Also, the industries have state-of-the-art research laboratories. In spite of such facilities available in the country and the highly skilled technical manpower, it is extremely unfortunate that we still keep on importing technology from abroad. It is about time that the government creates policies and provides intervention in order to facilitate commercialization of indigenous technologies, products, and technical services. As brought out above, it is suggested that appropriate incentives need to be provided to the industry in terms of tax-free import of equipments as well as exemption of excise duty on equipments which are used to commercialize indigenous technologies. It is also suggested that a high power committee chaired by the Secretary, Ministry of Petroleum and Natural Gas, may be constituted in order to oversee the need for import of any technology that is being done by the oil industry and to suggest how to avoid the same by commercializing indigenous technology in case there is an opportunity. The committee should also direct OIDB to create an escrow or a HEDGE fund to cover the perceived risk to new technologies which the industry might like to commercialize. Such funds have been created in the past and should again be revived across all sectors.

Non-support of indigenous technologies can be fatal for the economy in the long term and can also render the Indian hydrocarbon industry obsolete. Further, it will kill the innovation system of the country.

Hydrodynamic Cavitation Technology: Industrial Applications

A.B. Pandit

Cavitation is a physical phenomenon associated with three aspects: formation, growth, and collapse of vapor or gas–vapor bubbles within the body of a liquid due to variations in local static pressure. Decreasing the pressure over a liquid and bringing it to its vapor pressure at the operating temperature generate vapor bubbles in the liquid. When the pressure is brought back to normal, these vapor bubbles collapse with a bang to generate intense pressure and temperature at the point of collapse (Fig. 1). Such intense conditions (5000 atm and 12,000 K, intense turbulence) and resulting shock wave can bring about several physical, chemical, and biological transformations, even when the bulk conditions are ambient. This release of energy can be harnessed effectively for bringing about chemical, physical, and biological transformation. Mumbai-based HyCa Technologies Pvt Ltd., along with ICT, Mumbai, has developed the technology to create and collapse precisely tailored cavitation bubbles to modulate the pressure, temperature, and turbulence conditions by means of controlled variations in the static pressure of fluid. This article describes few case studies where the company's HyCator® brand of reactor systems was gainfully employed in effluent treatment plants, cooling towers, particle size reduction, and biogas production enhancement application. The objective of this article is to make reader aware of the potential of the tailored cavitation to bring about the change most effectively.

Applications in Effluent Pretreatment

The HyCator® brand of reactor system has been used to intensify various physical, chemical, and biological processes occurring in effluent treatment plants in an energy- and cost-effective manner.

The advantage of this reactor system is listed out in Fig. 2.

A.B. Pandit (✉)
Institute of Chemical Technology, Nathalal Parekh Marg, Matunga, Mumbai 400 019, India
e-mail: ab.pandit@ictmumbai.edu.in

P. Ghosh and B. Raj (eds.), *The Mind of an Engineer*,
DOI 10.1007/978-981-10-0119-2_43

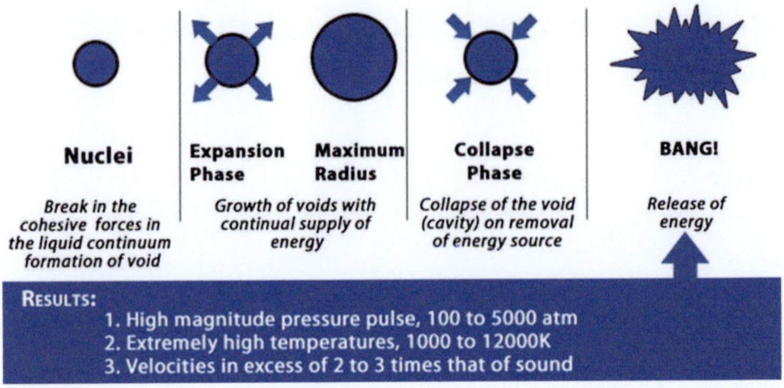

Fig. 1 Principle: hydrodynamic cavitation

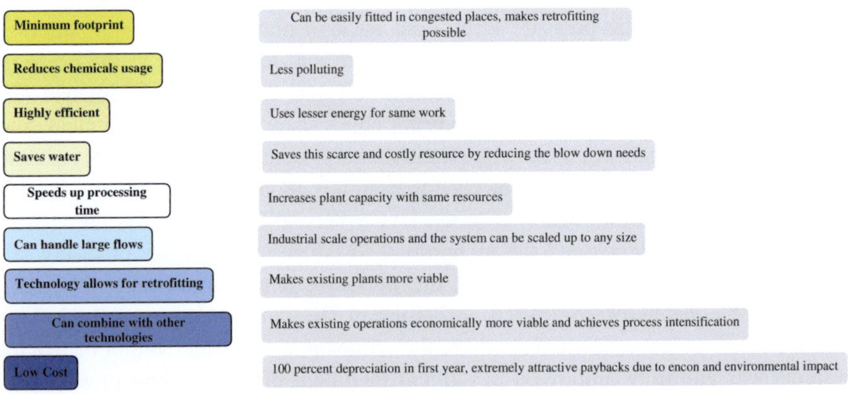

Fig. 2 Main features of HyCator® Reactor System

These reactor systems can be retrofitted to any existing effluent treatment plant to make the later more effective and efficient by reducing pretreatment time and costs, as well as by the reduction in the usage of chemicals in an environment-friendly way (Fig. 3).

The different types of reactor systems include the following:

Different reactor configurations have been developed to suit various applications, and the system can be customized to any scale of operation as targeted transformations. The HyCator®: OLM Reactor System generates tailor-made cavities for microlevel mixing. Length scales associated with cavitation are in the order of the diameter of a collapsing cavity, i.e., few nanometers to few microns, whereas in conventional mixers, the turbulent length scales are of the order of mixing element, i.e., few centimeters. Thus, cavitation is known to dissipate energy on the length scales required for micro- and molecular-level mixing which makes cavitationally induced mixing, a microlevel mixing. This makes cavitation-based HyCator®: OLM Reactor System much more energy efficient compared to conventional

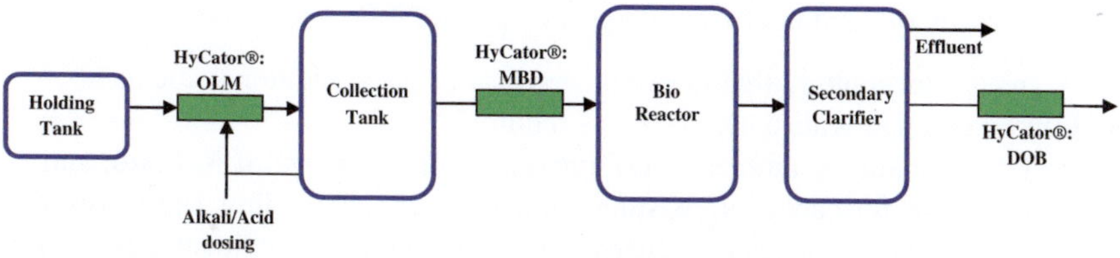

Fig. 3 Retrofitting stand-alone HyCator Reactor System into existing facility

methods of online mixing and can bring about the molecular-level mixing. Limitations do exit for highly viscous and viscoelastic fluids as the generation of cavities is more difficult in such liquids and the zone of effluence is limited.

In HyCator®: MBD Reactor System, the energy released during cavity collapse is harnessed for the breakage of the molecular bonds and generation of OH free radicals which are responsible for the oxidation of organic compounds in wastewater. The wastewater after passing through the reactor has much reduced COD levels, and biodegradability is also enhanced due to the breakdown of the larger biorefractory pollutants.

For HyCator®: MBD Reactor System used for effluent pretreatment, we first carry out multiple pilot trials on organizations' most representative effluent samples at our laboratory and set the trial protocol, and once achieving success, we request the organization to witness the trials. Conducting the trials as per the standard pilot trial protocol and conveying its analysis results will be the first step. Based on the results and observations, further course of action is decided. Without pilot trials, it is not possible to provide a full-scale solution as the effluent characteristics vary significantly.

HyCator®: DOB Reactor System helps to disintegrate (degradation and partial cell disruption) biomass in an energy- and time-efficient manner. This increases the activity of the biomass by defloccing the biomass making it accessible to oxygen and other substrates and nutrient, thereby reducing the generation of waste-activated sludge and speeding up further processing which are many times controlled by intracellular enzymes which are not easily accessible to the pollutants.

Advantages of Cavitation

Cavitation technology compares favorably with similar advanced oxidation processes (AOPs) like Fenton's process, wet air oxidation, ozonation and hydrogen peroxide treatment and ultrasonic/acoustic cavitation. These technologies require the addition of more chemicals, which in turn add to the effluent load that needs to be mineralized. Moreover, there is a requirement for higher bulk pressure and temperature as also longer processing times are needed on many occasions. Cavitation can also be combined with these conventional processes to bring about synergy between various processes.

Advantages of cavitation technology include the following:

- A greener technology that does not necessarily need additional chemicals;
- Can be coupled with other AOPs, if required;
- Bulk temperature is ambient; bulk pressure is in the range of 3–7 atm; and
- Enhances performance of existing effluent treatment facility (improves efficiency of aerobic reactor, increases biodegradability of effluent (BOD:COD ratio), reduces COD of effluent, etc.).

Similarly, cavitation-based reactor systems also compare favorably with other standard mixing technologies such as static mixing, jet mixing, and stirred tanks:

- It can operate with lower overall pressure drops and hence lower net energy consumption;
- It does not need a holding tank or static containers, since mixing is done online. Hence, it has low footprint;
- Mixing takes place on microscale, making it energy efficient;
- Can be designed and operated practically for any pressure and flow rate; and
- Can be fabricated in any material of construction for high wear and tear, corrosive resistance, and high-pressure and high-temperature applications.

Benefits of HyCator DOB Reactor System include the following:

- Increases the activity of microbes by partial disintegration and total deagglomeration of biomass resulting in high rate of reactions for acidogenesis, acetogenesis, and methanogenesis;
- Floc deagglomeration leads to better mass transfer;
- Minimum 8 % and a maximum up to 30 % increase in digester performance;
- Treatment of huge volumetric sludge streams;
- Continuous operation at varying sludge properties; and
- Stability against reactor blocking (sludge impurities).

Case Studies in Effluent Treatment

Improvement in COD Reduction Capacity of Bioreactor System

A Common Effluent Treatment Plant near Mumbai was operating two biotowers (A&B) for reducing the COD of partially treated effluent streams. COD reduction in biotower A was 40 % and in biotower B was 34 %, but even with this, the COD of the exit stream of ETP was not under the specified limits of discharge. Other alternatives to achieve this were to increase the size of the subsequent bioreactor or the residence time, i.e., reducing throughput or ozonation. All the alternatives required substantial modifications in the existing system or needed the addition of chemicals.

HyCator®: MBD Reactor System was installed in the inlet effluent stream of one of the biotowers (B) for increasing the biodegradability of the effluent. A detailed study was conducted on the biotower system to evaluate the performance of the installed device for COD reduction, biorefractory breakdown and oxidation, disintegration of biomass, and intensification of bioreactors. This was achieved by introducing air in the HyCator reactor, giving preoxygenated effluent to the biotower. The COD reduction in the biotower B increased from 34 to 54 %, at a mere additional operational cost of Rs. 0.32/m^3. The exit COD was reduced to within discharge limits. The feeding of the preaerated (preoxygenated) effluent to the bioreactor increased the biological activity of the immobilized microbes in the bioreactor without the need of the additional aeration vessel.

COD Reduction of Viscous, Partially Polymerized Glycerin Foot

A company was having trouble in treating a viscous, partially polymerized glycerin foots (distillation residue from glycerin distillation) stream. Although the stream was biodegradable, it would need extremely long hydraulic retention time (HRT) if treated in regular aerobic reactors, as it was partially polymerized and had high COD (170,000–50,000 ppm). Other options available were to use a bioreactor, but due to high viscosity and the presence of long-chain molecules, the residence time (volume) required in the bioreactor would have been very large.

HyCator®: MBD Reactor System with aeration (inside or outside) was recommended and designed for treating the same. Almost 70–95 % reduction in COD was achieved cost-effectively without any addition of chemicals such as H_2O_2, which would otherwise have been required to partially reduce the COD in a conventional ETP. The effluent stream was not required to be diluted before the subsequent aerobic treatment using the activated sludge; thus, considerable water saving was also achieved.

Conversion of Nonbiodegradable Ethylene Oxide to Biodegradable Glycols

Ethylene oxide (EO) is released during tanker unloading, which is arrested by scrubbing it with water. EO being highly soluble in water, antimicrobial, and poisonous cannot be taken to regular effluent treatment plant (ETP) as it will destroy the biomass present in the bioreactor. In this case, 2 m^3 of water containing 20,000 ppm of EO is generated and needs to be treated before it could be discharged. Conversion of EO (nonbiodegradable) to glycols (biodegradable) by conventional process requires very high temperature (>150 °C) and high pressure (30 kg/cm^2).

HyCator®: MBD Reactor System was recommended to treat this effluent stream. After successful pilot trials, a plant-scale HyCator®: MBD Reactor System was custom-made for reducing the EO content cost-effectively from 20,000 ppm to less than 3000 ppm in just 16 h. By using HyCator®: MBD Reactor System and no additional chemicals, EO was converted to a readily biodegradable material (glycols), which is further easily mineralized in a conventional bioreactor in the existing ETP.

Application in Cooling Towers

A reactor system has been developed for generating tailor-made cavities suitable for particular applications such as molecular breakdown especially useful in preventing biofouling in cooling tower water. Due to extremely high temperature and pressure and intense turbulence in the HyCator®: BFP Reactor System, shock waves are generated that are capable of destroying microbes and algae. The HyCator®: BFP Reactor System shown in Fig. 4 is a stand-alone unit, which will take its feed from cooling tower sump, and the treated cooling water will be discharged either to the line going for process or back to cooling tower sump as a closed circulation.

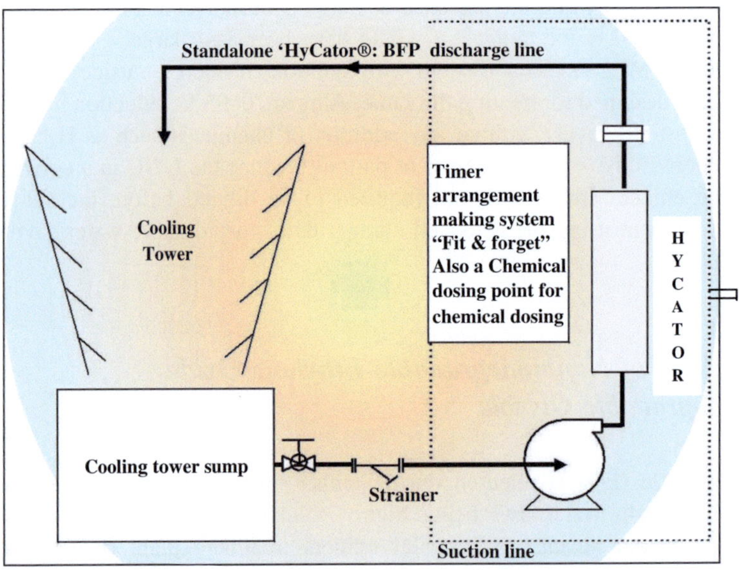

Fig. 4 Schematic of cooling tower circuit and stand-alone HyCator®: BFP Reactor System's installation

Potential Benefits

The reactor system prevents biofouling and reduces the blowdown to a very low value, which, in turn, will reduce the makeup water rearrangement. Once the frequency of makeup water is reduced, the addition of other chemicals will also reduce and other impurities due to the addition of chemicals will also reduce significantly.

In general, for normal feed water quality, the potential benefits include the following:

- Prevention of biofouling, corrosion problems, and scale formation;
- Environmentally safe: no chemicals added or unwanted residuals created by the process;
- No need of biocides;
- 40 % reduction in consumption of dispersants and corrosion inhibitors;
- Reduced blowdown of water due to operation at higher cycles of concentration, and the blowdown needs no treatment as there are no added chemicals.
- 40–80 % reduction in consumption of blowdown water.

Case Studies Involving Cooling Tower

Biofouling Prevention in Cooling Tower Water

Chemical treatment cost for biofouling and corrosion inhibition is high, and the plant management is always under pressure to reduce water usage and discharge. Cycle of concentration is also usually low. A hard scale often is formed in the summer season. The bacterial counts were 10^5 CFU/ml. HyCator®: BFP Reactor System was installed in the cooling tower circuit. A detailed study was conducted on the cooling tower system over a six-month period to evaluate the performance of the device for disinfection, scaling, corrosion, cycles of concentration, and heat transfer efficiency.

Makeup water consumption was reduced by 30 % and blowdown discharge reduced over 60 %. Bacterial microbial counts became nil, and cycles of concentration increased substantially. The results also indicated that the HyCator: BFP Reactor System treatment performed well compared to the chemical treatment without the addition of any chemicals. In this particular case, dosage of anticorrosion and scale prevention chemicals was also reduced significantly. Annual water saving exceeded 3600 m^3.

High Blowdown Water and Biofouling

The main problem with high blowdown water and biofouling is the requirement of the treatment of the blowdown water. The cooling tower is operated at low cycle of concentration. After the installation of HyCator®: BFP Reactor System, the biocides dosing was reduced to 10 % of the original and other chemicals (dispersant, corrosion inhibitor, and antiscalant) reduced to 40 % of the original. The bacterial counts came down to under the permissible limit. Blowdown was reduced by 40 %, and cycle of concentration was also marginally increased. Old scale was gradually removed, and no new scales were formed. The above cases clearly indicate the efficacy of cavitational treatment in the prevention of biofouling and reduced consumption of chemicals and makeup water due to reduced blowdown.

Application in Particle Size Reduction

In an innovative product HyCator®: PBD Reactor System, the mastered art of stimulating formation and collapse of such bubbles in the required manner and on the desired scale has been explored for size reduction. HyCator®: PBD Reactor System is fine-tuned for generating tailor-made cavities to grind the particles up to nanoscale level. Length scales associated with cavitation are of the order of a collapsing cavity, i.e., few nanometers to few microns, whereas in conventional size reduction equipments, the turbulent length scales are of the order of few mm. Thus, cavitation is known to dissipate energy on the length scales required for size reduction which makes cavitationally induced size reduction to nanoscale level. This makes cavitation-based HyCator®: PBD Reactor System much more energy efficient compared to conventional methods of size reduction.

As shown in Fig. 5, stand-alone HyCator®: PBD Reactor System can be fitted with minimal alterations in the existing system and be completely bypassed when needed. It requires minimal footprint and is ideal for place where safety issues are very stringent as an entire operation is concealed.

Comparison of Cavitation with Other Standard Size Reduction Technologies

Other similar size reduction technologies are as follows:

1. Colloid mills (e.g., ball mills, bead mills);
2. Disk mills;
3. Jet mills;
4. Rotor–stator mixers/high-pressure homogenizers.

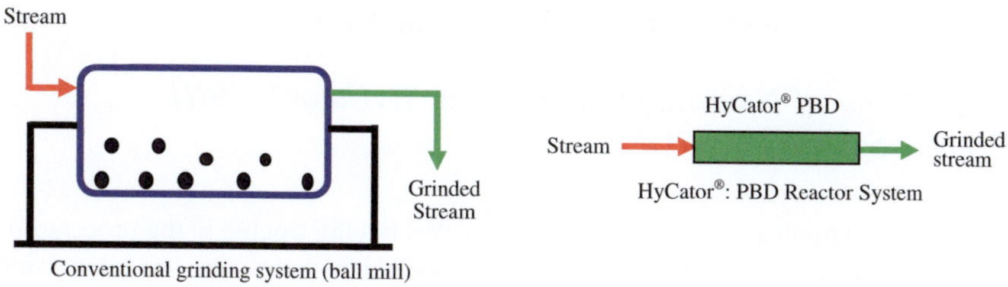

Fig. 5 Retrofitting HyCator®: PBD Reactor System into existing facility

These technologies face the following demerits such as

- Requires special equipment;
- Higher energy requirement;
- Operation is not cost-effective;
- Particle size may not be uniform.

Advantages of HyCator®: PBD Reactor System over above technologies are as follows:

- Online grinding possible;
- Can operate at lower pressure drops;
- Does not need any special equipment since size reduction is done online. Hence, it has low footprint;
- Need not have to use separate auxiliaries such as grinding media, etc., and hence, size reduction is energy efficient;
- Can be designed and operated practically for any pressure and flow rate;
- Can be fabricated in any material of construction for high wear and tear, corrosive resistance, and high-pressure and high-temperature applications.

Economic benefits of HyCator®: PBD Reactor System are as follows:

- Cost savings due to the reduction in energy consumption;
- No additional hardware requirement.

Other benefits are follows:

- Can be easily installed and needs only fitting it in bolted flanges at the required location.
- Requires no operational supervision and maintenance. But it may require slurry pump.
- Stream can be easily bypassed when required to bring in colloid/sand mill inline.
- Time saving.

Case Study of HyCator®: PBD Reactor System

Preparation of Nanosuspension Using HyCator®: PBD Reactor System

A renowned Mumbai-based FMCG company was having trouble in the process due to large nanosized (5900.0 ± 100.0 nm) suspended particles into the processing system and they wanted an absolute nanosuspension with particle size distribution as small as possible. Other alternatives to achieve this particle size distribution were to procure an extra grinding mill to decrease the size and to increase the residence time, etc. All the alternatives required substantial modifications in the existing system.

HyCator®: PBD Reactor System was installed in one of the units to prepare the nanosuspension. A detailed study was conducted on the existing system to evaluate the performance of the installed device for particle size reduction and intensification of process. By using HyCator®: PBD Reactor System, the particle size was reduced up to 300.0 ± 10.0 nm. The 90 % particle size reduction was achieved at an incremental additional operational cost and it did not need any maintenance.

Pigment Grinding Using HyCator®: PBD Reactor System

One of the Gujarat-based renowned dyes and pigment manufacturing company was using a ball mill for pigment grinding to reduce particle size from 0.66 to 0.53 μm. They were increasing substantially high manufacturing cost on account of huge power requirement for ball mill to achieve desired particle size as well as time

Table 1 HyCator®: PBD Reactor System versus bead mill for pigment grinding application

S. No.	Label	Cumulative treatment period required (h)	No. of passes	Mean dia. 90 % (μm)	Power consumed (Kw.h)	Operating cost (Rs. per liter) @ 5 Rs./kw.h
1	Sample (initial)	0	0	0.66	–	–
2	Particle size reduction using bead mill	144	9	0.53	8064	40320.00
3	Particle size reduction using HyCator®: PBD	59	20	0.53	3301	16506.00

Volume (batch size)—8844 l
Pump flow rate—300 l/h = 50 lpm
Motor power—75 Hp = 56 kW (at 60 % efficiency)

required for this unit operation was also high due to the requirement of multiple passes.

HyCator®: PBD Reactor System was proposed for grinding of pigments, and the organization achieved desired particle size reduction with 2.5 times lower power and treatment period that of previously required ball mill and substantially saved time and manufacturing cost as shown in Table 1.

Application in Biogas Generation Enhancement

HyCa Technologies has also developed HyCator®: BGG Reactor System to enhance biogas generation from anaerobic biodigesters. For an efficient utilization of feed in biodigesters, the disintegration pretreatment of digester feed process using advance technology is needed. HyCator®: BGG Reactor System has shown a positive effect on the degree and rate of digester's feed hydrolysis and ultimately on anaerobic digestion and has resulted in biogas production enhancement. By applying hydrodynamic disintegration, a controlled lysis of anaerobic digestive cells occurs in minutes instead of days. The intracellular and extracellular components are set free and are immediately available to supplement the biological degradation of the substrate which leads to an improvement in the subsequent anaerobic process. The cell of the activated sludge microorganisms ruptures and aids the digestion process leading to increased biogas production. Also, HyCator®: BGG Reactor System helps to disintegrate the larger-sized pollutant molecules into substantially smaller ones by the shear force, mechanical shock, and turbulence generated locally which help to degas the system so that pretreated feed gets easily digested further into digester, i.e., biodigestability of the feed is increased which also contributes to the enhancement in biogas production and reduction in the residual pollutant load leaving the digester.

Potential Benefits of the HyCator®: BGG Reactor System

- Increases the activity of microbes by disintegration of biomass resulting in high rate of reaction for acidogenesis, acetogenesis, and methanogenesis;
- Floc deagglomeration → better mass transfer of the substrates and the nutrients to microbes;
- Cell destruction → production of soluble chemical oxygen demand (SCOD) and proteins… → intensification of the anaerobic process;
- More biogas and less residual outlet COD;
- 8–30 % increase in the digester performance (VS degradation up from 42 to 54 %);
- Treatment of huge volumetric sludge streams;
- Continuous operation at varying sludge properties;

- Stability against reactor blocking (sludge impurities);
- Low maintenance;
- COD and color reduction in the outlet stream;
- Improvement in biodigestion and composting, etc.

Concluding Remarks

As it can be seen from the discussion and the details in the case studies, that cavitational transformation has a great future. The logic of delivering energy at the location of the transformation (be it biological, chemical, and physical) using the phenomena of cavitation has been exploited successfully. The mechanism of this energy delivery has been elucidated. The case study presented here does not limit the application of cavitation, and it has yet many more possible applications to explore.

The proposed techniques and the developed technologies have bagged the following prestigious awards:

- Gold medal at DST—Lockheed Martin Innovation Award for 2012.
- National award for the most innovative water saving product for 2011 from Govt of India/CII at the National Water Conclave, Jaipur.
- FE-EVI Green Technology Honouree from the hands of Hon Dr. A.P.J. Abdul Kalam on World Environment Day, 2011.
- Awarded the ET NOW/Bajaj Hindustan 'leap of faith' green entrepreneur of the year for 2012.
- Selected as a portfolio company of New Ventures India (a CII, USAID, UK Foreign and Commonwealth Office and World Resources Institute, Washington initiative) 2010.
- Showcased at Innovations India organized by IIT Bombay Alumni Association, Pune chapter and in Proto. in 2010.
- Showcased in the top five start-ups for the year 2009 in ET NOW.

R&D—Another Oxymoron?

Raja Manuri Venkata Gopala Krishna Rao

"R&D" is today's buzzword in the technological world, where these two activities are construed as inseparable. What I intend articulating herein, are my thoughts on this fondly debated topic today, engulfed by my experiences in the emerging areas of Composite Materials Science and Technologies that I pursued, in a highly demanding Aerospace Engineering Environment. It is a recall of what exactly these two terms meant to me, when I had to handle them in parallel (riding two horses at the same time), when I faced a dilemma (not a confusion) when to do the Research and when to do the Development. I also tried to figure out, what type of Research I was in, falling in between the Fundamental (postulative/hypothetical) and the frontier (Emerging/Futuristic). There I realized that, what I did could be most aptly termed the "Concurrent Research or Research Development Driven—RDD," running in parallel with the project objectives that I was mandated to meet for what I was hired and paid for. Implication was that we will deliver more than what was expected of us, for all that freedom and security we were blessed with. What resulted therein were a host of Scientifically Substantiated and externally funded Application Specific Indigenous Products, using Home-Grown Technologies. Publications, Patents, Academic Guidance, and Technical Training to Technology Transfer (TOT) were the by-products that came out in an "Engineering Sciences Laboratory," established under the prestigious CSIR Umbrella. Being a Chemical Engineer basically seems to have provided me the much needed "Multifunctional Character of an Interfacial Engineer," destined to work in a Multidisciplinary, if not an alien Environment. This article is dedicated to late Prof. R.B. Damania also a pilot, who initiated at NAL, the Program on "All Composite HANSA," the country's first DGCA-Certified two-seater light aircraft, for ab initio training of aspirant pilots. This program gave us the nuances of Polymer Matrix Composites for high-end applications also paving the way for many a subsequent nationally crucial programs such as Radomes and many.

R.M.V.G.K. Rao (✉)
Amritha Vishwa Vidyapeetham University, Coimbatore, India
e-mail: rmvgk@yahoo.com

R.M.V.G.K. Rao
"SAVITHRI", No. 1667, 16 'A' Main, 5th Cross, JP Nagar,
II Phase, 560 078 Bangalore, India

P. Ghosh and B. Raj (eds.), *The Mind of an Engineer*,
DOI 10.1007/978-981-10-0119-2_44

Passion and Purpose—The Two Horses

As I was trying to get over the *R&D Syndrome*, two questions stared at me with implicit answers, "Is Research for Research Sake?" With a "No," and is Development at the cost of Research? With a "Not at all." What dawned onto me was that these are like *two horses* one can ride with pride, when handled with both a passion to do so and a purpose in doing so, by walking that extra mile. Soon, it turned out that these two are in a way inseparable (when the deliverable was a visible Hardware, to be *scientifically substantiated*), but not interchangeable or synonymous (a gross *misnomer*, when this *combo term* is used casually/routinely). As a "Scientist Designate" in a Premier National Laboratory, mandated to take the Engineering Sciences to Industrial Applications, it then appeared a daunting task in a situation wherein the primary responsibility was to deliver the externally funded hardware, rather than pursuing a Passionate Scientific Research. And my conscience all the while haunted me, "Lo, you are the *Scientist First.*" So, there was a conflict between "Science and Conscience."

Research—Different Shades

"Research" as such is variously viewed as Pure/Fundamental/Basic, Applied/Application Oriented, the Forward Looking/Futuristic, or the Frontier, leading to Cutting Edge Technologies. So, where does one stand in this *galaxy of terminologies*? If "Research" was to mean mere exploration or a quest into the fundamentals of Pure Science (postulations/hypotheses for instance), the outcome remains dormant, but not necessarily meaning that it has no societal bearing. The immediate best it can do is to inspire many who venerate and value Research in its pristine form. The Researcher pursues what he enjoys passionately. It is for others to reap the benefits out of the resulting publications/seminal lectures, academic guidance, talks, celebrations, and so on, the hallmarks of his achievements. Also, it may have emerged out of curiosity to know why, as a cure for boredom. "There is no cure for curiosity" it is quipped. Such research can often lead to path breaking results of a Nobel Class. The best example of this that hogged the limelight in recent times is the discovery of "*The God Particle or Higgs Bosons that form very foundation of atoms,*" claimed to have taken us closer to the understanding of the origins of the physical universe, postulated in 1964 by Peter Higgs, a young physics teacher at Edinburgh and later Professor Emeritus after retirement (1996). Chance has it that in November 2013 (*half a century later*), Higgs an elusive professor who turned 84 learnt that he was awarded the "Nobel Prize in Physics," thanks to the Large Hadron Collider (LHC) experiments at CERN that confirmed existence of a particle consistent with Higgs Bosons. Happy and relieved at it, Higgs found "*this Nobel fame a bit of nuisance too.*" This passionate Higgs story (1964–1996–2013) can be likened to "Nishkama Karma" or "Disinterested Endeavor"—an anglicized term I picked up from my elder brother Prof. R.M.V. Raghavendra Rao, a litterateur par excellence and who directed my destiny toward an engineering profession. There are more parallels to Higgs discovery, as examples of "*frontier turned fundamental research.*"

The Applied Research with an implied end use has a different connotation in that the pursuer has an inherent interest in "Research," to apply his results remaining a driving force. This may even result in a Forward Looking Research (fanciful or purposeful again). The Researcher here has a priory a hope that someday down the line his results could be put to practical use. His research outputs are either widely published with adequate "*Experimental Data*" for posterity or patented for commercial exploitation (as in a Corporate R&D). This may even turn out to be a Frontier Research, for the future. Coming next is the Research Application Oriented, wherein the Researcher has clearly a process/product in mind, awaiting development, but delayed for want of decision makers and supporters (*Minds and Funds* typically with a phase lag). Here, when pursed with limited or no funds, his results will be "*Relevant*" *in* developing a Product with a purpose. In a way, this is *a vision*, since such a research output does form a precursor of the Product yet to be borne (*akin to a ready to use composite prepreg*), saving on effort, time, and associated costs (*like a straight line with an intercept*). Often, an in-house *pre-project funding* (PPF), however small it can be, provides well ahead, certain crucial inputs to an "*Impending Sponsored Project,*" also providing *continuity to the researcher's* penchant. The best example of this in the author's experience is the Development of the Large DWR-Radomes (Mark I/Mark II) for ISRO/IMD (more on this later). Thanks to the foresight/support from the then Director Dr. T.S. Prahlad, who while admiring researchers/publications, set the bottom line that scientists should not ignore *Lab's major sponsored programs*, to be executed in a la "*Mission Mode.*" "*Concurrent Research* (Research Development Driven—RDD) *presents* the case when a commitment is made to a funded project to develop and deliver a product, to be scientifically substantiated with well-identified/defined performance specifications, quality, timelines, and budgets (deadlines) hanging over the R&D Head like a *Damocles Sword*, with no room for excuses for delays". In a way, the promised deliverable sets the fixed boundary conditions (like in an unrealistic *idealized linear analysis model*). Thus, a passionate researcher has to put the *Product above the Paper*. By contrast, the so-called Sponsored Research, again from Government bodies, may not invariably have resulted in an end product.

Of all the shades of the research, the ***RDD*** showed the quick returns on balanced investments (intellectual and infrastructural), in typical government backed/run laboratories, as it is linked all the way, to an end product. Doing such RDD under the CSIR charter clearly demonstrated that "Research here is *not open-ended*, but *ends with a product.*" Here, the R&D appears on the same sheet (if not on the same page).

R&D Performance—Yardsticks

Doing the "R for R sake (*Publications*) and the D for D sake (*Products*)" is a thing with no ambiguities and misgivings. A genuine R&D in a Scientific and Industrial Establishment can be understood only through certain yardstick/performance indices, when it comes to evaluating one's performance. These constitute publications, presentations, patents, invited talks, seminar lectures, and academic guidance for the *R-Component*; while the Product—delivered with attendant progress reviews/deadlines, and compulsions of certification and acceptance from the sponsor's end—constitutes those of the *D-Component*. Then, how much the *R&D Scientist* needs to score on both these counts is a mute question. This turns out to be a gray area, governed by both subjectivity and objectivity, with no exact formula in sight yet. Different perceptions naturally emerge here. For some, it is difficult to publish because of the classified nature of their work, patenting becoming even more difficult when the sponsor's endorsement/clearances become an issue, thereby deterring them from doing enough in terms of meeting the yardsticks of the R&D put together. Also, time being a constraint in developing the *hardware in return to the price tag attached* to it, taking time off to "*indulge in publishing,*" of what they can and what they cannot. This is much like having variables more than the equations. *But,* for one, immensely smitten by the feeling that being a scientist with an innate obsession for research/publications (not for publicity sake, but for placing in public domain his experiences for posterity), riding these proverbial two horses becomes *inescapable*. Else, such a scientist can end up with a tendency of tilting toward an industrial type of activity, *losing his moorings with Science or blunting his Scientific Temper, when he has one*.

When an R&D Organization *leans gradually toward Hardware Development,* for whatever justifiable reasons, the evaluation of a Scientist Designate often if not invariably tends to be governed by what one delivers against External Cash Flow—ECF (another new buzzword). But for those, not in a position to bring in *ECF* by the very nature of their specialization in roping in the funds, the R-Component yardsticks come handy, but the question one must face is "*At whose cost the Research is done.*" In case of Hardware development, the ECF can take the precedence, for reasons obvious. But, for one whose conscience resonates between the R and the D, it is walking on a sword's edge with the risk of doing neither or sacrificing one for the other, even to the extent of permanently missing returns/rewards for his efforts. That is a price he must be prepared to pay without losing passion (heart), for doing the R&D in parallel.

R&D Management—Its Culture

Well, typically in public-funded scientific institutions, the R&D Managers evolve from within the organization. It is to be recognized that *Passionate R&D Professionals are usually born, but not created or made overnight*. In extreme cases, the criteria of how a R&D personnel is assessed can resonate in between, by what he develops and delivers against ECF and what research he does resulting in publications/patents, books, and so on. Here, two distinct classes of people emerge, by default, viz. the *Researchers and the Developers*, analogous to *Thinkers and Doers* (a once hotly debated issue in one of Country's Premier Organization, bogged down in a dilemma as to who should head and lead it?). Often, there can be little or inadequate overlap of the perceptions and philosophies driving respective activities within the same organization; the passions, commitments, and outputs however, and remaining undoubtedly visible in their own way. Herein crops up the issue regarding the *Manifesto or Charter of an R&D Organization* (meant to *bridge the gap between Academy and Industry*), as expounded in letter and spirit by its Founders, bringing into picture its attitude and environment it creates and provides for these two classes of people. The *Managers* should be the higher-order R&D professionals, having reached the level, meant to visualize and draw the *Road Maps* for the immediate, near, and far futuristic programs, in order to sustain the organization in a challenging and rapidly changing "*Techno-Scientific Environment.*"

The role of an R&D Manager in such an Organization has got to be versatile and unique, unlike what exists in an Industrial R&D (Return on Investments). *He has to lead from the Front/Rear, Plead when needed* (*generating funding*), *and bleed often* (*in a situation of do or die*), having committed to deliver a visible end product to the sponsor. It is a different matter that it is the organization which is ultimately accountable to the sponsor, no matter who stays and who leaves (in the spirit of Dr. S.R Valluri's statement, more on this later). The Manager has to take along all concerned at all levels in *synergizing their Strengths and normalizing the Weaknesses.* Building and Maintaining the "Research Reserves" are like keeping the over head (OH) storage tank full all the while (with a liquid level controller, if needed), in order that the taps do not go dry, one day. This analogy came off my hat over a decade ago much to the amusement of the *expert committee members*, when I was surprisingly invited for a "Discussion" for the highest position of my laboratory and the question close to my heart put to me was, "*How do you address the downward trends in publications in your organization?*" Well, it is a different matter that I suggested setting up a *Hypersonic Wind Tunnel Facility at my laboratory* (a dream I fancied in my mind, ever since I visited the NAL-Japan way back). In this case, the question refers to what unique contribution I could make sans composites, if I were to be selected. "Thank God," nothing of the sort happened.

The role of the *Captains of the R&D Organization* is vital as the watchdog all the time, in facilitating right outputs from groups of diversified disciplines, much like in a result oriented sport. But, what is missed out often is how to functionalize, align, and attach the research minded to the development savvy (like with the CNTs—*Carbon Nanotubes*) *to get the best of them, as* one cannot afford to have in an R&D, either of them exclusively. Both the R and the D groups have to work together and share the *Fruits of the R&D Success, for an inclusive growth of the organization.* The *Binding Matrix* undoubtedly is the R&D Manager, but the Captains are the Eye *in the Sky*. Often it is a testing time for the Manager, especially when he on his own volition or under compulsion accepts the R&D responsibility of *leading a well-groomed group of individuals with diversified potentials, disciplines, and attitudes*. Here, he should willfully become *an interface* among his team members with different inclinations, rather than being preferential to some and not being so to some. Hierarchy and ranks should not come in the way when it comes to giving due credit if not undue elevation. However, if some outperform in promoting *Academics* despite direct project responsibilities/pressures, the organization should *not miss spotting them* and reward them, by giving them independent responsibilities and empower them (recall Dr. T.S. Prahalad's vision here). As regards the technical staff (under employed, in particular) who acquires higher academics while contributing enormously to the sponsored programs, the organization should automatically upgrade them to suitable scientific cadres. If the system has limitations, it should "*Effect Bold Institutional Reforms*," *to reward in time and retrain them*. Preserving them ensures a Leadership-Bank for the health of the organization in the long run. Regretfully, not much has been done in this direction yet.

Often, he (the Manager) may need to even learn where it is not his domain or of no interest to him inherently. But having "*accepted the responsibility*" for whatever reasons, he has to *Lead, Plead, and Bleed*. Once he likes doing it, the results are fantastic, but first, the "*EGO must GO*." A Bottom-up Top-Down approach (willingness to learn even from below when needed with humility and managing the Equals and Superiors at the top with courage) enables the Manager rise like *a Phoenix,* pledging his long experience, maturity, wisdom, and foresight. Often, the *Ombudsmen* of the R&D Organization may need to help the Captains wake up and make up with mid-course corrections, if need be, to ensure smooth running of the R&D Band Wagon. If this is not done, it will lead to depletion of the *Research Reservoir* in the long run. At the National Scientific R&D Institutions, we are the Custodians of the Public Trust and should ask ourselves what we give back to the country and not what we get from it (*oft repeated by Dr. S.R. Valluri, the legendary former Director NAL, and later the Founder and First DG of ADA, and the visionary of the LCA program: thanks to Sri. G.N. Vittal, Chief Controller of Administartion, for timely reminding me on this patriotic statement of Dr. Valluri*). Respecting and Nurturing the High Science and High-technology Areas in contemporary and fast emerging fields is the prime responsibility of such Institutions. They should have strong Research Inclinations side by side the Product Indigenization Goals and Responsibilities, at the same time be the Beacon, or a hope for the Country's NEXGEN (e.g., the have-nots from the educational sector).

Concurrent Research (the RDD)

The Concurrent Research or the Research Development Driven (RDD) is need based and carried out mostly in parallel and occasionally in tandem with the Development programs, committed with the Sponsor's funding, timeline, and specifications clearly identified right from the beginning. The RDD thus is "*conjoint with a product*" from the beginning like in a mission mode, but not in Toto. Again, the task of the R&D Manager is to deliver what was promised, *standing under a Damocles Sword, rather than sitting pretty on a high pedestal.*

This is a very challenging situation, since the "Research here is Application linked" and *Contemporary*, to be carried out on well-defined and directed project relevant topics not necessarily linked to one's fancies or passions, with the expected results to feed into the end product, continuously. And, it is *time bound* since it has to run *Peggy Back* with on the development. But, such research *should not be coterminus* with the product delivery, paving the way to forward looking or the futuristic R&D Road Maps. That is when the R and the D represent two sides of the genuinely coined term "R&D," since the research results have to be compulsively fed into the Product under development, making it *scientifically substantiated and technically sound.* Yet it is a kind of Business and a win-win situation, both for the Donor (of challenge) and the Receptor (of opportunity). Here, while the deliverable is most often a visible hardware (a product), it can be a software too, if demonstrated/installed on an application platform (*Control Law* for a combat air vehicle—a priceless contribution by present Director Sri. Shyam Chatty and team). In a true R&D setup, the bottom line is that the *research component* is unambiguously visible, side by side the deliverable developed. Once this is done with least or no dependence on external sources, one can see a "Made in India" insignia, on every product *made in India and used elsewhere too.* It is a charm, ultimately to do this kind of research feeding its results into an end product, and the Researcher is at no one's mercy for funding, only rider being, "*Are you willing to ride two horses, and walk that extra mile*"? In this case, there is nothing like, someone questioning for accountability, the promised and delivered (promised version) speaking for itself.

Once again, it is worth recognizing here that R&D in its true sense is an outcome of "*intellectual and infrastructural synergy.*" Comprehensive Project Documentation covers technical and scientific reviews, prototype development processes, sponsor's acceptance/satisfaction indices, publications in open literature, patents, TOT records, and all stand as a testimony to the R&D prowess of the Organization as a whole. It is important to note that both these activities are *inherently dissimilar in nature,* resonating in between those of a "Research Passionate Scientist" and a "Hard Core Development Engineer." That brings us into the concept of doing both the R&D, in a hardware development-dominated (HDD) environment. In this case, it leaves no room for an attitude, "*I did my job, it is for you to do what you want to do with that.*" But then, these two horses need to be tamed and harnessed with both courage and commitment. Though this is a challenging and conflicting situation for a *Research Obsessed Scientist,* it is a win-win situation, as the Scientist gets the funding support (big or small) and the sponsor gets his end product (small or big)—sort of you can have the cake and bite it too. Question is how many wish to bite the bullet, taking them as Thrills in Chills and Charms in Challenges, with a will to put their neck on the anvil.

Thrill, Chills, Charms, and Challenges—Some Case Studies

The best example here is the development of an end product, with novel/new materials, new processes, and acceptance methods specific to the novelty (first time done, and none else can do). That was when my mind wandered in a wonder world called the "Composite Materials," especially when the products to be developed were as diversified as a 2-seater all-composite aircraft-HANSA (for pilot-training), and "Radomes" (Radar Domes for protecting ground-based Radars for disaster warning/management in coastal regions and for flight vehicles), both having a very close bearing on societal development and needs. These called for interdisciplinary participation (Materials, Processes, Technologies, Manufacturing, Mechanical Design/Structural Analysis, Fluid Dynamics and Flight Experiments, and Electromagnetics, including customized Engineering Services).

In essence, the HANSA demanded formulation and execution of Applied Research Programs on customized composite material formulations, their characterization, multilevel testing (coupon to component) till their certification/qualification and acceptance, concurrently with the development of inexpensive indigenous processes/fabrication techniques (*Appropriate Technologies*). Here, the emphasis was on very modest budget and easy absorption by a small-scale private sector for mass production, bringing together "*Basics of Polymer Science and different disciplines of Engineering and Engineered Materials and their Sciences*." This culminated in realization of a certificated light trainer aircraft of world class with a stringent weight budget (All Up Weight-AUW within 750 kg, vide the JARVLA regulations—not a gram exceeded, reminding the Act from The Quality of Mercy of the Shakespearian Drama, the Merchant of Venice). Many who worked in the program either lost or maintained their body mass indices (BMIs) in the process, *even as there was some skepticism about the* feasibility of such a weight budgeted aircraft. The indomitable courage and confidence imbibed in the "TEAM HANSA" by Dr. Siva Kumara Swamy, Program Director, and Dr. S.S. Desai, the CTFD Head and the Chief Quality Control Manager (CQCM-HANSA), in this regard were tremendous.

Coming to the Radomes, The "*DWR-RADOME*" provided yet another case of a challenging R&D venture, *standing up against imports prevalent at that time* to indigenously build the next-generation (NEXGEN) product for protecting the *Doppler Weather Radars* (the ground installed DWRs of ISRO) used for coastal disaster management and many AirBorne Radomes, in particular the ones for the Fire Control Radars on JAGUAR *Maritime upgrades* (of HAL/IAF) and the Weather Radars on NAL's-SARAS. Though a Radome is very complex confluence of multidisciplinary subjects and human interactions, the subject electromagnetics turned out to be a crucial design driver, and the last word, (Glorious tributes to Late Dr. Rakesh Mohan Jha, the EM-Emperor), for the final acceptance of the product by the sponsor/user.

The Bearer of the R&D Mantel

Here emerges the main question, how and who, should find the time or scope to do the *Research in parallel with the Hardware realization*. This is here that *a passion of a Scientist* (the Author) gave the clue that led to identification of myriads of project specific topics "*for short-term, medium–term, and long-term research opportunities,*" in training directly and indirectly, countless number of students/scholars/faculty both from within and outside the Organization. Everyone in the project teams was engrossed in the *technoscientific* significance of the product underdevelopment. Here, the policy of Prof. R. Narasimha FRS, the then Director of NAL, is that "the laboratory should also involve and benefit the young students and academic-aspirants liberally in its R&D pursuits reinforced by documentation of achievements," which gave a big Phillip and a new turn to the organization's Scientific Culture that persisted. Superimposed on all these are the stringent certification mandates imposed by the regulatory agencies such as the ISO, the DGCA (for the civil aircraft-HANSA), and the CEMILAC (for LRUs on the military Aircraft), as well as the continuous reviews by respective sponsors of these products —a situation not at all comfortable (most scary if at all), for a **Scientist**, wishing to be left alone as an *independent Researcher*, if there was a choice.

Project "HANSA" gave an impetus to a gamut of technical, scientific, and engineering aspects such as materials, choice, characterization/testing (coupons to component levels, invoking spectroscopic, thermal, mechanical, structural, and electromagnetics analyses), before and after rigorous exposure to application specific service environments, optimization of parameters for material formulation development and composite cure chemistry related to matrices and adhesives, in parallel addressing the scientific topics such as Synthesis and Cure kinetics modeling/optimization. For the first time in our country, the concept of the *thermodynamics parameter* "Glass Transition Temperature Tg" and its research took firm roots, as a crucial measure of the *composite cure as well as a service temperature index* for the Aerospace Composites. Throughout the program, Tg took the centre stage of material certification and acceptance. It is not hype if said that at one point of time, this parameter as one of the *design drivers* was among the most debated, *even capturing the curiosity and imagination of the Aerodynamicists, for a change.* That was when the Directorate General of Civil Aviation (**DGCA**) got convinced and certified a humble RT-Cured Epoxy Composite system with an enhanced Tg (by a novel step post cure process scientifically evolved in house) for the airframes. Selection and certification of a Flame-Retardant Resin System for the Engine Cowling and a composite fuel tank, both replacing the conventional Aluminum Alloy, were other excitements (Mr. Burt Routon, the legendary builder and flyer of world's first wet-winged all-composite experimental aircraft Routon-Voyager, endorsed the composite fuel tank, vide Dr. S.S Desai's Communications).

Research Studies on "hygrothermal effects" on the composites used (epoxies with glass and carbon reinforcements, adhesive joints, and sandwich panels) played a vital role in this program. This led to extensive studies on *Hot-Wet property*

Degradation trends, to generate the composite material Design Allowables (A Basis/B Basis), for the primary and secondary structures, so crucial inputs for the structural Engineers for their optimum Design and Analysis efforts to evolve a *safe design* for the composite parts. All these exercises demanded continuous inputs for the structural designers on the top floor and the design compliant process sheets for the fabrication teams on the shop floor. The all-composite (nonmetallic) HANSA using in-house groomed, inexpensive, and customized RTVBM Technology (*Room Temperature Vacuum Bag Molding*) made its maiden flight with an all up-weight of 750 kg *after rigorous weight reduction campaigns, and flight trials* made its maiden flight test on 11 May, 1998—The Technology Day, creating a great excitement among all those involved. The *CSIR newsletter* went viral with HANSA described as the "most-researched aircraft in its class" (Ref: Article of Sri. Hasan Jawaid Khan, its Editor). The certification for series production by DGCA in 2002/03 gave a sigh of relief as the TEAM HANSA dedicated to the project for more than a decade and a half **went hysteric**. Thus, HANSA provided the best example of a totally indigenous mission undertaken by a National Laboratory up to its certification, demonstration, and deployment in several flying academies. Today, HANSA has a silent and subtle message, "Hey, I am here as a great Made In India fully certified pilot-friendly aircraft, *rearing* to don the 50 and odd flying schools across the Country and beyond." *Any takers please?*

The DWR-Radome was an equally if not a lesser significant example that went beyond its indigenization as it culminated in the TOT to the End User/Sponsor, after about 5 years of efforts and more than 2 years of rigorous EM and Structural Tests on ground, before acceptance for its maiden installation (MARK I Version) at the Satish Dhawan Space Centre-SDSC, Sriharikota for ISRO-the sponsor, in 2003. Soon followed the MARK II version with less number of larger panels, the technology of which was finally transferred to the sponsor identified PSU (BEL). The research issues related to the Radome have been the selection and fine-tuning of a PUF-core and composite skins for the large sandwich panels, the incorporation of the cast in situ foam core in between the composite skins ensuring a strong bond, and the chemistry of foaming/curing to desired densities/pore architecture, for structural properties (strength and modulus) and EM Critical Properties (Epsilon, Tan Delta), to name a few. The Design Allowable established already for the HANSA Composite Air Frames came very handy here. On the EM front, a synergy had to be established between the development of the composite formulations, fabrication processes, and the EM Material Characterization for an optimally EM—designed and structurally integral Radome, to meet the stringent performance specs, on site. This is yet another example of a truly successful nationally important Inter-Institutional Program that literally gave yet another loud call, "No more imports please." These Radomes demonstrated at few Indian Coasts already, to withstand 250/300 kmph wind speeds, are all set to done the cyclone prone/ravaged Indian coast, replacing our own earlier generation Radomes (protecting the new generation DWRs). This indigenization effort will go a long way in strengthening our early warning capabilities, to ensure safety of vast coastal populations. *The need of the hour:* mass production of these radomes by PSU Licensee of this technology." The managerial support of late Dr. B.R. Somashekar and late Dr. S. Nagabhushana, during price negotiations for this program, was invaluable.

Well then, the Indigenization and certification (by CEMILAC) of a flight worthy Nose Radome for the Jaguar Maritime fighter aircraft, closely followed by its TOT to the sponsor (HAL, to be inducted very soon into the Jaguar Air Squadrons), was another example of achieving, "The Mission Impossible." The success story of this high-temperature resistance composite radome for a high-speed flight vehicle has many twists, excitements, and all the ingredients of a R&D venture, in a *filmy style*, in its execution in a record time that ultimately established a "*Template for Radome Indigenization Programs*," with in the country, to say the fact. Apart from several applied research activities addressing critical material and manufacturing issues, the radome's EM performance (near and far field) had to be evaluated on ground at LRDE, ISRO, and CAB facilities, including the radar supplier's facilities at ELTA-in Israel upon insistence. Finally, with the successful in-flight evaluation of the radome on the Jaguar Maritime aircraft platform by ASTE-IAF, ELTA (also our contender for these Radomes), gave a thumps up saying, "NAL Built Radomes are on par with the best available internationally." Thus, NAL had to get over the conflict of interests, in this program, to craft a strategy and establish its dominance in the development of this class of Radomes too. The solid backing of Dr. B.R. Pai, the then Director in this regard was very timely.

The aforesaid products and many of their predecessors/successors not covered herein for want of brevity were an off shoot of the Applications-Related Research Activities pursued concurrently with the product development encompassing composite materials, processes, and technologies with spin-off benefits. These are briefly different scaled versions of the Carbon Composite Air intake wind tunnel models of LCA-Tejas, the rigid composite Air Combat Simulators (ACS Domes) installed at the Hindon Air Base, the reusable Rocket Launcher Tubes for the MBRL's of ARDE/DRDO replacing the imported heavy steel tubes, the massive 300- and 500-KW wind turbine blades under a CSIR's NIMITLI enabled PPP Program, including the earlier porous composite support tubes for the RO-membranes used in water desalination plants (BRNS/BARC funded) and its TOT as well as a first time demonstrated composite Bus Body for the mobile speech and Hearing Unit, funded by All India Institute of Speech and Hearing-AIISH-Mysore. Thus, all these products firsts in the country *are capable of positioning Our Nation more self-reliant than ever before, in this field of Composite Technologies, in taking the* Chemistry to Component.

After Effects of a Passionate R&D and the Road Map

During my nearly three and half decades of stint at NAL (1973–2007- and 2008 on contract) and later over the past 7 years as a retired/not tired yet scientist from a prestigious National Laboratory, I tried very hard to understand the similarities and disparities between the R and the D. I struggled first as a scientist-in-charge of a *pilot plant*, with a number of unskilled/semiskilled/skilled workforce (masters in their own art), then getting an opportunity to elevate it to an *independent Unit* and finally to a full-fledged *scientific division*, which has today highly qualified human-power (technically/scientifically). Thanks to all my peers from within and outside, who gave me the mandates, that provided me and my Team a golden opportunity and *maneuverability like that needed for an unstable fighter aircraft*. The "FRP-DIVISION" that I was destined to create and groom has today many *Doctorate—decorated Scientists including enviable number of Women Scientists* and International Publications in its basket. (*For once, I could "discover an excuse" that there was "no time" to hit a century, for records sake*). All along, it had been my endeavor to strike a judicious balance between the R and the D, by interweaving these two in the warp and weft directions (akin to a planar bidirectionally woven fabric of glass or carbon reinforcement used in a polymer matrix of a composite material), and often by stitching them vertically (like in a 3D composite preform), to ensure the "*R&D Structural Integrity,*" invoking liberally the pressures from peers and demands of sponsors, and kindling passions of the research minded. In the process, a sandwiched I could create intentionally or unintentionally a platform to play my role at will in synergizing these two activities, research outputs supporting the product development, and products developed and delivered for a price, *nourishing, and nurturing* the former. The Division with the Road Map I left behind in 2008 had made its forays into many frontier fields of the S&T. These included Conducting Polymers/Composites/Nanocomposites, Radome paints, High Tg Polymer Matrices functionally tunable polymers by taking forward the *duel-Tg concept*, Radiation Cure Processes for rapid/energy-efficient green composites, Resin Infiltration Techniques, and Jip-Pregs (Just in time prepregs by resin film technology). Realizing a Jip-Preg machine after a long wait (2000–2008) was a great valuable addition to NAL infrastructure, thanks to the timely support of the then Director Dr. A.R. Upadhya. These are all set to replace the expensive Prepreg Technologies *if only harnessed further*. I am glad that even a tiny imported "Research Autoclave" that I revived from its coma state (a UNDP gift received long ago) serves as a humble reference for NAL-Autoclaves and that the wind turbine blades activity continues.

Tail Piece

I got a semblance (Saakshaatkaaram) of what an R&D could be made to look like, thanks to the organization which gave me the freedom to fire my imaginations into frontiers of realities. *All said and done*, "R&D" continues to remain an "OXYMORON," when the Research is done with mere Passion as the purpose, and the Development carried out sans Research. One thing I can say for sure is that for a scientist under the CSIR umbrella, "Research should be his soul and substance and Development his bread and butter." In this article, I tried to summarize my R&D Trials and Tribulations as a scientist, engineer, and technologist. Looking back,

I feel it would have been a missed Life Time Opportunity, had I not taken things as they came, willfully with a passion and bowing in all humility. At the end of the day, I have *some sense of fulfillment but with disillusionments too (for instance, missing another challenge of indigenizing the most complex Nose Radomes for the deadly IAF-*Sukhoi's, *despite our 3-year-long hectic proposals/presentations on request—alas for the simple reason that it was not commercially viable, even as we were ready to negotiate further). I wish* better Days are ahead, *for the Indigenous R&D.*

Acknowledgements Foremost the author thanks INAE, in particular Dr. Baldev Raj, the Hon President, and Dr. Purnendu Ghosh Editor in chief, for inviting me to pen my experiences on a topic of my choice, and (NMIT) Bangalore. I am indebted to my Academic Mentors (Prof. Manas Chanda, Dr. N. Balasubramanian, Prof. George S Springer of the Stanford University, Dr. A.K. Singh, and the late Prof. C.R.L. Murthy) and all my aforementioned Directors and former Director Dr. K.N. Raju for his very positive support. I thank many colleagues from other divisions/disciplines who were partners in my progresses, in particular, Dr. S. Viswanath, Sri. D.V. Venkatasubramanyam, Sri. Ningaiah, Dr. U. Ravindranath Nair, Dr. R.C. Majumdar, Dr. Chakrabborty, Sri. Satish Rohedekar, and Sri. C. Chandrasekhar; and Project Monitoring Heads Sri. S.C. Narayanamurthy, Sri. M.S. Ramachandra, Dr. Ranjan Moodithaya, and Teams. I place on record the invaluable support extended by my own TEAM-FRP, viz. Ms. N. Ankanna, C. Pragalathan, L.R. Chandramohan, Dr. Shylaja Srihari, Sri. T.H. Samiulla, Dr. A. Vanaja Sri. K. Mohan Kumar, Dr. B.S. Sugun, Dr. Sandhya Rao, Dr. L. Srikanth, Dr. Jyotsna Kiran Mathod, Dr. T. Padmavathi Sri. Rajendra Prakash M, Ms. Kavitha, S. Navaneetha, A. Revathi, and Sri. P.S. Sathyanarayana. My grateful acknowledgments to my colleague Sri. M. Subba Rao and his Team, for very valuable interactions I had. I specially acknowledge Dr. S. Bhogle, former Head Information Management Systems, for masterly articulating our achievements, through his Paste-Board. I recollect here the services of Sri. M.R. Narasimha Swamy, our former Head of Technical Secretariat, and his deputy late Sri. Md. Aslam, for their unique contributions in my student training and TOT programs. Last but not the least, I put on record tremendous support I received from my external sponsors/peers, in particular Sri G. Viswanathan, Director ISTRAC/ISRO, and his colleague Sri P. Sambasiva Rao, Prof. P.R Mahapatra, Dr. Acharya, and Prof. K.P. Rao all from IISc, Sri. S.K. Veluswamy from AGM-HAL, Dr. S. Christopher from LRDE/CABS and presently Chief of DRDO, and many from certification agencies. I have no words to quantify the enrichment I got from my timeless interactions with countless students/research scholars/trainees/contract staff. (*Missing acknowledging any is purely unintentional*).

I remain ever indebted to my Family (my wife Rathna, son Dr. Raja M.V. Pavan, and daughter R.M. Vauhini), for their decades of infinite patience and cooperation, in letting me achieve whatever I presented herein, never complaining of my unavailability to them when they were young and needed me the most. I am very proud of them. Above all, it is the Blessings of my Parents and the Elders of my large family in providing me the pedigree that made me what I am today.

Further Readings

"Engineering of Science-The Composites Way ", Annals of Ind. Acad. of Egg, Vole IV, April 2007, Raja Manuri Venkata Gopala Krishna Rao.

"Research Pursuits at the Fiber Reinforced Plastics Division, National Aerospace Laboratories (NAL-CSIR) Covering Journal Publications, 1976–2005 (***To be updated, still***), Special Publication Report SP 0502(2005), RMVGK Rao & Team.

"R&D-Riding Two Horses", Invited Lecture delivered for the Participants of Working Professionals Learning Project (AICTE Sponsored/TEQIP funded), Organized by NMIT, at Hotel ITC Gardenia, Bangalore on 18–20th Nov, 2013, RMVGK Rao.

NAL Archives (Presentations, Publications, Project Documents, Theses, weekly news pasteboards, etc…., etc……).

Many formal and informal Interactions I had with and inspirations I got from all my Directors.

"Higgs Finds Nobel Fame 'a bit of a nuisance", TNN, Times of India, June 2015.

Ten Levers of Smart Engineering

Aloknath De

Smart engineering is all about using insights to conceive, model, and scale an appropriate solution to a problem or an objective. During the process, scientific, economic, social, and practical knowledge is applied thoughtfully. This knowledge serves as an engine behind designing, building, and maintaining structures, machines, systems, materials, and even processes. Smart engineering aims at bringing 'wow' element in user mind and yielding an overall society benefit.

Through this article, we build up ten powerful techniques around smart engineering and analyze its associated motivations. We term these techniques as ten levers of smart engineering:

- **Need-of-the-Hour Engineering:**
  History records that different eras have laid stress on different aspects of engineering. In BC period, people have focused on primitive technology for agriculture—studying soil characterization, improving irrigation system, and finding means of land-plowing for harvesting. As civilization moved from Stone Age to Metal Age, society learnt to cook and prepare food. Agriculture to heavy engineering to electronics engineering—various themes have got emphasis during India's series of five-year plans. In present time, physical highway (roadways, airways) and information superhighway (internet 2 upcoming) are making our living world a global village. Mobile phone reaching to masses has enhanced our productivity. From world of voice and SMS, instant messaging and usage of context and social media are becoming prevalent. Need-of-the-hour engineering is about driving engineering efforts toward today's thrust areas. In present time, such need-of-the-hour engineering is around wide deployment and usage of broadband, and optimization of required connected infrastructure.

A. De (✉)
Samsung India Software Operations-Pvt. Ltd., #2870, ORION Building,
Bagmane Constellation Business Park, Outer Ring Road, Doddankundi Circle,
Marathahalli Post, Bangalore 560 057, India
e-mail: aloknath.de@samsung.com

P. Ghosh and B. Raj (eds.), *The Mind of an Engineer*,
DOI 10.1007/978-981-10-0119-2_45

- **Improvised Engineering:**
I call second lever of smart engineering as Improvised Engineering. This deals with how same or similar purpose is achieved by more sophisticated technology. For example, in early days, shadow from an anchored stick used to give relative time-of-the-day. Sun Dial and Jantar Mantar instruments are subsequent developments. Currently, we have watches of all types including high-precision instrument that captures split-second difference between winner and runners-up in Olympics 100-m race. Let us understand further with examples from printing and communication world. At some point, offset printing was almost the only way. Digital printers have seen journey from dot matrix to laser jet. 3D printing is bringing forth a paradigm shift from subtractive printing technique to an additive one. Similarly, Graham Bell's invention of telephone equipment to today's wireless connection is a steadfast progress. Smartphone has enriched voice communication but also has undoubtedly eased file sharing and multimedia data transfer.
- **Strip-Down Engineering:**
How many people have tried to unravel Coke's signature formula? Curiosity of knowing 'what goes inside a product design?' typically gets satiated through a process of *reverse engineering*. And then, there is very concept of *frugal engineering* which drives down the cost factor but at times misses to maintain durability aspect of product. I term the third lever of smart engineering as Strip-Down Engineering that combines the strengths of reverse engineering and frugal engineering. The engineering smartness here stems from applying Pareto's 80:20 principle and analyzing how to keep essential functionalities. The goal was to select top 80 % features from user perspective and implement them with 80 % cost reduction. For example, in a low-priced car such as Nano, question had arisen as to whether three nuts can tighten the tyre instead of four. Need for every element is critically questioned. Strip-down engineering constantly weighs cost-benefit trade-off and is a great vehicle to build product for bottom-of-pyramid (BoP) segment of population.
- **Performance-Boosting Engineering:**
Success of a product or service lies in its performance by relative as well as absolute measures. Performance-Boosting Engineering targets to enhance solution performance keeping constraints in mind. Let us take the example of mobile phone where we currently deploy octa-core processor. The evolution from single-core to octa-core has been a way to incorporate parallelization and increase processing power. In heterogeneous processing environment, appropriate partitioning of code across ASIC, DSP, CPU, GPU, MCU chips significantly drives up system's performance. This category of smart engineering also encompasses developing multiresolution systems, such as a spectrum of products from phone-with-full-connectivity-but-basic-camera to higher-resolution-camera-but-basic-phone-connectivity. This class of smart engineering facilitates introduction of more complex features, reducing the response time or boosting other system performance metric.

- **IntelliSys Engineering:**
  Intelligent devices employ functions of sensing, actuation, and control through sensors, actuators, and controllers. When the goal is describing and analyzing a situation, and taking decisions based on the gathered data in a predictive or adaptive manner, smart actions become essential to be performed. IntelliSys Engineering empowers conventional systems—promotes autonomous operation based on closed-loop control, energy efficiency, cloud-based solutions, and networking capabilities. Intellisys engineering helps us to measure how much energy is consumed in a typical home or how to save smartly when the tariff depends on the time-of-day. Similar to smart home platform, we can consider a connected car platform where tyre pressure sensors in each tyre sense a priori flat tyre situation. IntelliSys engineering for connected car enables cars to aid in navigation, cars togo driverless, cars to talk to each other by Wi-Fi technology, and even cars to fly!
- **Cross-Pollination Engineering:**
  By now, we have covered five categories of smart engineering and I have given illustrations from various faculties. Studying only one field dries up ones idea! Newer fields are emerging. And solutions for some problems require extensive knowledge of multiple faculties—either to tie ideas up, or to learn from one field and apply to another. This is what I term as Cross-Pollination Engineering. For example, knowledge of geology, soil engineering combined with biology helps to address problems pertaining to geo-microbiology. This helps us understand how bacteria and virus come to our food through soil contamination and what possible remedies could be taken up. Geography knowledge in conjunction with information system expertise paves way for geo-information system (GIS). Similarly, material science covers wide range of special materials targeted for nano-electronics to vehicle bodies. Cross-pollination engineering gives birth to new fields such as fiber optic communication that combines optical physics with telecommunication. Similarly, the field of music, coupled with acoustics engineering, opens a chapter of musicology by cross-pollination of subtleties in both fields.
- **Smart-Auxiliary Engineering:**
  Majority of times, engineering takes scientific discoveries forward in the form of products and utilitarian services. At times, engineering plays second fiddle to scientific projects—it helps in the next level of scientific discoveries through infrastructural support. Let us consider Large Hadron Collider mega-project recently conducted in CERN, Geneva. The very simulation of *Big Bang* has been an engineering feat—this has been a prerequisite to determine what happens after the big bang event. Supporting role of engineering should not be misconstrued as engineering trivia. One mouse can bother an elephant! One bird can hit a plane and knock it down. We recently found that Dreamliners have used, instead of Nickel Hydride, Li-ion batteries that produce twice the energy in half the space. However, pressure and heat wear out the insulation and may cause short-circuit resulting in fire on flight path. When we analyze how all Dreamliner flights were grounded worldwide because of batteries primarily, we start to appreciate the importance of Smart-Auxiliary Engineering.

- **Sustainable Engineering:**
  Benefits accrued from scientific discoveries are transpired through technology products in daily life. In order to promote *green* concept, we are reducing the use of thermal energy and going for solar and nuclear energy. Not just the power generation, smart grids are being designed for better power distribution. Sustainable Engineering encourages that we build products that consume less energy and make a zero-sum game with environment. Let us take example of electronic circuitry. Researchers have progressed to operate digital chipsets in 1.8 V instead of 3 V or 5 V. Energy-aware protocols have also been designed. These techniques substantially help in overall power reduction for electronics equipments. Sustainable engineering addresses concerns around energy consumption, electro-magnetic radiation hazard produced by cell towers and so on and so forth. How to reduce contamination as well as repurify natural resources for our well-being is the main concept behind sustainable engineering.
- **Nature-Inspired Engineering:**
  It looks gorgeous when city parks decorate vegetation in the form of horse or elephant! We can also design usable products inspired by flora and fauna. Cranes are built mimicking long-necked giraffe. Mercedes-Benz bionic concept vehicle has been designed understanding low coefficient of drag and rigid exo-skeleton of Boxfish. In Japan, the first design of bullet train running at 200 mph had been an engineering marvel, but had one problem: As it came from a tunnel, it produced a loud noise. This problem was resolved by studying beak of kingfisher as it swoops down to catch fish, touch the water surface, and then go back up. Look at our national flower lotus. The lotus leaves manage to remain free of contaminants as they possess a field of small bumps and dust is easily picked up by water drops. The same principle is used in the exterior paints, textiles, and so on. Nature-Inspired Engineering studies objects and phenomena in nature to understand how a fundamental scientific principle works in daily life and applies the notions in product design.
- **Forward-Looking Engineering:**
  Engineering foundations that are based on strong theory and driven by science can be quite forward-looking. Information theory to game theory, number theory to string theory all have been playing role in telecommunication, cryptography, and other associated areas. Let us think about how physical world and digital world are fusing! Technology is moving from reality to virtual reality, and then to augmented reality. User interactions are changing from touch-base to gesture-controlled. Integration of audio, visual, and haptic feedback is becoming part of next-user interaction. Why not consider quantum cryptography as another example of Forward-Looking Engineering? Quantum computing uses qubits with superposition and entanglement. Using these basic principles, quantum teleportation allows same entity to be in two places simultaneously, but observation decoheres. Forward-looking engineering aims to manifest scientific ideas or even science fiction concepts to reality.

Life with Experiments

K. Muralidhar

As a mechanical engineer specializing in fluid and thermal sciences, circumstances eased me into experiments for my doctoral research. Ever since, I have invested more than a fair share of my time in setting up apparatus, developing instruments, generating data, and analyzing results. It has been a remarkable journey with a promise of unending excitement.

Working in an academic environment, I have had a chance to set up experiments essentially from scratch. I could pick the geometry of my choice, the probes to use and appropriate measurement systems. I could debate on the use of mechanical gauges such as inclined tube manometers versus digital, single-channel versus multichannel, multiplexing, and instantaneous versus time-averaged records. I found myself developing a broad perspective on how specific decisions could be arrived at, in the process, getting grounded in the engineering profession itself.

When I saw myself building a fluid-thermal laboratory, it occurred to me that I was following the historical evolution of the subject itself, though on a compressed timescale. I could do simple experiments on formation of a boundary-layer over a flat plate and flow-induced oscillations. I could reproduce Moody's diagram. I saw alternate vortex shedding from a circular cylinder and, indeed, could perform Reynolds' laminar-turbulent transition experiment using dyes. Practically, all kinds of flows could be visualized using kerosene smoke. I realized that my understanding of the subject had greatly matured. I utilized this opportunity to set up undergraduate experiments that would accompany the main course on fluid mechanics.

Modern era is replete with examples where technology has been decisive in the way we live and work. Doctors give examples of medicine and surgery conquering disease. In experiments, several such developments have fundamentally altered the way data are collected—regarding questions we may ask and devices we can build.

I will narrate three examples here. First, the ability of computers to position and traverse probes, collect data, start and stop experiments, coordinate measurement systems, and analyze numbers has taken the subjective element out of research.

K. Muralidhar (✉)
Department of Mechanical Engineering, Indian Institute of Technology, Kanpur 208 016, India
e-mail: kmurli@iitk.ac.in

P. Ghosh and B. Raj (eds.), *The Mind of an Engineer*,
DOI 10.1007/978-981-10-0119-2_46

What started as a replacement of drudgery has led to creative experiments being performed, and very profound issues addressed. Data analysis and graphics capability of a computer now make it an oscilloscope, spectrum analyzer, voltmeter, temperature recorder, correlator, and many more rolled into one. Smaller and faster computers are now being embedded inside other instruments such as microscopes. Of course, computers are getting integrated with the human body and the consequence of this development is no doubt, curiosity, nonetheless, a source of alarm.

The second is the invention of lasers that can "see" everything as images that in turn are recorded by cameras and sent to large computers. Lasers now measure anything from particle velocity, to temperature, species concentration, and flames spreading in furnaces and engines. They do it in three-dimensional space and in real time. I find that the appreciation of what lasers can do is poor among the engineering community. I wish the issue is addressed.

Distinct from computers but connected to it is the virtual world of Internet and experiments that can be thus conducted. Internet brings in the possibility of collaboration among individuals and Institutes that are physically separated. Doing experiments in a laboratory elsewhere by sending commands over Internet is the simplest (but very useful) possibility. Of real value is when expertise from multiple sources are pooled together to conduct a joint experiment. Called synergy in the management context, cooperation has been a weak link in research, limiting us in many ways from hitting breakthroughs more often. This development awaits exploitation on a grand scale but has a transformative power of a paradigm shift.

Working in a laboratory environment forced me to ponder on the nature of research that engineers need to carry out. I had been struggling all along with the central purpose of my experiments, beyond niceties such as better understanding of the phenomena and significance of higher order effects. The obvious struck me later that the purpose was to examine cause–effect relationships in the framework of forces, fluxes, potentials, and energy. The goal was not to see things as they were. The purpose was to check what made a system work. This awareness led to several questions, for one, could predefined flow patterns be created? When I discussed the subject with my colleagues, I was told that creating new reality was a design perspective. Since then, I have come to view the goal of analysis, experiments, characterization, and indeed all of research as steps that yield a new process, or a device, or a system possessing certain desirable properties.

With computers, lasers, and cameras at our service, experimentalists now experience a new problem of plenty. Data are time-dependent and three-dimensional. Special devices now store the measurement records. Since the goal continues to be improving new processes or building new ones, the experimentalist is required to mine mountains of data to extract information of value. Fortunately, the subject of data analysis with large but 'gappy' information is gaining ground and we can identify patterns (or the lack of them) through numbers that exist purely in the virtual world. The mathematics needed for such studies is very sophisticated. Experimentalists had better not leave their analytical skills behind.

Historically, experiments have been way ahead of theory in driving human progress. Much of the present-day knowledge stems from controlled experiments, and in turn, reconciled with a newer theory or principle. The situation has turned topsy-turvy, ever since computers and numerical simulation have emerged on the scene. Computational experts now claim that their simulator can replace experiments. This claim is heavily discussed, but within the domain of fluid and thermal sciences, experts see it differently. There are strengths in each approach and the greatest gain is to be seen when the two work in tandem. A mathematical model getting triggers from an experiment running in parallel can simulate extremely complex phenomena. Conversely, an experiment may run with boundary conditions or geometry determined from a model running alongside.

My association with real functioning devices has had an unexpected benefit. I have been able to conceive of simple but new gadgets and configure new measurement systems that I feel are of use. The acquired wisdom is a payoff that all experimentalists deserve.

There can be no two opinions that theory, experiment, and simulation constitute the three pillars of the modern thought process. What could be debatable is the real perception that experiments carry the greatest load of them all.

The term *experiment* is very broad, context-specific, and takes on unique meaning to each group of professionals. Science experiments are highly focused with the primary aim to prove or disprove a theory. Engineers pursue a much broader goal, such as performance enhancement of a device. It may entail application of a collection of techniques and methodology to accomplish the overall purpose. Biologists record, characterize, and analyze species populations as a part of their experiments. Sociologists work with human populations and statistics to improve understanding of human behavior. There are great differences, division, and points of departure, depending on the subject matter. Yet, the importance of experiments is a common refrain, universally emphasized across disciplines; I am glad to be living in this world.

If experimentalists from various disciplines were to function together, they better develop a common language. Thematically, they may discuss interplay of experiments and theory, measurement on multiple length scales, and measurement on multiple timescales. Abstract communication will enhance cross talk and lead to fruitful interdisciplinary research.

I think my interest in experiments has rubbed off on students who have worked with me. Many who have graduated think experiments are important, can be done, and worth their time and effort.

Playing with Fuzziness and Ambiguity in Patterns—Challenges and Achievements

Sankar Kumar Pal

Background

I was born in Calcutta on September 13, 1950, the second of five siblings, to a very ordinary family of modest means. My father started as a clerk in Sree Saraswaty Press (a private concern at that time, now an undertaking of the Government of West Bengal) in Calcutta and retired, as far as I can remember, in 1978, as an accountant earning five hundred rupees per month. So I was well aware of the hurdles I would have to face in completing school education, let alone higher studies. I am extremely fortunate in that my parents supported me wholeheartedly in my pursuit of a Ph.D. degree, making great personal sacrifices in the process. I studied in four different schools. While the first three are in central Calcutta, the last one, Ariadaha Kalachand High School, is located in the northern suburbs of the city. I studied there for about four years and passed the Higher Secondary Examination from that school in 1966. Of those crucial four years, I spent three in my *mamar-bari* (house of maternal grandparents), as my parents' new residence, after relocating from central Calcutta, was rather inconveniently located for a safe daily commute. While the relocation from school to school and from an urban residence to a suburban one was primarily due to financial constraints, they did cause some amount of cultural, social, and environmental shock.

S.K. Pal (✉)
Indian Statistical Institute, 203, Barrackpore Trunk Road, Kolkata 700 108, India
e-mail: sankar@isical.ac.in

P. Ghosh and B. Raj (eds.), *The Mind of an Engineer*,
DOI 10.1007/978-981-10-0119-2_47

Education

I gained admission to the B.Sc. (Honors) program in Physics at a Ramakrishna Mission College, named Vivekananda Centenary College, Rahara, which is located further north. The college had strict discipline. As there was no one in my family to provide proper academic guidance, I turned to Prof. Ananta Kumar Sengupta (father of one my classmates) who was then the Head of Applied Physics Dept., Calcutta University. I am extremely grateful to him for his valuable advice. In those days, the qualifying cutoff in marks required for the admission to the B.Sc. (Honors) program was the highest for physics, among the pure science programs. I had initially been interested in studying chemistry, but changed my mind during admission. Some of my school teachers and relatives had discouraged/cautioned me against joining the physics program, thinking it to be too tough, but I took it on as a challenge.

After obtaining my undergraduate degree with honors, I had the option of joining either a two-year masters' program or one of the newly introduced five-year B.Tech. cum M.Tech. programs in Calcutta University. Despite the financial constraints, I opted for the latter, in Radiophysics and Electronics foreseeing its better opportunity.

Initiating Research

After completing M.Tech., I joined the Indian Statistical Institute (ISI), Calcutta, on March 01, 1975, as a CSIR Senior Research Fellow to work for a Ph.D. degree in the area of pattern recognition and man–machine communication by voice, a completely new discipline not related to any postgraduate courses we had studied earlier. I had no idea about pattern recognition, nor were there any textbooks on this subject; only a few edited volumes, mostly by Prof. K.S. Fu, were available in our library or in the market. Apart from these, the D.Sc. thesis of G.S. Sebestyen, titled "Decision-making process in pattern recognition," published as a monograph by McMillan Press in 1962, was available in our library. From these, I started to pick up the basics of sequential pattern recognition using statistical approaches. One day, my thesis advisor, Prof. D. Dutta Majumder, gave me a typewritten note, of about four pages, titled something like "Pattern classification with property sets," written by Prof. Ramesh Jain who had presented the concept in some seminar. There, I found a reference to Prof. Zadeh's famous paper, "Outline of a new approach to the analysis of complex systems and decision processes," which had appeared in the *IEEE Transactions on SMC* (vol. 3, pp. 28–44), in 1973. From these, I got some idea of multiclass belongingness of a pattern, which appealed to me very much as it seemed to be very natural for decision-making in real-life problems.

At that time, my institute (ISI) library was not subscribing to almost any IEEE transaction, as electrical sciences or computer science was not considered to be a core subject of research there. However, I collected a copy from the Institute of Radiophysics and Electronics, Calcutta University, and then subsequently got the other two seminal papers of Zadeh which had appeared in *Information and Control* (1965) and *Information Sciences* (1967). Apart from these, I bought an edited volume from my CSIR research grant, which I consulted often. It was titled "Fuzzy sets and their applications to cognitive and decision processes" by Zadeh, Fu, Tanaka, and Shimura, published by Academic Press in 1975. The Institute's library subsequently acquired another book on fuzzy sets, by A. Kaufmann, titled "Introduction to theory of fuzzy subsets: Fundamental and theoretical elements," Academic Press, 1975, though only in the later part of my doctoral work. These were the only information sources on fuzzy sets to begin with, in the early part of my research career.

Visualizing Fuzziness

Fuzzy sets are NOT *fuzzy!*

In the following, I shall explain how I encountered fuzziness/ambiguities in various problems of pattern recognition and image analysis, got motivated to handle them with fuzzy set theory, and subsequently continued to develop from time to time different hybrid technologies in the soft computing framework, as and when required to meet the then need. While doing so, I shall also describe the road map that I followed, acknowledging some persons concerned, and the different critical situations I faced.

A pattern recognition system (PRS) has basically three blocks, namely measurement space, feature space, and decision space. Uncertainties arise from deficiencies of information available in a situation. Deficiencies may result from incomplete, imprecise, ill-defined, not fully reliable, vague, contradictory information in various stages of a PRS. For example, vagueness can occur in the measurement space due to experimental error, limitation in instrument/measurement to go for finer details, and availability of input in linguistic form. Sometimes, it may be convenient and appropriate to express the input feature value in interval form, or having one side of the interval unknown or even both sides fuzzy. In case of handwritten characters, for instance, vagueness comes from badness in writing, not from randomness. Accordingly, the resulting classes in the decision space may become nonconvex, elongated, and overlapping, thereby making them intractable. These necessitate the design of classifiers with the capability of generating linear to highly nonlinear boundaries, or modeling overlapping class boundaries efficiently. In case of overlapping boundaries, it is natural and appropriate to make a multivalued or fuzzy decision on an unknown pattern, i.e., a pattern has the possibility of belonging to more than a class with a graded membership. Depending on the distribution of membership values over different classes, the output decision on a pattern with respect to a class may therefore be *soft* and linguistically quantified as—"definitely belongs," "definitely does not belong," "combined choice," "second choice," etc. For a doubtful pattern, it is always better to say "doubtful," rather than misclassifying it. In that case, the aforesaid multi-valued decisions have at least an opportunity to get the pattern correctly classified with some higher level information (e.g., syntactic, semantic), if available. This, in turn, dictates that a good classifier should be able to restrict the misclassified samples within less number of classes.

I was interested in working on speech and speaker recognition problems. Speech, being patterns of biological origin, their characteristics depend greatly on speakers' health, sex, age, temperament, spirit, and mind, thereby resulting in considerable amount of fuzziness in them and overlapping among the classes. For example, the same word uttered by a speaker at different times in a day may have different characteristic features. Accordingly, I started developing methodologies for nonparametric classification and recognition, and published my first IEEE paper—"Fuzzy sets and decision-making approaches in vowel and speaker recognition," *IEEE Trans. SMC,* 7, pp. 625–629, 1977. Subsequently, I published on plosive recognition and self-supervised adaptive recognition systems, and submitted my Ph.D. thesis to Calcutta University in 1978 titled "Some studies on pattern recognition and man-machine communication by voice with fuzzy set theoretic approach." The foreign examiner of my thesis was Prof. K.S. Fu, the father of pattern recognition, Purdue University, USA. While appreciating the work as a pioneering contribution, he envisaged a future problem to study the sensitivity of different fuzzifiers and hedges, which were used in the distance function and similarity measure, on recognition performance. Prof. Zadeh mentioned in the Foreword of my 1992 IEEE Press book, coedited with Jim Bezdek, that "S.K. Pal first applied fuzzy sets to the speech recognition problems in 1977."

Meanwhile, I started realizing that the processing of gray images could be another good candidate area for fuzzy set theory application. Since it is gray, the basic concepts of image regions, segments, edges, skeletons, relations among them, etc. do not lend themselves to precise definition. For example, a question like—"Where is the boundary?"—has no precise answer. Whatever hard decision that one may make for extracting those features/primitives would always lead to an uncertainty. In other words, it is appropriate and also natural to consider the various tasks of processing of a gray image to be fuzzy, NOT hard, to manage the associated uncertainty in processing as well as in recognizing the content. Again, in an image recognition or vision system, once an uncertainty is caused in edge detection, segmentation, skeleton extraction, etc. on account of the application of hard decisions (0 or 1) at the processing stage, it is likely to propagate further to the primitive extraction stage and finally affect the decision-making process in identifying the image contents. This further justifies the significance of fuzzy processing whereby the uncertainty can be minimized at the final stage of a vision system by retaining the gray information in the preceding stages as much as possible, and the ultimate output will not then be biased/affected much by lower-level decisions. One may note that gray information is very informative and expensive to collect too; once they are made crisp by a threshold, the information is lost and can no way be retrieved. At the point of final decision-making at the highest level, one can always make them binary.

I was then looking for an opportunity to work in image processing. At that time, laboratories with complete software and hardware facilities for working in (gray) image processing were not readily available in many universities/institutes, not even in the developed nations. Luckily, I got a Commonwealth Scholarship to study at Imperial College London in 1979. (Though there was a possibility to work at Purdue University, USA, as a Postdoc Fellow with Prof. K.S. Fu, I chose to go to Imperial College London for a second Ph.D. To me, doing good research and publishing high-quality research papers is the primary concern, Ph.D. is a by-product.) Some digitized image data (in paper tape) were collected from Philips Research Laboratory, Redhill, Surrey. Then, I started developing various algorithms for enhancement including image definition, edge detection, primitive extraction, and image entropy measures using fuzzy sets, and publishing them in IEEE Transactions and Electronics Letters. Subsequently, I obtained another Ph.D. in electrical engineering in early 1982 in the area of fuzzy image processing. In a gray image, there are two types of ambiguities, namely grayness ambiguity and spatial ambiguity. The former is concerned with whether a pixel can be considered to be black or white, and depends only on the gray value, whereas the latter is concerned with both the gray level and location of pixels characterizing the geometry of image subsets. In the process of formulating the algorithms, Zadeh's contrast enhancement operator (INT), S and π membership functions, max and min operators, index of fuzziness, and entropy of fuzzy sets were used. With contrast enhancement of a fuzzy image around a fixed crossover point, the difficulty in deciding whether a pixel is black or white reduces and accordingly the values of its index of fuzziness and entropy decrease (*IEEE Trans. PAMI,* 4, pp. 204–208, 1982). Similarly, given a set of fuzzified versions of an image, the one with minimum index of fuzziness or entropy gives the best segmented output for object extraction.

As an application, we choose the problem of identifying different stages of skeletal maturity (growth) with age from X-ray images of radius and ulna of wrist. The problem is significant from the point of determining the various stages of malnutrition of babies. We collected the image data from Prof. L.F. Turner, Institute of Sick Children, London. Here, the shapes of radius and ulna at several stages of growth have overlapping character, i.e., they look alike. Accordingly, these were handled with fuzzy syntactic recognition approach, where both the primitives (e.g., vertical, horizontal, and oblique lines, and curves) and the relations among them were considered to be fuzzy in developing the unambiguous grammars using production rules. Since the same set of production rules with different membership functions characterizes more than a class, the number of rules required is less, as compared to those using deterministic rules (*IEEE Trans. SMC,* 16, pp. 657–667, 1986.)

It may be mentioned here that my thesis advisor Dr. Robert A. King, Department of Electrical Engineering, was basically an expert in the area of signal processing and had not worked earlier on fuzzy set theory or image processing, till I joined him. However, he was convinced about my ideas and allowed me to work independently to develop the subject. One may further note that there was another pioneering group on fuzzy image processing led by Prof. Azriel Rosenfeld, University of Maryland, College Park, father of image processing, working since late 1970s, particularly in fuzzy geometry, connectedness, and topology on image subspace, among others.

After returning to India in May 1983, I started developing, with my students, multivalued recognition systems with linguistic input, fuzzy syntactic recognition methods, and various entropy measures and image segmentation algorithms, among others. Detailed theoretical analysis is provided for the said multivalued recognition system (*IEEE Trans. SMC,* 24, pp. 1001–1021, 1994). Problems like estimating the entire class from a set of few sampled patterns, selected randomly, were dealt with. We have defined correlation between fuzzy sets and fuzzy operators using ordinary sets (*IEEE Trans. SMC,* 17, pp. 840–847, 1987) to enrich the theory. In the area of image analysis, we have given various definitions of image entropy based on exponential gain function and other quantitative indices for image processing tasks. The exponential gain function relies on the fact that a better measure of ignorance is $(1 - p_i)$ rather than $1/p_i$ (as used by Shannon), where p_i is the probability in receiving the *i*th event (*IEEE Trans. SMC,* 21, pp. 1260–1270, 1991). Accordingly, we have defined higher order fuzzy entropy, image entropy, and hybrid entropy. As the order of image entropy increases, the validity of the segmented outputs, with respect to minimizing uncertainty, becomes more meaningful and valid. Hybrid entropy takes care of both probabilistic and fuzzy entropy and has significance in digital communication, particularly in noisy environment, where the concern is whether a bit is transmitted or not in a noisy channel and if it exceeds a threshold or not.

During 1986–1987, I visited the University of California, Berkeley, and the University of Maryland, College Park, as a Fulbright Fellow. That was the first time I met Prof. Lotfi A. Zadeh and Prof. Azriel Rosenfeld in person. Among the several characteristics of Lotfi, two features that appeared to be unusual and thus impressed me are—when I wanted to write a paper with him, Lotfi told me that he loves to work alone (showing his list of publications) and advised me not to put his name, and he never discussed fuzzy sets when we were together, whether in a car or a restaurant or at his house.

My reminiscences and road map would remain incomplete, if I do not mention the criticism that I received often, like many other fuzzy researchers, from my colleagues when delivering lectures or seminars within my Institute and outside. The situation can be felt easily considering that I have been in an organization named Indian Statistical Institute, surrounded by probabilists and statisticians. However, we have always viewed it as follows:

- Fuzzy set theoretical approach supplements the probabilistic approach and it is not a competitor, rather providing enrichment.
- We find a better solution to a crisp problem by looking at a larger space at first, which has different (usually less) constraints and therefore allows the algorithm more freedom to avoid errors by commission to hard answers in intermediate stages—*notion of embedding*.

Neurofuzzy and Rough–Fuzzy Computing

In the late 1980s, I got interested in artificial neural networks (ANNs) and started developing with my students various neural and neurofuzzy models mainly for classification, clustering, rule generation, feature selection, and connectionist knowledge-based systems. The idea of synergistic integration was to enable ANNs to accept linguistic input (*low, medium, high, missing features*) in addition to numerical input; exploit the ANN characteristics like adaptivity, robustness, ruggedness, speed via massive parallelism, optimality and capability in generating highly nonlinear boundary; and uncertainty handling capability of fuzzy sets in the input, output and during training. This greatly enhances the application domain of ANNs. Our article "Multi-layer perceptron, fuzzy sets and classification" (*IEEE Trans. NN,* 3, pp. 683–697, 1992) received the *Outstanding paper award from IEEE Neural Networks Council*. We have developed a series of generic models and demonstrated their applications to noisy/overlapping fingerprint identification, speech recognition, atmospheric science, image processing, etc. Through integration, it has also been possible to make a layered network, which is usually used as supervised classifier, act as an unsupervised classifier using the fuzziness measures as error detectors (*IEEE Trans. FS,* 1, pp. 54–68, 1993). Efficient unsupervised feature selection models were also developed accordingly (*IEEE Trans. NN,* 11, pp. 366–376, 2000) along with some application-specific models, e.g., mixed category perception (*IEEE Trans. NN,* 6, pp. 1091–1108, 1995) and fuzzy clustering network for hardware realization (*IEEE Trans. NN,* 11, pp. 1174–1177, 2000).

To enhance the computational intelligence characteristics of the said fuzzy networks, particularly for mining large data sets, we then started integrating the merits of rough sets and genetic algorithms into them. I had become interested in rough sets (RSs) and genetic algorithms (GAs) when I was visiting the NASA Johnson Space Center, Houston, TX, during 1990–1992 and 1994 as an NRC Senior Research Associate. Dr. Robert N. Lea was my official advisor in NASA JSC. Bob was more than a friend who not only gave me free hand to carry out research whatever I wanted without interfering, but also helped in my personal/family matter whenever required. I used to attend several seminars on rough sets organized in the Software Technology Branch, Information Technology Division. Two features of rough sets, namely granular computing with information rules and uncertainty analysis with lower and upper approximations, drew my attention. Since RS has the capability in extracting the domain knowledge, whether supervised or unsupervised, with reduced dimension in the form of information granules/rules, these can be encoded as initial network parameters for reducing its learning time significantly. Similarly, GA-based learning, with chromosomes defined in terms of the network parameters and modified genetic parameters (*IEEE Trans. SMC-B,* 28, pp. 816–828, 1998), can replace the traditional gradient descent search technique which is slow and often gets stuck at local minima. The aforesaid synergistic integration of the four tools in the soft computing paradigm results in gain in terms of performance, computation time, and compactness of the network, among others (*IEEE Trans. KDE,* 15, pp. 14–25, 2003). So, it has wide application in mining data sets with large dimension and size, and in knowledge discovery. In this context, the convergence of GAs, which was a long-standing problem, was also proved theoretically (*Int. J. Patt. Recog. & Arti. Intell.,* 10, pp. 731–747, 1996).

Meanwhile, I also realized that since both fuzzy sets and rough sets provide algorithms for different kinds of uncertainty, why not we integrate them to have a much stronger paradigm for uncertainty handling than either of them. In 1997, I visited Prof. Andrzej Skowron, Warsaw University, Poland, under an INDO-POLISH collaborative project; there I met Prof. Z. Pawlak, father of rough sets. Andrzej and I edited a volume—Rough-Fuzzy Hybridization: A New Trend in Decision Making, Springer, Singapore, 1999, which is the first of its kind.

One may note that Pawlak's rough set theory is based on the concept of crisp set and crisp granules, and provides a framework of handling uncertainty arising from granularity in the domain of discourse or limited discernibility of objects. However, in real-life problems, one or both of them may be fuzzy. A gray image is such an example where the set (e.g., object region) can be fuzzy and the granules (e.g., pixel windows) may be overlapping. In order to model this, we have defined generalized rough sets (*IEEE Trans. SMC-B,* 39, pp. 117–128, 2009), where the set and granules could be crisp as well as fuzzy. Accordingly, one could use "granular *fuzzy computing*" or "*fuzzy granular* computing" depending on the application.

Again, in an image, nearby gray levels have limited discernibility, i.e., nearby gray levels *roughly resemble* each other, and the values at nearby pixels have *rough resemblance*. Therefore, in the rough–fuzzy computing framework, image ambiguity may be viewed as resulting from fuzzy boundaries of regions + rough resemblance between nearby gray levels + rough resemblance between nearby pixels. Accordingly, we defined generalized rough–fuzzy entropy based on lower and upper approximations. Its merits over fuzzy entropy and the significance of fuzzy granules have been demonstrated (*IEEE Trans. IP,* 18, pp. 879–888, 2009) as well as have shown the superiority of unequal-sized granules to equal size (*Applied Soft Computing,* 13, pp. 4001–4009, 2013).

Merits of rough sets and fuzzy sets have also been integrated judiciously in clustering problems where rough sets deal with vagueness and incompleteness in class definition, and fuzzy sets enable handling of overlapping partitions. Each cluster here is represented by a cluster prototype, a crisp core (lower approximation), and a fuzzy boundary. Membership values are unity for the objects in the crisp core region and are in [0, 1] for those in the fuzzy boundary region. In other words, rough–fuzzy clustering provides a balanced mixture between *restrictive partition* of hard clustering and *descriptive partition* of fuzzy clustering. Therefore, it is faster than fuzzy clustering and is capable of better uncertainty handling/performance (*IEEE Trans. SMC-B,* 37, pp. 1529–1540, 2007). Thus, wherever fuzzy c-means or c-medoids have been found to be useful in the past four decades, rough–fuzzy clustering would be superior. Merits of rough–fuzzy integration have also been observed recently for designing a granular classifier with least dispersion index (*Pattern Recognition*, 45, pp. 2690–2707, 2012), case-based reasoning classifier (*IEEE Trans. KDE,* 18, pp. 415–429, 2006), granular neural computation for feature selection (*Neural Networks,* 48, pp. 91–108, 2013), and spatiotemporal outlier detection (*IEEE Trans. KDE,* 26, pp. 194–207, 2014).

Other salient contributions made are in data condensation (*IEEE Trans. PAMI,* 24, pp. 1–14, 2002), unsupervised feature selection (*IEEE Trans. PAMI, 24,* pp. 301–312, 2002), and active support vector learning (*IEEE Trans. PAMI, 26,* pp. 413–418, 2004) in mining problems; stemming (*IEEE Trans. SMC-B,* 37, pp. 350–360, 2007) and page ranking in Web intelligence (*IEEE Trans. KDE,* 21, pp. 21–34, 2009); and gene selection (*IEEE Trans. SMC-B,* 40, pp. 741–752, 2010) and gene function prediction (*IEEE Trans. BE,* 56, pp. 229–236, 2009; *IEEE Trans. BE,* 59, pp. 1162–1168, 2012) in bioinformatics. Among the several research collaborations performed, mention may be made of the Department of Computing, Hong Kong Poly University, Hong Kong, and the Department of Applied Mathematics, University of Naples, Parthenope, Italy, specially the former, for giving very good academic dividends.

Current Research Work

Let me now mention two of the research problems we are recently dealing with:

Fuzzy Granular Social Networks and Big Data Issues

A social network is viewed as a collection of relations between social actors and their interactions. These actors form closely operative groups, which are often indistinguishable in the process of problem solving. This resembles the concept of granules, i.e., clumps of objects in the universe of discourse, drawn together, for example, by indistinguishability, similarity, proximity, or functionality. One may further note that the basic concepts of "conceptual similarities" "between nodes," "cluster of nodes," "relation between nodes and their interactions," etc. do not lend themselves to precise definition, i.e., they have ill-defined boundaries. So, it is appropriate and natural if a social network is represented in terms of fuzzy granules.

Accordingly, we have developed a fuzzy granular social network (FGSN) model which provides a unified framework to represent social networks effectively and efficiently. Here, a granule is constructed around a node with fuzzy boundary. The membership function for computing the degree of belonging of a node to the said granule is determined depending on the problem in hand. Accordingly, some of the popularly known network measures are newly defined under granular space, including the entropy of the network to measure the uncertainty arising from fuzziness (*Inform. Sci.,* 314, pp. 100–117, 2015).

The FGSN model is now being used to solve various problems of network analysis such as target set selection and community detection, and deal with the associated big data issues. For example, one may refer to an investigation for detecting fuzzy-rough *communities* (*Pattern Recognition Letters,* doi:10.1016/j.patrec.2015.02.005).

Computing with Words, Z-Numbers, and Machine–Mind Development

Here, I explain our work on computing with words (CWWs) and the significance of the Z-numbers, as recently explained by Zadeh (*Inform. Sci.*, 181, pp. 2923–2932, 2011). The CWW paradigm is inspired by the astounding ability of the human brain to perform tasks on the basis of concepts encoded in the words and phrases that frame natural language statements. It aspires to induce this amazing decision-making ability in the computer—a step toward evoking machine–IQ (M-IQ) in a computer. CWW is imperative when (a) the information to be conveyed lacks numeric precision; (b) the situational imprecision can be exploited to arrive at robust, low-cost solutions; (c) numeric computing principles cannot be applied; and (d) words express a lot more than numbers. Potential areas of application of CWW are in semantic web searching, linguistic summarization of text samples or complex phenomenon, and subjective decision-making.

The concept of the Z-number, proposed by Zadeh in 2011, depicts it as to being able to provide the perfect environment of amalgamation of CWW and behavioral computing. We have started investigating this issue (*Fundamenta Informaticae,* 124, pp. 197–229, 2013; *Theoretical Computer Science,* 448, pp. 2–14, 2013), have developed *Z**-numbers as augmented Z-numbers for machine subjectivity representation (*Inform. Sci.,* 323, pp. 143–178, 2015), and have utilized the study to design a machine–mind framework for subjective real-world comprehension (*Natural Computing*, Springer (doi:10.1007/s11047-014-9478-x).

Summary

The aforesaid research in pattern recognition and machine intelligence *has led to the emergence of several modern disciplines* involving fuzzy sets with other computational paradigms, as evident from the literature. This has also led to the *incorporation of rough sets as a component of soft computing*; the original definition of Zadeh had components such as fuzzy logic, neurocomputing, GA, and probabilistic reasoning. This augmentation by rough sets has enhanced the computational intelligence property of soft computing and triggered its multifarious applications.

I graduated seventeen Ph.D.s, till now, in India and abroad who are well placed in premier institutes and industry. Out of the twelve Ph.D.s from ISI, one is an SS Bhatnagar awardee, two are FNA, two are IEEE Fellow, four are FNAE, and three are FNASc, so far.

Prof. Lotfi Zadeh visiting the Center for Soft Computing Research, Indian Statistical Institute, Calcutta, in February 2006: (*L* to *R*) Dwijesh Dutta Majumder, Lotfi Zadeh, and Sankar Pal

The impact of the research contributions is characterized by *h-index* = 68 and *g-index* = 135 with more than 22,000 citations (Google Scholar) and several prestigious national and international awards/honors received from various dignitaries such as three prime ministers of India, president of India, and president of Iran.

For furtherance of research in the said topics in a consolidated manner, our Institute created a new unit in March 1993, called Machine Intelligence Unit (MIU), under my headship. As recognition of our research contributions, the Department of Science and Technology, Government of India has established the nation's first research center in soft computing in October 2004 at ISI, Calcutta, under my leadership. The international conference Pattern Recognition and Machine Intelligence (PReMI) is the brainchild of MIU. The conference is unique in the sense that it provides a platform to exchange ideas regularly between these two communities for mutual benefit. The first edition was in Calcutta in 2005, the fourth one in Moscow in 2011, and the sixth edition in Warsaw in 2015.

"Pattern recognition and soft computing" have not only led me win several prestigious awards and honors in India and abroad (including *SS Bhatnagar Prize* and *Padma Shri*), and attain the highest administrative and academic position, namely Director, of Indian Statistical Institute, a premier research institute of international repute, but also gave me opportunities in visiting about forty countries for academic purpose and more importantly, to know the common people and culture there and make new friends. It will remain incomplete if I do not acknowledge my Ph.D. students and young colleagues/collaborators with whom I have done most of the work. They keep me always young.

For further references on this, one may go through the following monographs: (i) S.K. Pal and D. Dutta Majumder, *Fuzzy Mathematical Approach to Pattern Recognition*, John Wiley & Sons (Halsted), N. Y., 1986; (ii) S.K. Pal and S. Mitra, *Neuro-Fuzzy Pattern Recognition: Methods in Soft Computing*, John Wiley, N.Y., 1999; (iii) S.K. Pal and S.C.K. Shiu, *Foundations of Soft Case-Based Reasoning*, John Wiley, N.Y., 2004; (iv) S.K. Pal and P. Mitra, *Pattern Recognition Algorithms for Data Mining*, CRC Press, Boca Raton, Florida, 2004; (v) S. Bandyopadhyay and S.K. Pal, *Classification and Learning Using Genetic Algorithms: Applications in Bioinformatics and Web Intelligence*, Springer, Heidelberg, 2007; and (vi) P. Maji and S.K. Pal, *Rough-Fuzzy Pattern Recognition: Application in Bioinformatics and Medical Imaging,* Wiley—IEEE, N.Y., 2012.

A View of India Through Kolam Patterns and Their Grammatical Representation

Kamala Krithivasan

Kolam is a form of decorative design that is drawn on the front courtyard of a house in India. Drawing of kolam patterns is more prevalent in South India, especially in the rural parts. With big apartment complexes coming up in cities, the front yard concept is scaled down, but still small patterns are drawn in front of the front door of the apartments. In North India, the patterns are called Rangoli, and they are drawn with color powders.

In South Indian villages, the women members of the house get up early in the morning and prepare the surface in the front courtyard by sprinkling water. Sometimes, cow dung is also used to wax the floor. This is because cow dung is supposed to have antiseptic properties. Even when the floor is still not completely dry, decorative kolam patterns are drawn, so that they remain there for sometime.

The drawing on the courtyard is carried out by women who deftly design with pinches of rice flour held between their fingers (Fig. 1). Nowadays, chalk powder is also used. The use of rice flour is motivated by the idea to help ants to get their food easily. The rice powder is said to invite birds and other small creatures to eat it, thus showing how to have a harmonious coexistence. Throughout the day, the drawings get walked on, rained out, or blown around in the wind and new kolams are drawn the next morning. In some places, kolams are drawn twice a day, early in the morning and late in the afternoon.

Kolams are thought to bestow prosperity to homes. It is like a welcome board placed in front of the home. Goddess Lakshmi, the Goddess of prosperity is invited by this action. If kolam is not drawn in front of a house, it amounts to mean that the people in the house are observing grievance. If a death occurs in a family, kolam is not drawn for thirteen to sixteen days depending on the clan to which the family belongs. On the death anniversary of elders also, kolam is not drawn in front of the house. In contrast, when a child is born into a family (the birth may be in another town/village), irrespective of the time of the day, the front yard of the house is decorated immediately with kolam. It is to welcome the new entrant in the family. On happy occasions and festivals, the patterns are large and elaborate.

K. Krithivasan (✉)
87, Sripadmanaba Nagar, 5th Street, Adyar, Chennai 600 020, India
e-mail: kamalakrithi2002@yahoo.co.in

P. Ghosh and B. Raj (eds.), *The Mind of an Engineer*,
DOI 10.1007/978-981-10-0119-2_48

Fig. 1 Kolam drawing

In wedding halls, kolams are drawn one day prior to the wedding or function using rice flour paste. For this, rice is socked in water for more than an hour and ground nicely. It is made into a watery paste. A small piece of cloth is soaked in it and held between the ring finger and the middle finger, and as the hand moves, the thumb presses the cloth so that white liquid flows uniformly and patterns are drawn. After letting it dry for a while, the design looks beautiful and stays for a long time. On the borders and at the central places of the pattern, red powder (kavi) paste is used to make the design more attractive. Kolams are drawn in the pooja room of a house which is a place of worship. In temples also, large kolams are drawn. Some volunteers do it as it is considered as a kind of service. Nowadays, white paints are used in temples so that the patterns stay for a longer time in spite of people walking over it. It is also believed that the lines must be completed (closed curves) symbolically meaning that it prevents evil spirits from entering the shape and hence entering the inside of the home.

It used to be a matter of pride to be able to draw large complicated patterns without lifting the hand off the floor, standing up in between, something like Eulerian circuits. On occasions like Navarathri—meaning nine nights (in South India) (in the north, it is called Dasara—meaning ten days)—large patterns are drawn. It starts on the new moon day falling between 16 September and 15 October. In South India at the same time, dolls (called Golu) are kept in a decorative way inside the house. It is quite common to have competitions. A group of people visit the houses in a particular area and award prices to the best Golu and the best kolam patterns.

In the month of Marghazhi (Dec 15–Jan 15), people get up very early. Different large designs are drawn everyday. In villages, in some places of the pattern, a particular type of flower is kept with a small mud base. It is the flower of the pumpkin plant. Mostly, young girls get enthused and learn new designs and show their caliber. In olden days, men folk join together in that month early in the morning and go around the streets singing songs in the praise of God. In quite early times, young girls used to get up early, join together go to the river, and take bath and while returning singing songs in praise of God. It was thought that such practice would get them good husbands and also it was supposed to be good for the country.

Decoration is not the sole purpose of a kolam. It involves good physical and mental exercises. Bending and drawing the kolam gives strength to the muscles. Drawing properly a pattern needs focus and a good deal of concentration.

Kolam is an interesting tool to stimulate the mind of a person. It develops the memory and concentration powers of a kid when she tries to learn this. It can be practiced on a piece of paper with a pencil or on the floor or board with a chalk.

There is a lot of room for creativity. New designs can be drawn. In many magazines in India, there will be a separate section for women, and in that section apart from giving new recipes, etc., new kolam patterns drawn by readers are published.

There are different types of patterns in kolam. Usually, they are drawn in two ways. One is to use dots and draw curved lines around them. The other one is to draw the dots and join them in a proper manner. Figures 2 and 3 illustrate the two methods.

Some kolams are single patterns (hanging lamp) (Fig. 4). One method of covering a large area of floor space is to combine some of the smaller design to make a bigger one. Figure 2 shows how four of the first one is combined to form the second one and four of the second combined to form the third one and so on. Here, the lines are straight. Making them curved, we get a pattern like Fig. 5. They may be looked at as patterns having exponential growth.

Fig. 2 Anklets of Krishna

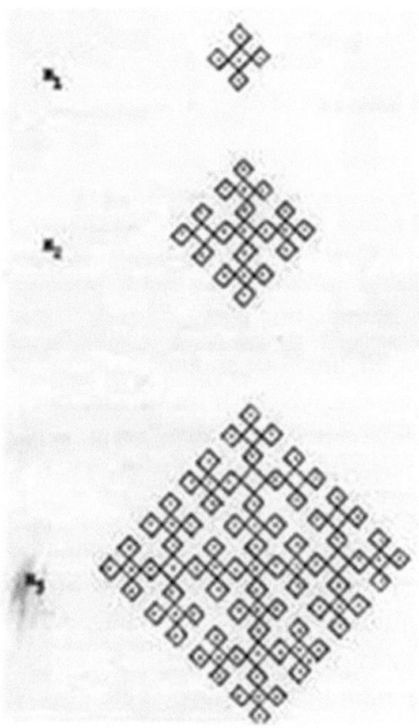

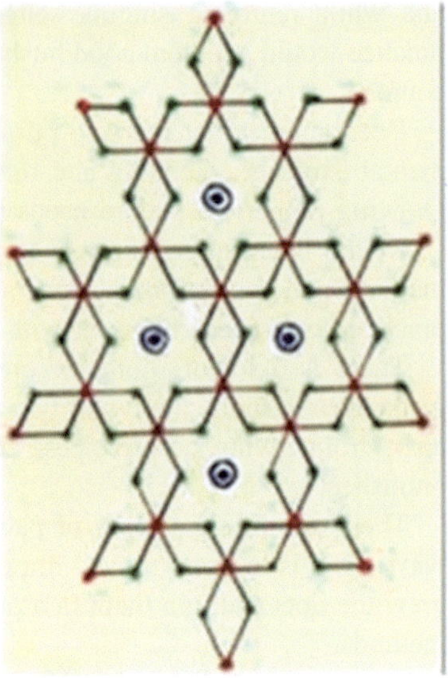

Fig. 3 Vilvathalam

Fig. 4 Hanging lamp

Figure 6 is another example of how similar designs can be drawn. How big patterns are got from smaller ones is shown in Fig. 7. The other type is by drawing dots and joining them. One such kolam vilvathalam is given in Fig. 3. Another design is Kooja (Fig. 8). They may be looked at as kolam having polynomial growth. Some kolams are circular in nature. One example is the kolam Hridhaya Kamalam (Fig. 9). It is sometimes interesting to find out how many lines (closed curves) make up the pattern. Some of the kolams are drawn by a single closed curve (Eulerian circuit). Examples are Krishna Salangai and Hridhaya Kamalam.

There are certain beliefs associated with kolam patterns. It is also believed that certain patterns have some effect on the health and wealth of a family. There is a

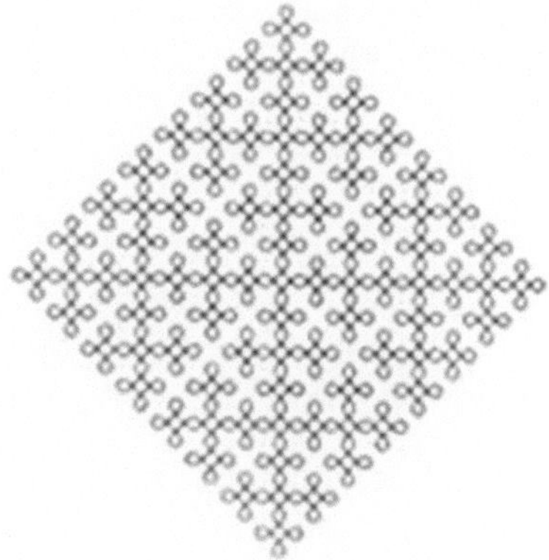

Fig. 5 Anklets of Krishna with curved lines

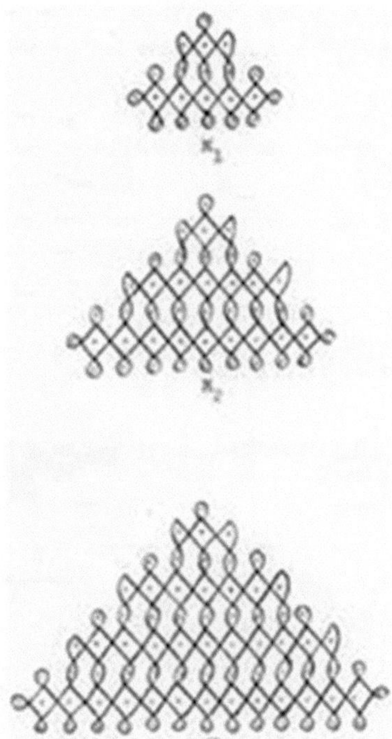

Fig. 6 Mountain

poem called 'Soundharyalahari' consisting of hundred verses. They are in praise of Goddess Adhi Parasakthi and were written by a saint Adhi Sankara during the tenth century. Each song is believed to have some particular power for ailing or bringing wealth, etc. For each verse, there is a particular design and it is believed that if one recites the verse for a specified number of times for a specified number of days after drawing the pattern and placing the idol of Goddess on it and offering a specified dish prepared to the Goddess at the time of finishing, he/she will be relieved of the ailment or get what they wanted. There is also a design called Meru or Srichakra

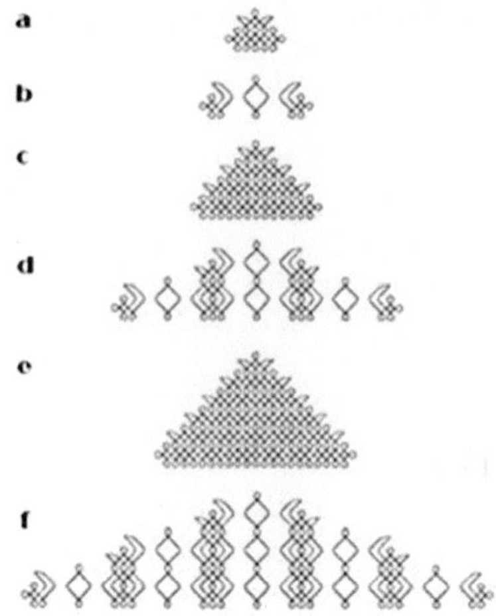

Fig. 7 Curves making up the mountain kolam

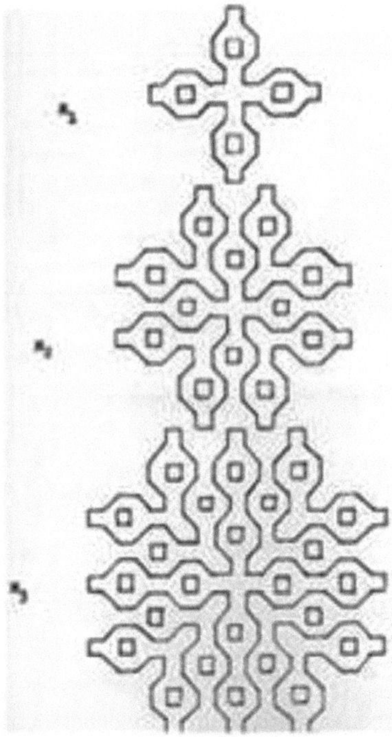

Fig. 8 Kooja

which is kept in temples and in the place of worship in homes. The design is given in Fig. 10. This design is said to have some divine power. There are nine levels of closed contours in that, and two composers, Sri Muthuswami Dikshitar and Sri Uthukadu Venkatasubbayyar, have sung nine songs describing the powers of each level. These songs are sung during the nine-day celebration in September/October.

Fig. 9 Hridhaya Kamalam

Fig. 10 Srichakra

Not only Hindus (main religion in India) but also people from other religions draw kolam patterns. Figure 11 gives a pattern which is drawn on Christmas day.

There have been some attempts to generate kolam patterns using formal grammars [1–5]. This idea can be used to generate the kolam patterns using computers. Formal language theory was started by the need for a formal grammar for natural languages and programming languages. The grammar usually generates a language consisting of strings of symbols from an alphabet. But later, array grammars to generate rectangular arrays of symbols were defined. There are two ways to generate kolam patterns using array grammars. In one method, basic patterns are treated

Fig. 11 Christmas kolam

as symbols and an array grammar generates rectangular array made up of these symbols, and the arrangement is made properly to connect the basic patterns in a proper manner. Another method is to generate labeed dots and give simple instructions to join them [1]. For example, to generate kolam vilvathalam, an array of green, red, and blue dots is generated using array grammar (Fig. 3). Simple instructions are given as follows:

1. Join a red dot to the nearest green dots (the six nearest green neighbors for the interior red dot and two nearest green neighbors for the exterior red dots).
2. Draw small circles around blue dots. This kolam with green, blue, and red dots and lines joining them are given in Fig. 3. Patterns which are basically circular in nature such as Hridhaya Kamalam are generated in [2].

Another way of generating Kolam patterns by a computer is to generate strings using L-systems and interpret the symbols in the system as moves of a cursor. This approach is followed in [3–5]. L-system is a formal model where parallel rewriting of symbols is considered, and the start point is an axiom. The reader is referred to [6] for basic details on grammars and L-systems.

L-systems are used for a number of applications in computer imagery. It is used in the generation of fractals, plants, and for object modeling in 3 dimensions. Applications of L-systems can be extended to reproduce traditional art and to compose music. We describe below how L-systems are used to generate kolam patterns.

Here, the description of the pattern is captured as a string of symbols. An L-system is used to generate this string. This string of symbols is then viewed as commands controlling a LOGO-like turtle. The basic commands used are move forward, make right turn, make left turn, etc. Line segments are drawn in various directions specified by the symbols to generate the straight-line pattern. Since most of the patterns have smooth curves, the positions after each move of the turtle are taken as control points for B-spline interpolation. We see that this approach is simple and concise. First, we see an example of a fractal and then a kolam pattern.

Many fractals can be thought of as a sequence of primitive elements. These primitive elements are line segments. Fractals can be coded into strings. Strings that contain necessary information about a geometric figure can be generated by L-systems. The graphical interpretation of this string can be described based on the motion of a LOGO-like turtle.

A state of the turtle is defined as a triplet (x, y, A) where the Cartesian coordinates (x, y) represent the position of the turtle, and angle A, called the turtle's heading, is interpreted as the direction in which the turtle is facing.

Given the step size d and the angle δ, the turtle can move with respect to the following symbols.

f Move forward a step length d. The state of the turtle changes to (x', y', A), where $x' = x + d * \cos(A)$ and $y' = y + d * \sin(A)$. A line is drawn between the points (x, y) and (x', y').

F Move forward as above but without drawing the line.
\+ Turn the turtle left by an angle δ. The next state of the turtle will be $(x, y, A + \delta)$. Positive orientation of the angle is taken as anticlockwise.
− Turn the turtle as above but in clockwise direction.

Let S be a string and ($x0$, $y0$, $A0$) be the initial state of the turtle, and step size d and angle increment δ are the fixed parameters. The pattern drawn by the turtle corresponding to the string S is called the turtle interpretation of the string S.

Consider the following L-system.

$$\text{Axiom}: w: f+f+f+f$$
$$\text{production}: f \rightarrow f+f-f-ff+f+f-f$$

The above L-system is for quadratic 'Koch island' (Fig. 12). Here, we see how successive patterns are drawn using this L-system. The images correspond to the string generated for different derivation steps n are shown in the following figures. The angle increment δ is 90°. The step size d could be any positive number. The size of the 'Koch island' depends on the step size and the number of derivation steps.

Koch constructions are a specific case of L-systems. The initiator corresponds to the axiom in L-system. The generator is represented by the single production. Consecutive positions of the turtle can be considered as control points specifying a smooth interpolating curve. B-spline interpolation is used for most of the kolam patterns.

Kolam patterns are drawn in a similar way using L-systems. An axiom and rules are given. They are used to generate strings, and the symbols in the string are interpreted as moves of a cursor and the kolam is drawn. We have given below the L-system for the kolam candies (Mittai Pottalam). The pattern is given in Fig. 13. Axiom: (− D − − D)

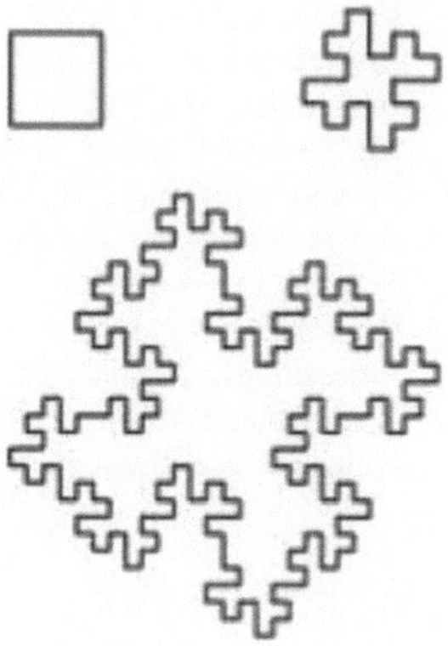

Fig. 12 Fractals generated by L-system

Fig. 13 Kolam patterns generated by L-system

Productions:

$$A \rightarrow f + +ffff - -f - -ffff + +f + +ffff - -f$$
$$B \rightarrow f - -ffff + +f + +ffff - -f - -ffff + +f$$
$$C \rightarrow BfA - -BfA$$
$$D \rightarrow CfC - -CfC$$

Angle increment = 45°.

Kolams are part and parcel of Indian culture. They are not only decorative designs but give good exercise to the brain too. Indians are proud of this tradition and hope this will be appreciated by people from other countries too.

References

1. Siromoney G, Siromoney R, Krithivasan K. Array grammars and kolam. Comput Graphics Image Process. 1974;3:63–82.
2. Siromoney G, Siromoney R. Radial grammars and radial L systems. Comput Graphics Image Process. 1975;4:361–74.
3. Prusinkiewicz P, Krithivasanand K, Vijayanarayana MG. Applications of L-system to algorithmic generation of South Indian Folk Part Patterns and Karnatic Music. In: Narasimhan R, editor. A perspective in theoretical computer sciences commemorative volume for Gift Siromoney. World Scientific, pp 229–247 (23).
4. Prusinkiewicz P, Hanan J. 1989. Lindenmayer systems, fractals. In: Lindenmayer A, Fracchis FD, Krithivasan K. Lecture notes in bio-mathematics (vol 79). Springer, Berlin.
5. Prusinkieurcz P, Krithivasan K. Algorithmic generation of South Indian Folk Art Patterns. In: Proceedings of INCOCG, Singapore, September 1988; pp. 323–335.
6. Krithivasan K, Rama R. Introduction to formal languages, automata theory and computation. Noida: Pearson Education India; 2009.

Further Readings

Traditional customs and practices-Kolams (http://www.indian-heritage.org/alngaram/kolams/).
Dr. Gift Siromoney. KOLAM (http://www.cmi.ac.in/gift/Kolam).

A Logical Perspective of Formal Verification: A Narrative on the Genesis and Evolution of the Formal Verification Group at IIT Kharagpur

Pallab Dasgupta

On 13 Aug 2009, a relatively private condolence meeting took place at the famous IBM TJ Watson Research Centre, USA, in memory of the departed soul of a former computer science professor of the Indian Institute of Technology, Kharagpur. Among those present were some of the most senior architects of IBM's state-of-the-art processors, who fondly recollected having learnt the basics of computer architecture from this professor. Over the next month or so, literally hundreds of obituaries poured in from his students all over the world. As his son, I learnt that a teacher lives on in the hearts of generations of students and that is one of the most gratifying aspects of this profession.

Having grown up in the campus of the oldest IIT, I was fortunate to witness an era where the size of computers shrunk from the size of a room to the size of a button. The cost of the transistor defied all laws of inflation, and in the process, the semiconductor industry empowered mankind with unbelievable computational power. Behind this, success story lays fascinating outlooks in computer architecture, scalable optimization algorithms and tools, and fundamental innovations in device physics.

What is often not perceptible at the surface is that the entire process of building an integrated circuit from its concept is based on the foundations of symbolic logic, namely the ability to capture the behaviour of electronic devices in terms of mathematical logic. Electronic design automation, which has equipped the chip designer with semi-automated tools for developing a chip having billions of transistors, would not be a reality had it not been based on logic and automated deduction.

My father was not a logician by training, but his Ph.D. made several fundamental contributions in circuit minimization based on logic optimization. By the time I studied the subject of logic optimization as an undergraduate student of computer science, the circuit industry was already using these and more advanced techniques ubiquitously. The history of mathematics has never witnessed a more

P. Dasgupta (✉)
Department of Computer Science & Engineering, Indian Institute of Technology, Kharagpur 721 302, India
e-mail: pallab@cse.iitkgp.ernet.in

P. Ghosh and B. Raj (eds.), *The Mind of an Engineer*,
DOI 10.1007/978-981-10-0119-2_49

rapid adoption of the advances in logic-based techniques as it has happened in the chip design industry.

The type of logic commonly used for digital integrated circuits such as processors is called Boolean logic or alternatively propositional logic. In simple terms, the circuit description defines the functionality of the circuit and its components in terms of logic formulae. Given the end-to-end functionality of a circuit, a synthesis tool automatically deduces the optimal combination of elementary circuit components (called logic gates) that together realize that functionality. The designer specifies the functionality of concurrent communicating circuit modules using high-level languages such as Verilog or VHDL, and the synthesis tools perform the underlying logic optimization automatically.

As the complexity of the circuits evolved from a few hundreds of gates to billions of gates, as is prevalent today, the task of verifying that the circuit has been designed correctly has become the most dominating task in the chip industry. Verification has not only become very complex, often accounting for more than 70 % of the design cycle time—it has also become a very critical task considering the use of electronic chips in a wide range of safety critical embedded systems, including medical instruments, automotive/avionic control systems, and atomic reactors. With the emerging ubiquity of embedded software-based control, the verification task now needs to address both the hardware and the software running on it.

The task of program verification, that is, whether a software program performs its intended task correctly is a much older and fundamental problem in computer science. Programs are also logical specifications, but the underlying logic is richer than Boolean logic and therefore more complex. In a seminal work presented in 1931 that shook the foundations of logic and deduction, Kurt Godel had shown that any logical system capable of expressing all relations between natural numbers was necessarily incomplete. In simple terms, this means that if a framework based on logic had the power of capturing arithmetic relations between natural numbers, then there will be formulas in that logic whose truth cannot be determined by logical deduction. Since programs use arithmetic, it follows that in general we can have no automated system of deduction that conclusively proves the correctness of any given program. Mathematical verification techniques for programs therefore primarily rely on abstractions and conservative approximations.

Digital circuits, unlike programs, are finite state systems, and therefore, the task of deducing whether a circuit will ever reach a bad state is always decidable. An automated system for deciding whether a circuit behaves correctly has two inputs—the *specification*, which defines the correct end-to-end behavioural requirements of the circuit, and the *implementation*, which is the Boolean logic of the circuit and its components. If the specification is also defined in terms of formal logic, then the automated system is called a *formal verification system*, and the task of deducing whether the implementation models (or implies) the specification is called a *model checking* problem.

The formative years of the formal verification research group at IIT Kharagpur coincided with the early years of adoption of this technology by the chip design and electronic design automation companies. The intuitive appeal of formal verification is quite significant. Circuits and programs have been traditionally verified through simulation and testing; however, such verification is not complete unless it is carried out with all possible inputs. The input spaces for most real-world systems are typically so large that exploring all input combinations are infeasible in practice. In a complex system, this leaves the system vulnerable to errors that may happen on those inputs that were left out. For example, in a complex railway yard, the number of ways in which trains may arrive and leave is so large that it is impossible to test a signalling system for all possibilities. On the other hand, formal verification proves a specified requirement (such as mutual exclusion among the conflicting routes in the yard) through logical deduction, and therefore, the coverage of behaviours is exhaustive leaving no scope for error.

The foremost challenge faced by the formal verification community was in choosing the language for defining the specification. The language should be expressive enough to capture the design intent but at the same time should not trade off the decidability of the underlying logical system in favour of expressive power. Specifically, the specification involves the notion of logical behaviours over time, which is not explicitly captured by Boolean logic. For example, the requirement that a traffic signal must turn yellow for some time before it turns red has a temporal connotation that cannot be expressed in Boolean logic (Fig. 1).

In 1977, Dr. Amir Pnueli proposed the use of a specification language for formal verification which extends Boolean logic with temporal operators. Thus, *linear temporal logic* (LTL), for which Dr. Pnueli was awarded the Turing award in 1996, enables the formal specification of properties over time within the syntactic fabric of a decidable propositional logic. LTL remained within the ambit of program verification, until the circuit community pushed for the development of formal language standards for writing assertions or formal properties of circuits. Today, it is believed within sections of the verification community that one of the good things that came out of the Pentium FDIV bug of 1994 was the rapid adoption of assertion-based verification in the semiconductor industry. Today, assertions are used extensively by all chip design companies, and formal verification of functional units in integrated circuits has become standard practice.

We started the Formal Verification Research Group at IIT Kharagpur at a strategic point of time, when the leading chip design companies (such as Intel,

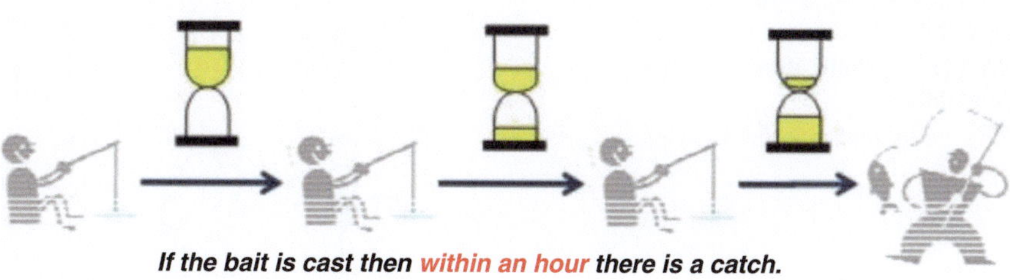

Fig. 1 Illustration: temporal logic allows us to specify timed behaviours

IBM) and electronic design automation companies (such as Synopsys) were collaborating through the Accellera consortium to merge their individual assertion languages (Forspec of Intel, Sugar of IBM, and Open-Vera of Synopsys) into a uniform language standard called SystemVerilog Assertions. This paved the way for designers to express the design intent through formal properties—possibly the widest use of logic among non-logicians.

While the task of developing language standards for formal property specification was being pursued, computer scientists and logicians were working overtime to address a very important bottleneck in formal verification called the *state explosion problem*. Circuits are inherently component-based concurrent systems, and hence, the number of states of a circuit is a product of the number of states of the components. To understand the problem of state explosion, consider a three aspect railway signal, which has only three states, namely red, yellow, and green. The railway yard of Howrah station near Kolkata has more than 150 signals, and therefore, the number of possible combinations of the signal states is 3^{150}, which is more than the Avogadro constant!! Some of these combinations are bad because they allow trains on conflicting routes, and the verification task is to determine whether such combinations are allowed by any sequence of execution of the signalling circuitry (Fig. 2).

The state explosion problem introduces an engineering dimension to the problem of formal verification. A theoretically complete proof procedure becomes infeasible because the state space defined by the consequents of the logic is too large to be stored or traversed explicitly. Contributions by Prof. Edmund Clarke, Prof. Alan Emerson, and Dr. Joseph Sifakis, for which they received the Turing Award in 2007, established symbolic techniques which allow the proof procedure to work on a succinct representation of the state space, that is without making the state space explicit. While this and other engineering innovations have helped in scaling the technology significantly, there is a still a very large gap between the requirements of formal verification and the feasibility of existing technology.

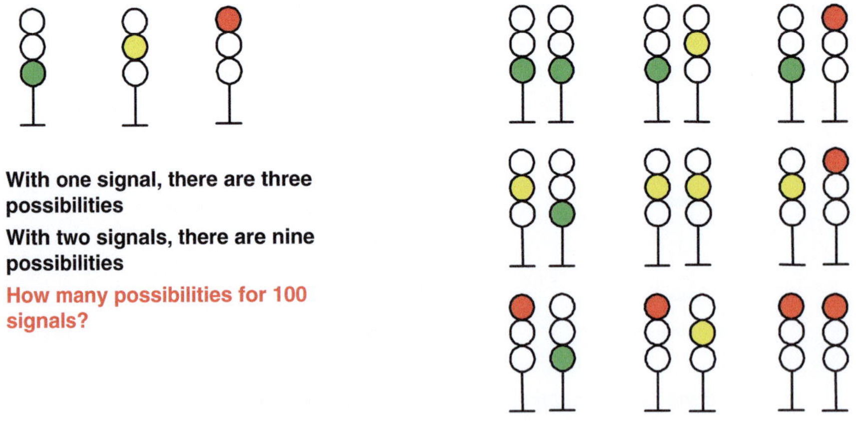

Fig. 2 Illustration: the state explosion problem

The focus of our research group has been shaped through the understanding of the practical difficulties in leveraging formal verification technology in industry. The first challenge came from a group within Intel, which posed a simple question with a complex answer—*Can the task of verifying a formal property on a circuit be decomposed to proving local properties on the component modules of the circuit?* The question essentially asks for a divide-and-conquer approach to formal verification. To us it was clear that attempting to automate the division of the functionality would not be practical in the absence of domain knowledge, since the state explosion problem would seriously affect the applicability of the technology. Hence, we proposed a bottom-up approach where the existing practice of writing assertions for the components could be leveraged to prove architectural properties through automated deduction. This new paradigm of verification was named design intent verification, since it involved a meta-level of verification where the design intent expressed in terms of architectural properties was being verified in terms of component properties (Fig. 3).

In principle, the design intent verification paradigm introduces a logical basis for engineering component-based designs. We realized very soon that this is a powerful concept which can be applied in a wide variety of domains. In the years that followed, we applied this technique on a wide variety of problems—each requiring proof procedures of different types, but with the same underlying philosophy.

One of the most interesting engineering applications of design intent verification was in the automotive domain. Complex features in automotive systems require coordinated behaviours involving multiple distributed components. In order to achieve the desired timing in an end-to-end functionality, the designer must budget the time among the component actions. Since components are typically outsourced

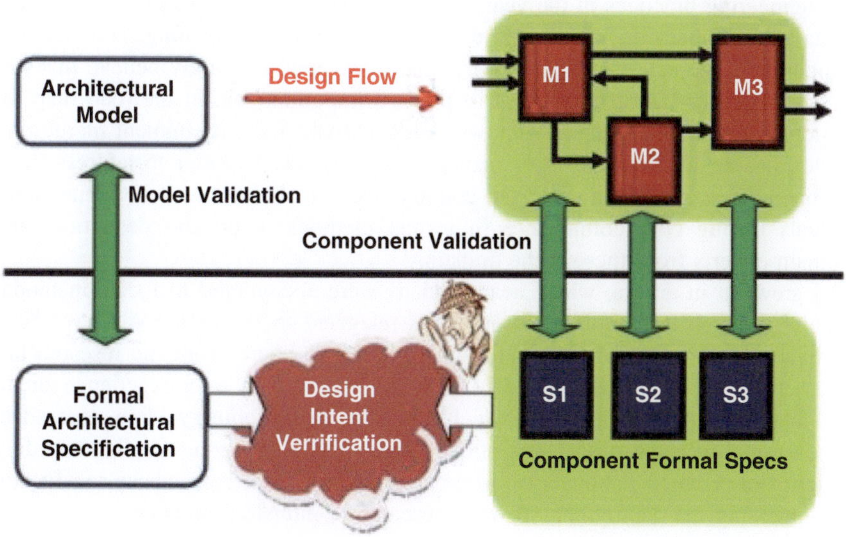

Fig. 3 Illustration: design intent verification

and a component may participate in multiple features, time budgeting is a complex global constraint satisfaction problem, which essentially justifies the specification of timing requirements in sub-system technical specifications. In collaboration with General Motors, we developed a framework which extended our earlier framework to real-time specifications, thereby adding fundamentally new decision procedures.

The core principles of design intent verification have more recently been used for formally verifying the architectural power intent in complex low-power digital integrated circuits. In simple terms, to save power consumption (and thereby extend battery life), most processors and system-on-chip designs regulate the voltage, frequency, and power on multiple islands of the chip (called power domains). The architectural power management strategy is implemented through a global controller which orchestrates the power controllers of the various domains. In collaboration with Synopsys, we developed a tool flow for verifying this orchestration at a feasible level of abstraction, by extracting out properties of individual power domains.

Yet another enriching experience was in developing formal specification languages and verification techniques for analog and mixed-signal circuits, which are essentially infinite state systems because the time and value domains are dense. We applied our techniques on circuits from the power management domain (power regulators, battery chargers, etc.) and used them to develop the frameworks for verifying the integration of cell phone power management units. This was done in collaboration with the National Semiconductor design centre at Scotland (now acquired by Texas Instruments). We have also been working with companies such as Freescale Semiconductors and Texas Instruments on analog verification through the prestigious consortium, Semiconductor Research Consortium (SRC).

Symbolic logic and automated deduction will play an increasingly important role in engineering practices in the years to come. New formal approaches will have a remarkable impact on knowledge discovery, meta-level reasoning, and decision support in large data-intensive cyber physical systems. I strongly believe that formal methods and AI techniques will play an important role in the road map for systems such as smart electrical power grids, climate and environment monitoring systems, and integrated disaster management systems. In order to address these problems in a structured way, the country needs to build teams involving individuals having core competence in formal methods, logic and deduction, and domain experts from these niche domains.

I grew up in an era, where academicians were encouraged to focus on fundamental research and applied research was relegated to the industry. I have been fortunate to work in the area of formal verification where academic research has been driven by serious requirements from the industry. My own experience taught me that the types of problems which come from the industry often necessitate fundamental research of the highest complexity and is therefore equally rewarding in terms of research publications. More significantly, there is immense satisfaction in witnessing the deployment of one's research in industrial practice.

Engineering and Innovation in Manufacturing and Production of Akash Missile System

Prahlada

Introduction

Akash Mobile Air Defence Weapon System has been designed for defending and protecting the vulnerable areas (VAs)/vulnerable points (VPs) from aerial targets penetrating from low, medium, and high altitudes. The system developed by DRDO has completed all development flight tests, user trials, and transfer of technology for production.

This paper brings out various engineering issues/problems that were faced in producing the first indigenous surface to air missile system in the country and how successfully these issues were resolved, industry got benefited, and production targets are being met.

Engineering Improvements and Innovation in Akash Weapon System During Production

During the development, certain functional as well as operational features were identified for the improvement by the users over and above what are indicated in ASR/GSQR. For example, the preparation time of missile had to be minimized, number of vehicles to be reduced, thus reducing manpower and time for launching. The improvements carried out in production version equipment and their benefits are indicated in subsequent paras. These modifications have already been verified through user operated validation flight trials on production version equipments. The general production engineering measures such as improving ergonomics, minia-

Prahlada (✉)
442B, Flat No. 2A, Naman Apartment, 2nd Cross, Ideal Homes Township,
Raja Rajeswarinagar, Bangaluru 560 098, India
e-mail: prahlada.ramarao@gmail.com

P. Ghosh and B. Raj (eds.), *The Mind of an Engineer*,
DOI 10.1007/978-981-10-0119-2_50

turization by adapting latest technology, and standardizing certain hardware across all subsystems are not discussed here. Only some of the solutions that fall under innovative engineering and managerial schemes have been brought out.

Improvements Implemented in the Command Control Center After User Trials

(a) **Inter-visibility computation to aid deployment**: Provision has been built in command control center to compute line-of-sight inter-visibility between any two stations given the latitudes and longitudes. This aids the commander in deciding the location for the deployment in order to achieve radio communication between them at the earliest. This has been achieved with the help of available digital terrain map.

(b) **Automatic Bias estimation and correction**: Provision has been made to estimate the bias offset of the reporting radars. This has been achieved by providing for the selection of one of the radar as reference radar and correcting all others with respect to that.

(c) **Manual backup for Multi-Sensor Data Fusion** (**MADF**): control center contains a proven, state-of-the-art, and automatic multi-sensor data fusion software for accurately synthesizing the real air situation picture. But under unforeseen conditions where data of any radar cannot be automatically correlated, manual intervention has been provided for association and disassociation with which the operator can provide information to the system for foolproof data fusion.

(d) **Clutter zones processing**: Provision to control false tracks has been made by creating zones in control center where no automatic tracks will be created.

(e) **Integration with Higher Echelons**: Work is at an advanced stage of realization to integrate control center with Integrated Air Command and Control System (IACCS) of the Indian Air Force for exchange of track data and control information. Control information includes the rules of engagement, automatic query for identification, and response and clearance to engage.

Improvements in Flight Level Radar

Programmable Signal Processor

Upgradation to programmable signal processor, using of SHARC processors employing digital Fourier transform (DFT), has resulted in providing greater flexibility for fine-tuning certain ECCM features and facility for future improvements without hardware changes. Radar signal processing in coherent mode has resulted in improved sensitivity (minimum detectable signal) and doppler filtering.

(a) **Digital Radar Scan converter as replacement of Analog CRT display**
The analog display of the radar has been replaced with a radar scan converter (RSC), which uses a touch panel and a 21 in. LCD interface for creating and viewing of targets in real time.

(b) **Automatic PRF Selection Algorithm for ensuring high reliability of tracking**
A new algorithm for PRF (Pulse-repetition frequency) selection based on velocity estimates of the targets has been implemented.

(c) **New tracker algorithm for improved accuracy for "High-g" maneuvering targets**
Upgradation of tracker with the incorporation of "Interactive Multiple Model" for tracking has been completed. This feature has also been validated with sorties of "High-g" maneuvering targets.

(d) **Modular and Compact receiver**
Receiver Mark II has been configured as a modular and compact solution to bring down size and power consumption.

(e) **New algorithms for weapon control and automation**
The decision support software, namely weapon control computer (WCC), which is a part of the FCC computer subsystems, has been upgraded with new Launcher Selection Algorithms. This has improved the reaction time significantly due to reduced human interaction. human–machine interface (HMI) for the radar computer has also been developed which enables remote operation of the radar.

(f) **Highly secure operating system based on Linux for display system of Control Center**
The erstwhile operating system on Microsoft Windows suffered from the drawback of its susceptibility to virus attacks. The Linux operating system is more secure and robust.

(g) **Replacement of TWT (x-band Command Link transmitter) with Indigenous Microwave Power Module (MPM)**

   (i) The indigenously developed MPM (Microwave Power Module) based Transmitter weights only 12 kg compared to 100 kg imported transmitter and consumes lower power, occupies less volume and costs less than three times the imported one.
   (ii) The new indigenous MPM provides for greater compactness, higher stability, and increased reliability. Moreover, due to its compact design, it has been mounted on the rear side of phased array antenna itself.

Innovations in Production of Phased Array Antenna

Pickup and radiating Plates: Following improvements were carried out to reduce machining of pickup and radiating plates which used to take six months per set:

(a) Formulae 62 stress-relieving machine were used for in process stress relieving during machining of the plates. Tooling was developed for clamping the plates during stress relieving.
(b) The total process of manufacturing of these plates has been optimized to the extent that one set of plates is now made in one month.
(c) Inspection of the plates is carried out on the machine itself, and the clamping of plates on the machine is done using the same holes with which the pickup and radiating plates are clamped in the actual array (hence simulating the actual condition).

Array PCBs: The initial PCBs were made through Micropack, Bangalore. Since the size of the machine on which the holes were drilled (for PCMs location) was not big enough, the requisite tolerances were not met and a lot of rework was required to be done in the assembly stage. A vendor was found in Ahmednagar with the requisite size of the machine, and the PCBs are now manufactured through them meeting the required tolerances. The vendor was developed by BEL to meet the overall requirements.

Continuity checks on PCBs: There are four pins in each PCMs (phase control modules), which provide connection to the PCB. Hence, there are about 4300 × 4 (17,200) connections. Continuity checks have to be performed to confirm correctness of all these connections. Special jigs have been made to reduce the total cycle time of checking these connections.

Slow cycling of Array: To remove the infant mortality in the PCMs, the array is subjected to multiple cycles of high power testing (for varying duration). After each cycle of full power, each PCM is checked for performance. Initially, it was all done manually, and hence, the time taken for each cycle was more than two days. However, now automated test setup has been designed to bring down the cycle time. Also multiple tools have been made so that this activity can be simultaneously performed in multiple arrays and hence bringing down the overall time cycle.

NFTR testing: characterization of first array in NFTR took six months. The following changes were brought about to reduce the cycle time:

(d) Initially, the array was tested without radome (since no reference surface was available when radome was put in the front side of the array). Hence, tests had to be repeated with radome. However, now the testing is done with the radome, and four machined plates are put at the four edges of the radome. Hence, the cycle time got reduced.
(e) The application software for collimation was optimized to reduce the total iterations required for collimation of the array.

(f) It is proposed to upgrade the NFTR facility. In this upgrade instead of one probe, there would be an array of probes which would substantially bring down the overall testing time of the array. It is envisaged that after hybrid upgrade, the total cycle time for testing of the array at NFTR would become about two weeks.

Transportation of Array from BEL-GAD to Bangalore: Special fixture has been designed which simulates the actual container top on which the FLR/TLR antennas are mounted. The antennas after customer clearance at BEL-GAD are transported to Bangalore on these fixtures. The fixture has four ISO corners and can be bolted on to standard trucks/trailers having ISO corners for clamping of standard containers. Multiple fixtures have been made so that simultaneously multiple antennas can be transported to Bangalore.

Improvements in the Akash Launcher System

(a) **Improved Operation and Maintenance features**

(i) Provision of identical motors and drive for azimuth. Elevation and outriggers to achieve standardization and better maintainability.
(ii) Provision of the entry of three taboo zones, both in azimuth and elevation for manual mode of operation in order to avoid inadequate slew of launcher to the restricted zones.
(iii) Improved layout of mechanical subsystems and electronic module on launcher to get easy access for maintenance and replacement.

(b) **Human–machine interface (HMI)**

(i) Provision of standby local console to give redundancy in operation.
(ii) Provision for health monitoring of subsystems remotely from BCC.
(iii) Enhanced security against unauthorized operation of the equipment through the provision of software access control for the operator's console.

(c) **Power supply system**

(i) Provision for automatic switching over to the alternate power supply source to enable uninterrupted operation of the launcher. The choice of power supply is automatically made from among various sources such as DG set main, DG set standby, commercial power supply, or batteries.
(ii) Provision for silent operation by using battery bank.
(iii) Battery monitoring unit (BMU) to be provided for continuous online health monitoring of each battery in battery bank.

(d) **Transportation**

(i) The minimum ground clearance of launcher has been increased to 400 for 300 mm without adversely affecting the feasibility of transportation by air, rail, and road.
(ii) Provision of two different canopies for transportation of launcher without missiles and with missile.

Missile Hardware Improvements

Hardware Changes

(a) **Sensor package unit**
The sensor package unit has been changed from free gyro-based system to roll rate-based system. This change has been carried out as technology upgrade as free gyro-based systems have become obsolete. Fiber optic-based sensor package has been indigenously developed and flight tested. Production process for this unit is also established. About Rs. 12.0 lakhs of foreign exchange has been saved per item, and the dependency has been reduced. The weight of this unit also got reduced from 3.2 to 2.5 kg.
(b) **Pilot tube**
The pilot tube has been removed from the configuration. The dynamic pressure calculation is being done onboard using sensor package unit and ground-based commands using radar estimates of height and velocity. This provides enhanced reliability and improved aerodynamics of missile by reducing its weight and drag.
(c) **Wings and fins**
The wing and fins with reduced hinge moment have been designed in order to improve the maneuverability and controllability of the missile.
(d) **Delivery of "Ready to Fire" Missile**
The following changes have been carried out in order to provide ready to fire missile to the user:

(i) The Pyro Relay Unit has been modified to provide extra safety for ground fired pyros. With this, the pyros will be fitted in the factory itself, thereby providing improved safety and no preparation in the field.
(ii) The air bottle is charged in the factory itself and need not be charged in the field.
(iii) Wings and fins need not be removed during transportation. The container has been modified to carry the missile with wings and fins duly assembled.

(e) **Improved reliability of command guidance unit**
The bias-T unit which was outside the CGU has been integrated in the system itself to provide improved reliability.

(f) **Additional safety during checks**
The firing pulse from radio proximity fuse to SAM which was earlier going directly has been routed through a relay to provide improved safety during the checks.

(g) **Thrust on Indigenization**
The following subsystems which were earlier imported have been indigenized in order to reduce foreign dependency.

   (i) Thermal battery
   (ii) Umbilical connector and push-pull connector.

(h) **Improved shelf life of missile**
The Section-III Airframe has been changed from magnesium alloy to aluminum alloy to provide better corrosion resistance and hence improved shelf life.

(i) **Improved Command Destruct Logic**
The software has been modified to carryout command destruct based on a new logic for initiating maneuver after the cross over. Accordingly, the warhead is exploded at a safe height. The change has been successfully carried out and validated as per the requirement projected by Indian AF.

(j) **Improved reaction time at launch**
The auto launch time has been reduced to 3 s from 5 s in order to provide better reaction time for the Akash weapon system.

Management Policies Adopted

Quality Management Methodology Through Production Phase

Since for the first time quality assurance of such a complex multi-discipline engineering work is attempted in the country, there was a requirement of ensuring high quality and reliability. Considering this requirement, Master Quality Assurance Documents were created to indicate the policies for inspection and acceptance of items at module level, subsystem level, system level, and weapon system level. This document also consists of the inspection and procedure to be followed for variety of equipments and their environment test requirements. The responsibilities of each stake holder (BEL, BDL, manufacturing agencies, DRDO, Service agencies, etc.) were listed, and stages were defined. Constitution of number of boards such as local waiver boards, standard design review boards, failure analysis boards, and configuration control boards were defined to bring in clarity to production and inspection procedures. The QA responsibilities were assigned to the agencies spread across the country.

Issues Resolved During Production of Akash Missile System

In order to ensure efficient and troublefree production normally associated with the increasing rate of production, quality, reliability, new environment, re-engineering, validation of performance of production equipments, etc., some of the issues addressed during production are listed as follows:

1. Flow forming of managing steel rocket motor casings instead of roll bending
2. Paintings, surface preparation on maraging steel rocket motor casings
3. Plating of En 24 components, magnesium alloy casings
4. Electronic stress screening (ESS) tests on all electronic packages
5. Sealing and rain proofing of missile
6. EMI/EMC testing of missile
7. Corrosion of studs—(pitting/white marks)
8. Air intake bent tubes—Indigenization
9. Reconfiguration on trailers
10. Pre-flight validation of modifications
11. Infrastructure establishment for missile storage/preparation buildings
12. Stage inspection clauses and testing classes
13. Weapon system integration documents
14. Phased flight trials of weapon system
15. Validation flight testing of first batch production equipment (trials by IAF)
16. Vetting of all QA plans, BOMs, TS, ATP
17. Amendments through SDRCs/acceptance of deviations through waiver boards
18. Batch qualification testing of propulsion systems
19. Modified missile containers
20. New missile check out facility
21. Configuration finalization for Army equipment on Tatas with Rail transportability
22. Roadability trials on systems mounted on HMVs
23. Second vendor development for fuel-rich grains (FRG)
24. Inspection constraints of MSQAA—manpower/quality/delay
25. SOP to prevent damage of missile container during transportation
26. Air bottle—monopoly issue—new manufacturers—alternate vendor qualification
27. Alternate to Tatra chassis—re-engineering requirement
28. "O" rings/gaskets—rubber components—additional vendor-aging qualification tests
29. Revised sampling plan, batch qualification plans based on production stability

30. Widening tolerances on drawings after studying manufacturing spreads over 50 sets
31. Weapons system integration site—joint investment by production agencies
32. Proper storage of items/electronic units after receipt storage at lead production agency
33. Microporosity issues of booster grains
34. Divergent nozzle—process improvements

Conclusion

Engineering and production challenges faced during serial production of this sophisticated missile system have been indicated in detail. The innovations and improvements undertaken for various weapon system elements have been brought out along with the benefits obtained. Various issues that were encountered and resolved during production have been explained. Having resolved all engineering and management issues, the weapon system is under continuous production and induction into Indian Defense Services with all these innovative features and benefits. The complete set of Indian industry (large-scale, medium-scale, and small-scale industries) are busy on production; product development and lifetime product support for the next 3 decades till the system gets phased out.

Acknowledgments The authors acknowledge the contributions made by all the system managers, project managers, teams from various DRDO laboratories, public sector undertakings, ordnance factories, industry, academia, quality agencies, etc., who are engaged in development, system integration, testing, evaluation, and production of Akash missile system. Our special thanks to all the project managers, system managers from various DRDO laboratories, engineers from various industries, and their teams. Our sincere thanks to late Dr. R.R. Panyam, who as project director of Akash, have led various teams and made significant contributions in taking the indigenous Akash missile system (AMS) into production.

The First Steps Toward Self-reliance in Solid Propellant Rockets

Rajaram Nagappa

India's standing in solid propellant technology today matches with the best in the world. Some of the largest solid propellant motors in the world power India's launch vehicles. And a still larger solid propellant power plant will power the GSLV Mk III experimental flight later this year. The seeds for this level of achievement and growth of this essentially homegrown technology were sown in the 1960s.

I was always fascinated with aeroplanes and enrolled for aeronautical engineering course at the Madras Institute of Technology (MIT). MIT, which was established in 1949, was the only institution in the country those days to offer a course in aeronautical engineering at the graduate level (the course at the Indian Institute of Science was at the master's level). MIT was a compact institute with an intake of 100 students every year, and the strength of the aeronautical engineering course was only 10 per year with the result the faculty–student interaction was rather strong. The added advantage was the aero-faculty boasted of some outstanding teachers, who instilled in us sound theoretical and practical knowledge. The syllabus included a course on rockets and missiles.

The Indian Space Program came into being with the establishment of Thumba Equatorial Rocket Launching Station (TERLS) and the launch of the first sounding rocket, the USA supplied two-stage Nike-Apache on November 21, 1963. Dr. Homi Bhabha and Dr. Vikram Sarabhai, founders of the Indian Space Program, foresaw the necessity of indigenous capability in the development of sounding rockets followed by capability in the development of satellite launch vehicles. In 1966, the Atomic Energy Commission had approved the establishment of Space Science and Technology Centre (SSTC) for this purpose. I joined SSTC as senior technical assistant in January 1967—I was the thirtieth member in the newly found organization. The people who had joined ahead of me had specialized in branches of engineering relevant to aerospace and some of them had been trained for brief durations in Japan, in the laboratories of the Institute of Space and Astronautical Science (ISAS). One of the ideas the propulsion and chemical engineers had picked

R. Nagappa (✉)
National Institute of Advanced Studies (NIAS), IISc. Campus, Bangalore 560 012, India
e-mail: r.nagappa@gmail.com

P. Ghosh and B. Raj (eds.), *The Mind of an Engineer*,
DOI 10.1007/978-981-10-0119-2_51

up was to test the initial designs in a small rocket motor[1] called the pencil rocket. Back at SSTC, this idea was utilized for the initial development of propellant systems.

The solid propellant motor has no moving parts. The propellant consists essentially of an oxidizer—ammonium perchlorate commonly used—in a matrix of fuel binder. Quite often, metal powder like aluminum is added to increase the propellant density and energetics. The propellant has the consistency of hard rubber (shore hardness in the region of 70). The burning surface at any given instant of time influences the chamber pressure and thrust produced by the motor. The propellant is therefore cast with different core shapes such as star, wagon wheel, dendrite, fins, and slots to provide the desired burn surface regression. The other controlling parameter is the propellant burn rate, in which the ammonium particle size plays an important role.

In SSTC at that time, there was no organized structure; the reorganization into discipline-based divisions and groups happened much later. The engineers had taken up development task in their respective areas of interest and consequently, there was some level of duplication. I joined S.N. Prakash, who was engaged in propellant development and propulsion system design. Propellant formulation was a trial-and-error approach. Laboratories were yet to be set up and only rudimentary equipment was on hand. The laboratory was an old house inherited when the fisher folk of Thumba were relocated to establish the rocket launching station. Early experiments involved composite propellants comprising ammonium perchlorate (AP) in a polyester resin binder. Pestle and mortar was used for hand grinding the AP; a laboratory stirrer powered by a small motor was used for blending the resin and plasticizer; an egg-beater served the purpose of mixing the resin and AP; finally, the thick paste was filled and hand rammed into the rocket chamber in which the central core had been earlier located followed by curing in an oven. Most of the propellant formulations used polyester resin as the binder along with minor quantity of plasticizer. Variations in the percentage composition of the fuel and oxidizer were tried. A few compositions were tried with epoxy resin. If the propellant preparation setup was rudimentary, propellant property measurement and inspection was mostly absent.

The rocket chamber was a thick mild steel tube of 50 mm diameter designed with a large factor of safety. It was closed at the fore-and-aft ends by flange with openings for placing the igniter and nozzle block, respectively. The assembly was held together with four tie rods with a notch in the center. These motors were referred to as Rohini 00 motors or RH00 for short. (Dr. Sarabhai had chosen to name the indigenous development program as Rohini rockets—in fact, the engineers in SSTC were referred to as Rohini Engineers.) The idea of using a heavy-walled tube was from considerations of (a) withstanding pressure excursions beyond the design pressure and (b) refurbishment and reuse of the hardware. The notch in the tie rod was designed to fail if abnormal pressure excursions occurred.

[1]The power plant of a solid propellant rocket is generally referred to as 'motor'.

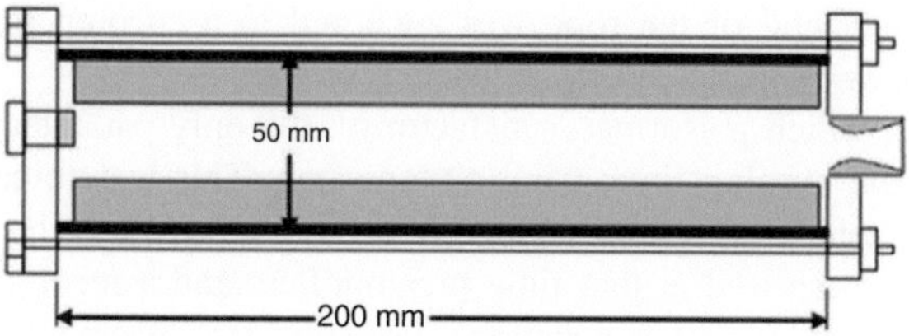

Fig. 1 RH-00 motor

This was a handy pressure relief system as in the initial days of development there were many cases of overpressure and failure. The motor is schematically shown in Fig. 1.

A number of trials had to be done before a propellant grain without visible surface cracks could be obtained. Buried cracks and debonds went unnoticed. Visual examination was the only technique, and internal cracks were obviously not always visible to the eye. A strand burner had been designed and fabricated and was used to measure the burning rate of the propellant samples. The igniter was characterized by doing a closed bomb test. For this, the setup was taken to the NASA supplied DOVAP trailer for recording the pressure transient.

A small test stand had been put in place at Veli hill. I used to carry the assembled RH-00 motor, a 12-V lead–acid battery, an igniter tester and the firing console in a jeep to Veli hill every time the motor had to be tested. The approach road to the test stand was not motorable and the motor and the test paraphernalia had to be carried by hand for the last 200 m. The precautions taken were to earth the rocket motor on the test stand, keep the igniter wires shorted till the final connection was made, and keep a safe distance from the motor while it was tested. The manual filling of the propellant into the chamber resulted in blowholes and gaps. This in addition to the cracks not visible to the naked eye resulted in quite a few explosions during the static tests. There were instances of non-ignition as well as hang fire. The latter was quite unpredictable—as one was not sure, if the grain would ignite after some time or not. In such an event, a healthy waiting period of 30 min was observed before approaching the test article. In addition to these precautions, the humid weather of Trivandrum was perhaps helpful in preventing buildup of static charge or we were plain lucky—during this period, there were no accidents.

The hardware though simple took some effort to fabricate. In-house fabrication facilities were just coming up and Trivandrum did not boast of any major industrial support infrastructure. The industrial estate in the Pappanamcode locality had some workshops and the fabrication of the hardware was carried out there. This demanded conservation and reuse of hardware. The notched tie rods came in handy for pressure release in many cases. It was the aft-end flange with the nozzle which used to be ejected away either due to overpressure or due to the bolts giving away and would get lost in the dense bushes in the neighborhood of the test stand. An ingenious method of retrieving the flange and the nozzle assembly was adopted by drilling an extra hole in the flange. A nylon rope was threaded into this hole and knotted. The other end of the rope was anchored to a coconut tree. The ejected nozzle after the test was hauled in, inspected, and reused.

For the tests, which performed satisfactorily, the only parameter that could be measured was the burning time using a stopwatch. This was remedied when the instrumentation setup was commissioned in a room adjacent to the test stand. The instrumentation comprised at that time preamplifiers and a recorder to record the chamber pressure and rocket motor thrust. Pressure transducers were commercially procured along with a dead weight tester for calibrating the transducer. A load cell designed in-house and fabricated in the workshop was used for measurement of thrust. A tensile testing machine was used for calibrating the load cell. As the transducers were available in limited numbers, their deployment was done with great deliberation, after testing a few RH-00 motors without instrumentation and ensuring a level of certainty and repeatability of the stopwatch measured burning time. One can still recall the excitement when the first measurement of pressure and thrust was done. Analyzing the data for the exact burn duration, the achieved pressure, and thrust provided a level of validation to the design calculations.

RH-75: India's First Indigenous Rocket

It was around mid-1967, when Dr. Sarabhai decided that the training and experimentation had provided enough knowledge and should be put to realization of a system. The engineers had shown him some tests and presented their ideas of the laboratory and facility requirements to proceed further with the tasks. Sarabhai decided that two teams would independently build a 75-mm-diameter rocket called the Rohini 75—RH75 for short. The rockets built by the two teams were designated as Mk I and Mk II. The exercise, if one may term it so, involved the design and development of the complete rocket systems and its flight test. RH75 would be an unguided rocket; meant for learning; the requirement was it should be able to take off and perform a stable flight; and no payload or performance target was set.

The Mk I team was led by Y. Janardana Rao, an aerodynamicist who had worked extensively with NASA prior to joining SSTC; A.E. Muthunayagam would take care of the propulsion and the mechanical engineering aspects of hardware design and realization. C.V. Ouseph would work on the propellant development. M.C. Mathur, also an aerodynamicist, who had worked with Bristol Aerospace in Canada, led the Mk II team. S.N. Prakash supported him for the propulsion and propellant systems. R. Vivekanandan would provide the aerodynamics support to both the teams. Both the teams started off in right earnest chalking a development and rocket realization strategy.

The enunciation of the RH75 development provided opportunity for fresh recruits like me to get involved in the design and development activities. By then, I had applied against a post in the Mk I project and had been selected as Engineer SC. Besides me, two more engineers had joined the Mk I propulsion team.

The key element was the propellant and both the teams concentrated on this aspect. This was a slow process as the propellant facilities were still being set up and some facilities had to be shared between both the teams. Also, though computer program for propellant combustion had been developed, the formulation per se was still a trial-and-error approach and the yield was poor. At this point, we came to know that the Cordite Factory at Aravankadu (CFA) in the Nilgiris was producing extruded double-base propellant grains, which was 67 mm in diameter. The diameter of the propellant grain was close enough to the design diameter of the RH75 rocket. It was felt engineering the rocket around this grain would save precious time and the Mk I team could chalk up an early success. The propellant could be substituted once the indigenous development came through.

An expedition to Aravankadu indeed confirmed the availability of solventless cordite blocks with the factory designation SU/K. The propellant grain had a seven point star inner geometry and length of 550 mm. The factory had a stipulation of a minimum order quantity that was far in excess of our development requirements. In the absence of an alternative, we ordered nearly a tonne of the propellant. When the batch was ready, we requisitioned a jeep with trailer from the transport section and drove up to Aravankadu to collect the propellant. There were no explosive vans available in SSTC/TERLS complex—even the sounding rockets received from the USA and France were unloaded at Cochin port and transported by trucks to Thumba with a pilot escort. Safety features enforced during transport were the speed limit and an earthing chain on the vehicle for grounding any built up charge.

For the project team, speed of accomplishment was as important as a good design. Rather than design and fabricate the rocket casing, a survey was done to see what was readily available in the market. The Indian Aluminum Company (INDAL) was manufacturing extruded seamless aluminum tubes and aluminum rods. It was decided the tubes could be used for the rocket chamber, and appropriate machined lengths of the aluminum rods could be used for the fore-and-aft closures for the chamber with appropriate openings for the nozzle and igniter. A check showed that the wall thickness of the tube was adequate for the design maximum pressure with necessary factor of safety. INDAL agreed to fabricate a die and extrude the tubes to our requirements, but insisted that a minimum order quantity of 1 tonne had to be placed. Orders were placed for aluminum tubes of 75 mm diameter with wall thickness of 2 mm and aluminum rods of 75 mm diameter. The alloy offered by INDAL was B51SWP, a solution-treated and precipitation-hardened alloy whose ultimate tensile strength was around 280 MPa.

The rocket motor length was obviously dictated by the propellant grain in this instance. It was felt that a single cordite grain of 550 mm would be too short and to have some meaningful thrust and performance, a longer length was desirable. The engineers at CFA had informed us that the grains could be bonded by applying acetone on the surfaces to be bonded and maintaining slight pressure. Bonding two lengths of the cordite grain to achieve a grain length of 1100 mm was an obvious conclusion. The grain ends were chamfered and bonded to realize the full length grain.

In a solid propellant rocket, the chamber pressure is dependent upon the propellant burning surface area at any given instant of time and the nozzle throat area through which the combustion gases are exhausted. In the RH-75 design, a design issue came to the fore. The nozzle throat had been arrived at to maintain the maximum pressure within the capability of the case material. This resulted in a throat diameter which was larger than the initial port diameter of the grain, which was contrary to normal design practice. My calculation showed that this will lead to overpressure as sonic velocity would be reached at the end of the grain rather than in the nozzle throat. We discussed this, but I could not convince the team that the pressure achieved will be beyond the case capability. We went ahead with the first test using a heavy-walled mild steel proof motor case. The test result showed overpressure as predicted but did not end with catastrophic result as the proof motor had a large factor of safety. The maximum pressure could not be recorded as the pressure transducer showed a flat value corresponding to the saturation limit of the transducer.

In the discussion following the proof motor test, it was decided that the grain port area had to be increased to a value higher than the throat area to ensure that the sonic velocity occurs in the nozzle throat. The center port of the grain was a 7-point star. It was noted that the diameter of the outer star point was slightly greater than the throat area and it was decided to enlarge the propellant bore equal to the outer star diameter for a length of 300 mm. The grains were taken to the workshop and radial drilling machine was used to machine off the star lobes and render the port circular for a length of 300 mm.

Thankaiyyan was one of the two machinists in the workshop. A well-built person, he found drilling into the cordite grain was like cutting butter and we had to constantly remind him to do the drilling slowly. Even while using the coolant, the grain became quite warm. The other precaution we took was to carry out the work after the main shift was over. Thinking back, I feel the exercise we did was risky, but the spirit and adventure of accomplishment was so great we just went after the job. We managed to drill the cordite grains without any mishap.

The modified grain was tested in the proof chamber and provided good result. The maximum pressure and thrust as well as the burn time matched with the prediction. With this successful achievement, it was time to take up tests with the flight chamber. The flight chamber construction was quite simple. The aluminum tube formed the chamber. Even at the initial stages of design, we realized the importance of thermal management and sealing gas ingress to undesired areas. This called for ensuring not only good bond between the propellant grain and the casing, but also good sealing at the grain termination points and at the hardware joints. For the RH75, this was achieved essentially through resin systems. The outside diameter of the grain was smeared with polyester resin and then inserted into the tube. Polyester resin was then gravity-fed into the annular gap between the grain and the chamber to the point of overflow.

The fore-end closure was fashioned from the aluminum rod and housed the igniter. The aft-end closure also used the aluminum rod of a longer length and housed the nozzle. Both the closures were attached to the chamber with twin rows of screws. Araldite applied to the closures prior to assembly ensured gas tightness of the joints. The nozzle used a graphite block for the throat and some part of the divergent. The remaining part of the divergent was made of mild steel and functioned as a heat sink for the hot gases. For the 2 s of burning of the motor, this proved to be adequate. A basket-type igniter with a charge of black powder and SR 371 was used to ignite the motor. The rocket motor parts are shown in Fig. 2.

The choice of the bonding/filling resin and curator ratio was done with abundant caution. Resin samples were prepared in paper ice cream cups with different resin/curator ratios. The temperature due to the exothermic nature of curing was measured and a ratio where the cure temperature remained below 60 °C was chosen. After successful static tests, the motor was integrated with the fin shroud and nose cone. The nose cone was filled with 1 kg of lead—euphemistically called inert payload—to maintain the center of gravity of the rocket forward and to provide an adequate margin of static stability. The first flight of the motor took place on November 20, 1967, and was tracked by the range radar. The height achieved was 10 km. This was a historic event as this symbolized the first indigenously built rocket to be flown from Indian soil. This would prove to be the stepping stone to further accomplishments in solid rockets. The RH-75 Mk I rocket can be seen in the following picture.

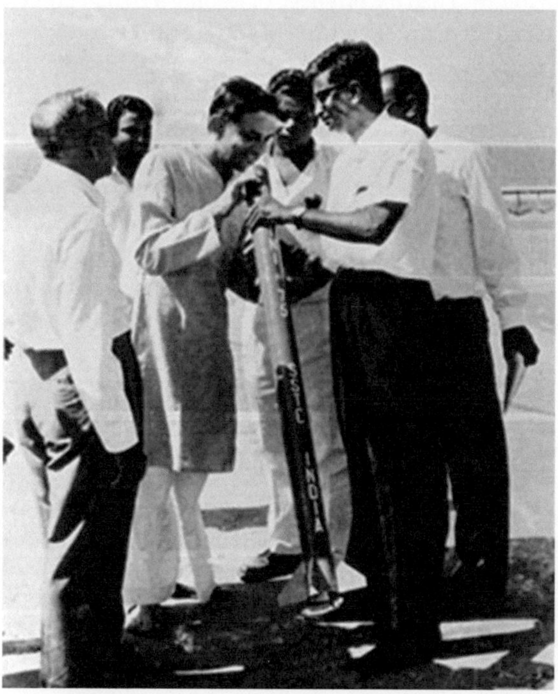

Dr. Y.J. Rao showing the RH-75 flight rocket to Dr. Vikram Sarabhai. The others in the picture are Shri H.G.S. Murthy (extreme left) and Dr. A.E. Muthunayagam (partially hidden by Dr. Rao). Shri Murthy was at that time director, TERLS

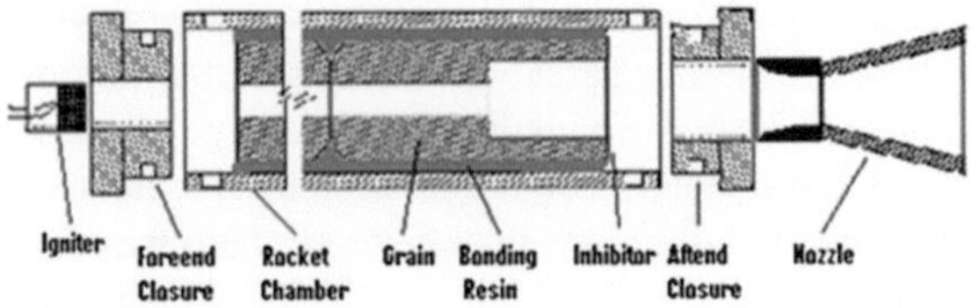

Fig. 2 RH-75 assembled motor

The environment at SSTC was informal, competitive, and knowledge seeking. Engineers were encouraged to come out with new development ideas and propose technology development projects. New propellant compositions (at a time three agencies were separately conducting propellant formulation trials), different case materials and manufacturing techniques, use of ablative materials and composites all came about during these early years. Experimentation included multistage rockets and boosters employing multiple rocket grains. It was a good learning experience, which paved the way in the later years for systematic and organized approach to realizing solid propellant motors for the sounding rockets and launch vehicles.

In the years to follow, I had the privilege of leading the solid propellant rocket team and the team chalked up many significant events. Realization of case-bonded motors for the SLV program was followed by motors with higher energy propellants for the ASLV motors. An improved fourth-stage motor with poly aramid (Kevlar) fiber motor case was another achievement. Major technology development in terms of hydroxyl terminated poly butadiene (HTPB)-based propellant systems, maraging steel for booster motor case and filament-wound Kevlar motor for the third stage, gimbaled nozzle construction for the upper stage were all developed and realized for the motors of the Polar Satellite Launch Vehicle (PSLV). When the PSLV booster with a diameter of 2.8 m and containing 125 tonnes of propellant was static tested in October 1989, it was third largest motor in the world. All elements of the motor were made in-house or fabricated/processed by Indian industry. Only few items like fasteners and resin curator, which were not economical to make in view of limited quantity requirements or freely available due to usage in other industries were procured. The present set of solid propellant rocket engineers at the Vikram Sarabhai Space Centre have contributed to further refinements and new systems. They realized and tested in January 2010 a solid motor of 3.2 m diameter and employing 200 tonnes of solid propellant, longer motor segments and employing a submerged gimbaled nozzle for vectoring the thrust for control of the launch vehicle in pitch and yaw directions. With the retirement of the space shuttle and Titan IVB solid rocket motors, this motor is next only to the 230-tonnes propellant solid motors used in the Ariane 5 vehicle.

This has truly been a remarkable and satisfying journey.

Note: The development of solid propellant rockets in the Country is brought out in more detail in my recent book "Evolution of Solid Propellant Rockets in India." The book is published by DESIDOC under the DRDO Monographs/Special Publication Series.

Challenges in Developing Indigenous Technology

R.D. Kale

Until late 1970s, there have been in the Departments of Atomic Energy and Space, little emphasis and effort placed on developing critical technologies though know-how transfer was already in existence in these Departments for sometime. With the advent of an ambitious programme of nuclear power reactors, that of satellite launch vehicles and the well-known integrated guided missile development programme or IGMDP of DRDO "indigenisation" became a key Mantra in all above departments pursuing cutting edge technology areas. While this was fraught with many problems and challenges, it was yet inevitable in key technological areas such as sodium-cooled fast breeders where there was little choice if at all for importing such technologies. Having been associated with such projects as prototype fast breeder reactor from early 80s, I would like to present my views on the indigenisation efforts on the then emerging high-tech areas based on my own experience and that of my colleagues in IGCAR. The challenges to overcome in the indigenisation process can be broadly classified under three categories, viz., technological, industry limitation, and procedural; the last one refers in particular to tendering and contracting procedures under government regulations. The nature of these challenges can be best understood by discussion of some examples in the technologies encountered.

Technology of Liquid Sodium Pumps

Fast reactors use liquid sodium as heat transport medium to convey the nuclear heat to steam generators (SGs) by circulating sodium in their primary and secondary heat transport circuits. For the Prototype Fast Breeder Reactor or PFBR, gigantic sodium pumps are used to circulate 15,500 CuM/h sodium flow consuming 3400 KW power at a high operating temperature of 400 °C in their primary circuit.

R.D. Kale (✉)
Flat 105 A-8, Planet Millenium, Near Rakshak Society,
Pimple Saudagar, Pune 411 027, India
e-mail: ravindrakale42@gmail.com

P. Ghosh and B. Raj (eds.), *The Mind of an Engineer*,
DOI 10.1007/978-981-10-0119-2_52

Similar pumps are also needed in the secondary circuit. Such pumps are made by few select Western countries and not available for import being deployed in sensitive/strategic technology. The major challenge in the pump hydraulic design lays in achieving high cavitation resistance performance while keeping pump diameter as compact as possible. Here was a challenge never faced by Indian pump industry till then. A development project had to be conceived in collaboration with a reputed pump industry and hydraulic tests were initially conducted on smaller models initially to study parametric effects to arrive at a suitable impeller and diffuser combination before constructing a 1/2.75 scale larger model pump. For studying cavitation, erosion effects not only visual observation of cavitation in the impeller was necessary but also a suitable paint had to be developed for the assessment of impeller erosion in short-term water tests. Again as the suitable paint developed in the Western world could not be accessed, a separate collaboration was entered into BHEL who were familiar with such processes for cavitation study in hydraulic turbines. Vapour cavity lengths developed in the impeller were observed and later measured under stroboscopic light in transparent shroud of the model impeller. This was carried out for varying suction pressures and was later correlated with eroded paint areas of the impeller to determine the point of maximum erosion and its exclusion from the operating states of the pump.

Another challenging aspect of the pump design was its ability to accommodate differential thermal expansion during its high-temperature operation. A unique single point tilting pump rotor concept was evolved in which the vertical pump rotor supported by a spherical bearing undergoes a small inclination of 0.4° as the pump passes through a temperature range until it reaches its operating temperature. A hydrostatic sodium lubricated bearing at the bottom of the shaft guides the rotor safely. The concept had to be validated on a near full scale rotor assembly measuring 7.5 m before finalising the pump mechanical design. The complete development of the primary and secondary pumps spanned over two five-year-plan periods with major participation from IGCAR, M/s KBL, FCRI Palghat and among others.

Steam Generator Tube to Tube Sheet Welding

A very different technological challenge was faced in procuring 23-m-long and slender seamless tubes in 9Cr. 1Mo ferritic steel. These seamless tubes are must for the fabrication of sodium-heated SG to avoid a weld in the middle of the tubes for improving the reliability of the unit against a violent sodium water reaction. The Nuclear Fuel Complex was till then producing shorter tubes but they successfully came through this challenge and produced these tubes which had to pass Eddy current examination as one of the important specifications to exclude any non-metallic inclusion. The tube material was produced by vacuum induction melting and further refined by electroslag process. Then, there was a further challenge of achieving high-quality butt weld between the tubes and the tube sheet provided with machined spigots for the purpose. This weld had to meet 100 % volumetric examination which could be carried out only by micro-focal rod anode X-ray equipment. The welding itself had to be performed by internal bore welding with pulsed TIG welding. The acceptance criteria specified for these 2.3-mm-thick weld were isolated porosity of 0.4 mm maximum and a cumulative pore diameter of 5 mm over a weld circumference of 53 mm. The weld concavity and convexity both were limited to 0.3 mm, very challenging indeed. Special welding sequence with preheating and different high pulse levels of current had to be followed to achieve the first ever 25-m-long 19-tube model SG. The unit has been thoroughly tested over nearly 10,000 h in a specially constructed Steam Generator Test Facility.

Industry Limitations

When the programme of sodium pump development was launched, it was clear that no Indian pump industry could take up the pump design and development in its entirety. This was because the pump design involved high-temperature mechanical design and heat transfer considerations from hot sodium, besides a finely tuned hydraulic design. No pump industry was prepared to take up the responsibility for an integral design and development. Attempts were also made by opening a dialogue with a reputed pump industry working with NPCIL projects as well as with the then SERENA/FASTEC combination of European industries in the fast reactor field but to no avail. Hence, the development programme was launched with Indian industry taking up hydraulic design to meet our specifications coupled with in house efforts (IGCAR) for the rest of the pump design and development. The in-house efforts were devoted to design the ten-metre-long pump shaft for high operating temperature, the bottom hydrostatic sodium lubricated bearing among other components, and developing manufacturing technology for the long shaft of composite solid and hollow forgings, hard facing of bearing parts and a special float type non-return valve, the last one being designed in collaboration with FCRI, Palghat.

Limitations were also recognised in the manufacture of the SG test model measuring 25 m in height and 0.3 m in diameter. Here was necessary a close co-ordination in procuring long seamless tubes from NFC, thick 9 Cr 1 Mo tube sheets and plates from Midhani, in order to supply in time all free issue materials to L&T, the main SG supplier. Tremendous efforts on the part of designers were called for to coordinate with industrial groups as all above raw materials were being made to the stringent specifications for the first time in the country. The efforts placed in the SG did not stop here as considerably more work with industry was required to put in place a successful design of a conventional oil-fired heater; the major challenge being the heat transfer medium which was liquid sodium necessitating extraordinary material surveillance and inspection, e.g. helium tightness of all the welds..

Challenges Faced in Awarding Contracts

In case of sodium pump, a dialogue was initiated with pump industries both Indian and foreign for evolving suitable hydraulic design through a two-stage process. The first step involved in constructing a 1/3 model and the second consisted in building a full scale "hydraulic prototype" to cover adequately the scale-up from successful model results. Considerable efforts were devoted in convincing the Department (non-technical) authorities to accept such a tender as the actual full-sized pump was to be separately built for the project later. Positive replies to the communication from IGCAR were treated as pre-qualification bids after satisfactory bidder assessment. Following a three-stage technology, contract tendering (first time in the Department) bids were formally invited on a limited tender basis. The stages included (a) technical; (b) commercial without price; and (c) price bids. The order placement was preceded by a detailed techno-commercial dialogue and negotiation over a period of one-and-half year, the order being approved by a high-level committee that included secretaries in the department, in addition to Director and senior technical personnel. The whole process from the initial technical query to ordering stage took nearly four years, whereas the two-stage development took another five years. A further period of 5 years became necessary for additional development contract subsequent to a revision in the pump design that had to be evolved with the number of pumps being halved and made as compact as possible to suit a revised primary circuit design from economic considerations. Much more canvassing and convincing act had to be performed to pilot the project further.

In case of SG technology, it was equally important to design and construct in a time bound manner a test facility simulating all operating conditions of the actual plant. Towards this, a 5.7 MW test facility was constructed on war footing in order to freeze the SG detail design for PFBR. The 25 m tall, 0.25 m diameter 19-tube unit was accommodated in a 36-m-tall high bay with rest of the systems located in a smaller building. The vendors for the fabrication of SG model, the fired sodium heater and the feed water pump of unique high head (170 bar) and low flow of 10 CuM/h were selected after considerable pretender technical dialogue and careful deliberations during the tendering stage. Vendors' reputation and their capability to affect timely deliveries were given enough importance. EPC packages for steam—water system, fired heater and control-instrumentation proved extremely useful even in this project of R&D nature. An important feature of the periodic project review was the inclusion of accounts officer in every meeting to keep them abreast of importance and urgency of certain procurements and to effect expeditious release of payments in all works contracts. Emphasis on fast communications was a hallmark in running this project especially e mail communications which were introduced then recently (1998 onwards). The SG facility was a success story of the task forces set-up at different levels in IGCAR. All above considerations helped immensely in achieving this project without cost overrun though the project took six months more for its completion than the allotted 5-year period.

This article has attempted to bring out the technological and other challenges met with in developing difficult technologies indigenously. The author would like to acknowledge gratefully the contributions of many of his former colleagues both junior and senior that have been used in preparing this article.

The Challenge of Developing a High-Speed Rotor Assembly—How It Was Met

Ram Kanwar Garg

Preamble

The development of a high-speed rotating machine commonly known as 'gas centrifuge' and used in enrichment of uranium is considered a difficult and guarded technology, but strategic for the nuclear energy programme. This was achieved in the Bhabha Atomic Research Centre (BARC) during the period 1980–1990 by entirely indigenous R&D efforts. For me, as a leading participant in this development, right from conception to the final stage, it was both a challenge and unique opportunity which not many engineers get throughout their professional career. I take pride in sharing with fellow engineers and scientists my experience and my belief that for a dedicated team, no task is impossible if the right working environment is provided.

Being a multidisciplinary project, the R&D team was comprised of about twenty engineers and scientists in the disciplines of chemical, mechanical, electrical, electronics and instrumentation engineering, metallurgy, physics and chemistry. One unique feature of this team was that except two of us, all others had joined directly from the training school run by BARC to work on this project. The credit for the success of this project goes to all of them who worked with dedication and team spirit and employed several innovative ideas.

Glimpses of some of the materials, components and systems needed for this programme and the technological and managerial approaches that were adopted for this development, leading to the setting up of a demonstration facility, are provided in this article.

Right from the inception of the nuclear energy development programme started immediately after independence under the leadership of the visionary

R.K. Garg (✉)
4, Vikram Jyoti Cooperative Housing-Society, Deonar, Near Telecom Factory, Mumbai 400 088, India
e-mail: gargrk@gmail.com

P. Ghosh and B. Raj (eds.), *The Mind of an Engineer*,
DOI 10.1007/978-981-10-0119-2_53

Dr. H.J. Bhabha, the focus has been on achieving indigenous capability for harnessing this new source of energy. Towards this end steps were initiated for training of manpower, production of special materials required and development of relevant technologies. For those not familiar with the basics of nuclear power, it may be worthwhile to mention that whereas in a thermal power plant, coal, oil or natural gas is burnt in the boiler for raising steam which runs the turbine and in a nuclear power plant, uranium is used as fuel in the nuclear reactor to generate heat. Uranium, thus, becomes the essential material for the nuclear power programme. Moreover, unlike coal, oil or gas, uranium as it occurs in nature is mined as an ore and has to go through a number of processes to make it usable in a nuclear reactor. After use in the reactor, the spent fuel has again to be processed or managed with great care due to very high level of radioactivity in it. The processing of the spent fuel also gives plutonium, a man-made material which again can be used as fuel in a nuclear reactor. Uranium as it occurs in nature has only 0.7 % of U^{235} isotope which is fissionable and serves as fuel in the reactor, the bulk (99.3 %) being U^{238}. Nuclear power reactors can use uranium in natural form (but need heavy water as coolant) or enriched in U^{235} to the extent of 3–4 % where ordinary water can be used as coolant. For some special types of nuclear reactors or for use in nuclear weapons, much higher enrichments are required. All the steps for producing fuel for nuclear reactors constitute what is known as the 'nuclear fuel cycle'. Needless to say, the development of technologies for the fuel cycle is the most important prerequisite for a successful and hurdle free nuclear power programme. These technologies are not easily available and some of them, viz. uranium enrichment and plutonium extraction from spent fuel, not available at any cost.

By about 1970, India had in operation two nuclear power reactors imported from USA fuelled with enriched uranium. Two natural uranium heavy water-cooled reactors were under construction with collaboration from Canada. Two research reactors, one based on natural uranium and the other on enriched uranium, had been operating for over 10 years. Based on R&D efforts in BARC, processes for extraction of uranium from ores, conversion of the uranium concentrate to metal or oxide of nuclear purity, fabrication into fuel elements for use in research as well as power reactors and for extraction of plutonium from spent fuel had been developed and production plants had been setup or were in an advanced stage of construction. Processes for heavy water production were also under development. Thus, indigenous technologies had been developed for all the steps of the fuel cycle except uranium enrichment. Even preliminary studies had not been initiated. It was thought that uranium enrichment was a difficult technology and highly capital and energy intensive. At that time, enrichment plants were operating only in the USA, Russia and France. These plants were based on the 'gaseous diffusion' process, using uranium hexafluoride as the gas for isotopic separation.

Around this time, some reports appeared about another process, viz. 'gas centrifuge' being developed jointly by UK, the Netherlands and Germany. In Germany, 'nozzle separation' process was also being pursued.

Difficulties in the separation of isotopes of an element arise due to the very small difference in the physical and chemical properties of the isotopes or their compounds more so for heavy elements like uranium. A very large number of stages are required for meaningful separation. Studies on separation of some of the isotopes of light elements like hydrogen (for heavy water production), boron and nitrogen were already in progress in BARC. These processes were based either on difference in the physical properties or on the chemical exchange behaviour. A small plant for separating deuterium and hydrogen by electrolysis of water had been operating at Nangal as an adjunct to the fertilizer plant.

Initiation of Studies on Uranium Enrichment

With the status of development summarized above and with the confidence gained in development of technologies of the other nuclear fuel cycle processes, when the plans of the department of Atomic Energy were being prepared for the decade 1970–1980, it was proposed that preliminary studies may be initiated on uranium enrichment as well. By this time, two other projects in which I was the leading participant were nearing completion. I, therefore, got the opportunity to take up R&D studies on this topic. To start with, a project team was constituted in 1970 with three fresh engineers from the training school of BARC. Over a period of 6–7 years, the team size increased to about twenty engineers and scientists in various disciplines, all fresh from the training school.

It was decided to simultaneously carryout literature (whatever little was available) survey and feasibility studies of all the processes reported to be in use for production and those under development in other countries of the world. The processes in these categories were the 'gaseous diffusion', gas centrifuge, nozzle separation and chemical exchange. Separate groups were constituted for study of these processes. For the first three processes, the feed material has to be in the gaseous state, viz. uranium hexafluoride. So a separate group was given the task of developing the technology for its production. It involved production of fluorine gas as well, since it was not available in the country. The technology for production of uranium tetra fluoride was already established.

On the analogy of deuterium enrichment (for heavy water production) by chemical exchange, experimental work on the chemical exchange process for uranium enrichment could be started immediately. However, after trials with a few systems over a period of about 2 years, no meaningful results could be obtained. The separation factors are perhaps quite low, requiring a very large member of stages to get enrichment which could be detected by the mass spectrometer available in BARC at that time. This approach was, therefore, not pursued further.

The other three processes based on the small difference in the physical properties of the molecules of the isotopes of U^{235} and U^{238} require, first, the development of the separating element (diffusion membrane, nozzles with very small opening or very high-speed rotating cylinder). Each one requires different materials and fabrication technology. Other equipment and components required for these processes include compressors/vacuum pumps, valves, instrumentation, piping and heat exchangers. All items have to be compatible with uranium hexafluoride (UF6) which is a corrosive gas. Moreover, in view of the properties of UF6, the whole system has to work under sub-atmospheric pressure and is required to have high level of leak tightness. Studies on the three processes continued over a period of 7–8 years to evaluate and assess the technical feasibility of making the separating elements based on laboratory scale work, survey of availability of required materials and equipment in the country or abroad freely, special fabrication facilities, if any, to be established, etc. The costs and the time frame for the first prototype and the possibility of scale up subsequently were also taken into consideration. Based on the outcome of these studies, the 'gas centrifuge' process emerged as the choice for further development and all efforts were directed towards it.

Development of the 'Gas Centrifuge' Process

After preliminary studies, the following plan of action was prepared:

(i) Development of the rotor assembly (centrifuge) consisting of

   (a) a tube of suitable length and diameter capable of rotating at a peripheral speed of at least 300 m/s—which means made of material with high specific strength, and having other suitable mechanical properties, resistance to corrosive action of UF6 and thin enough to reduce the power required for its rotation.
   (b) high-speed motor for long continuous run and a stable high-frequency power supply source for the motor.
   (c) a bearing and suspension system to support the rotor and to ensure dynamic stability at high speeds.
   (d) a casing to protect against crash of the rotor and to maintain high vacuum to eliminate drag loss.
   (e) provision of facilities for electron beam welding, heat treatment, passivation, plating, dynamic balancing, etc.
   (f) Arrangements for feeding the UF6 gas and removal of the separated streams.

(ii) Sourcing or providing special materials and manufacturing facilities to produce the above components, in large numbers—initially in hundreds and subsequently in thousands.
(iii) Interconnection of the units in parallel as well as in series mode through a network of pipes and valves and sourcing of other equipment such as vacuum pumps, condensers and instrumentation for monitoring and control.

From the survey carried out, it was apparent that none of the components of the rotor assembly would be available off the shelf and everything would have to be developed ab initio. Each member of the team was assigned the task of developing a specific component or system. He was given the freedom to select his line of action and to obtain guidance or help from any source he considered necessary, within BARC or outside. I acknowledge the full support of various groups in BARC and my seniors in the execution of this project.

The first important decision to be taken was on the material to be used for the rotor tube, taking into consideration the possibility of its availability indigenously or from outside sources as unrestricted supply and facilities for fabrication into tubes of the required dimensions. The candidate materials having high specific strength (tensile strength/density) were identified as certain aluminium alloys, titanium alloys, high-strength steels, maraging steel and fibre–resin composites. It was learnt that Midhani was in the process of developing certain grades of maraging steel for application in defence and the space research programmes. Maraging steel was, therefore, selected as the material of the tube. To get the tube of the required dimension, the product from Midhani had to go through two fabrication steps, viz. extrusion, to make the blank tube of a higher thickness followed by flow forming to the required dimensions and wall thickness. It was found that at that time only one flow forming machine was available in the country, with a government organization and a private party was intending to import a similar machine for making cans. The machine imported by the government organization had not been installed. The private party was, therefore, encouraged to import the machine as early as possible. Thus, three organizations were identified and contact established to ensure procurement of the tube—Midhani for the maraging steel material, Nuclear Fuel Complex for extrusion of the blank and the private party for flow forming. Facilities for other operations like welding the top and bottom discs to the tube by electron beam welding, dynamic balancing, heat treatment, passivation, etc., were set up in-house. The EB welding machine was developed by a separate group in BARC.

The motor to drive the rotor at a constant speed of 40,000–45,000 rpm was the next major component which had to be developed. The power rating was arrived at through design calculations and experimental studies. Motors with wound rotors were ruled out due to mechanical reasons. The choice narrowed down to hysteresis-type solid rotor motors. Further, to eliminate any radial loading due to radial magnetic flux, axial flux motor was envisaged. Thus, a unique motor design was developed with the stator in the shape of a pancake and rotor disc directly attached to the bottom end of the rotor disc. The material of the motor rotor had to be optimized taking into consideration mechanical strength and hysteresis properties. The motor stator had also to be potted to guard against corrosion. A high-frequency drive was also needed to convert the 50 Hz supply to 700–800 Hz to run the motor at 40,000–45,000 rpm. For initial trials, various sources were rigged and finally low-power switching inverters were developed. One of the groups in BARC greatly helped in this endeavour.

Non-contact pickups and associated instruments for measuring the speed and vibrations were also developed, since none of these were available in the market.

The other critical component which had to be developed was the bearing to support the long rotor. The bearing had to work in vacuum, was subjected to axial as well as radial load of a few kilograms, had to run at very high speeds, continuously for years without any maintenance or lubrication. No available bearing could meet such stringent conditions. Design and fabrication of the bearing had, therefore, to be taken up in-house. After studying various systems, a hybrid design of a pivot and jewel bearing incorporating hydrodynamic action was adopted. Spiral grooves were made on one of the bearing surfaces which would generate a fluid film between the two surfaces and ensure a long life. A stable hydraulic fluid with very low vapour pressure had to be selected and procured. The problem of machining of the pivot and the jewel to the required sub-micron tolerances and engraving the spiral grooves had also to be tackled. Some enterprising entrepreneurs were roped into undertake this job. To reduce the load on the bearing and for radial stability, a magnetic suspension system was incorporated.

By 1983, the above efforts enabled us to make the first prototype unit which could run at speeds close to the target and on which some separation trials could be conducted using some gas mixtures. This was followed by isotopic separation using the process gas, viz. UF6, which by that time could be made on a small scale. A few more units were then assembled and connected in series to partly simulate the conditions of a cascade. A semi-pilot-scale facility was, thus, available for component and system testing and for conducting separation trials. Concurrently, a decision was taken to set up a demonstration plant with a few hundred units. A new site was selected and civil construction work started in 1984. Along with experimental work in the semi-pilot plant and construction work of the demonstration plant buildings, activities connected with sourcing of materials, fabrication and machining, procurement of bought out items had to be intensified.

The demonstration plant was planned to have a cascade consisting of a few hundred rotors to start with, facilities for testing of components and assembly of rotor components as well as for UF6 production. By 1989, all the construction activities and installation work were completed and the plant was commissioned with UF6 process gas, and enriched uranium was produced in the country for the first time in kilogram quantity. A venture which appeared to be fraught with many uncertainties became a success story and a matter of pride for all of us who were participants in it. It was possible due to the full commitment and dedication of the team, their untiring efforts and innovative approach and cooperation of many groups in BARC.

With the successful completion of this project, I retired in 1990 with a sense of satisfaction and fulfilment and with the firm belief that determination to achieve the goal, confidence in oneself and the team members and hard work are the keys to the success of any venture.

My Tryst with Indigenous Armour Development

T. Balakrishna Bhat

Born little after India's independence, as a child, I used to imagine and feel that I owned the whole great country. My father, Madhav Bhat, was a respected school headmaster who used to get respectful salutations from members of nearly every house situated all along a 7-km-long path of walk to his school through hills and fields. He would reciprocate with appropriate affectionate words without stopping his walk. It was a thrill to walk with him. My mother, Indira, would often be cheerfully singing tunes and hymns while taking care of the small farm, all the workers, children, cows and guests with infinite patience and love. It was a great joy to do every type of work to help her and receive her blessings. Early in the morning, every day, father would gently wake me and teach me Shlokas, words, spellings and grammar while sitting by my bedside even as I lay on the bed with closed eyes, and end the day similarly at night. This process, I believe gave me many things, including a habit of not wasting a single moment from the time one wakes up to the moment one falls asleep.

At the age of fourteen, I joined Sri Ramakrishna Mission Balakashram at Mangaluru, where the rigorous discipline of perfectly doing all the chores along with studies toughened the mind–body system. Here, I had the chance to tutor (free) classmates and juniors, which increased my grasp on the subject and also strengthened my self-confidence.

In 1967, I joined the B.Tech. programme at IIT Madras. Here, I concentrated totally on understanding the principles and deeper aspects, rather than on securing grades. This quest made me read a large number of books and to some extent journals available in the Institute library where I would often sit up until it closed late in the night. After B.Tech., I secured admission with scholarship to pursue Ph.D. at Washington State University, but, when I learnt that USA had sent its 7th fleet in readiness to attack India during the Bangladesh conflict, I changed my mind. Instead, I went to IISc Bangalore to study M.E. from where DMRL recruited me.

At DMRL, for the first six months or so, I visited all the groups and glanced through all the books and journals and generated hundreds of research ideas.

T. Balakrishna Bhat (✉)
Defence Metallurgical Research Laboratory, Kanchanbagh, Hyderabad 500 058, India
e-mail: tbalakrishnabhat@yahoo.co.in

P. Ghosh and B. Raj (eds.), *The Mind of an Engineer*,
DOI 10.1007/978-981-10-0119-2_54

Initially, I worked on TEM and intermetallic alloy systems. One day, Dr. V.S. Arunachalam, who had joined as our new director, called me aside and in his characteristic excited way, asked me whether I prefer to work on pure science and may be hope to get a Noble prize someday, or work on an important development work. Because of the training at IIT, proud to be an engineer, I immediately chose the latter path. Next day, Dr. Arunachalam called me to his office and excitedly explained the scattered notes in his little diary about the Chobham armour trials shown in a hazy way in England to the visiting Chief of Army Staff and asked me whether we can quickly develop and demonstrate a similar one. Though I knew nothing, I sensed that every atom in my body was excited.

First I made a quick dash to TBRL, ARDE and weapon-related laboratories studying all available reports and papers to understand the nature of the threats and their operating principles and mechanisms. To find some solution, I went into a contemplative enquiry mode and scanned the rather difficult journals such as "Journal of Applied Physics" at the libraries in IISc and TIFR to look for sound principles based on which one can construct appropriate armour materials on our own, ab initio. Various ideas such as Konda's effect, deflection of shockwaves, splitting of the jets, avoiding momentum multiplication, using extremely high viscosities of glass like substances, facilitating lateral dispersal of momentum and energy, breaking up the projectiles or deflecting the projectiles were conceived. Appropriate tailor-made materials and structures were thought of. It was realised that while in most engineering materials and applications, we need to maximise strength, sometimes strength and toughness, in armour we need to maximise the product of strength, ductility and the volume that participates in energy absorption. Increased speed of plastic wave and increased homogeneity of strain that accompanies it is critical. These are unique requirements. Further, it was observed that while homogeneous deformation is key for maximising energy absorption, inhomogeneous flow is desirable for momentum absorption such as in the case of HEAT and for turning or breaking the shots. For dissipating or absorbing shocks, layered structures should be preferred. Accordingly, many new materials and structures were conceived and made.

The first results of the trials on the HEAT rounds came within a few months and, may be for the beginner's luck, were truly fantastic. Soon, larger samples were made and tested at PXE Balasore. The plates not only defeated the HEAT rounds but also withstood the KE, APDS rounds. The rounds were trapped inside. To see what happened to it, the plate was brought to DMRL. It was cut open the same night to see what actually happened to the shot. I and Dr. Arunachalam walked from Laboratory Quarters to DMRL at well past midnight to examine the plate from inside. To our surprise, the shot was not inside, hiding, but had actually broken into to fine dust! It was an exciting beginning. A comprehensive and confident programme thus began at DMRL.

Using a variety of starting materials such as ceramics, hard steels, tough composites, and energetic explosives, the armour programme advanced in many directions to meet a host of challenging requirements. Success after success came in the form of armour system for MBT Arjun and its continuously improving features.

For T-72 Ajeya and for T-90, the required armour technology was developed indigenously. Armour for light vehicles, helicopters and many other applications such as Lancer helicopter, Vijayanta tank, ICV-Abhay and Mi-17 helicopters also emerged out of the programme to meet the requirements. It gives great satisfaction and excitement to me and my research team.

More than 20,000 tonnes of various armour materials have been produced to meet the various requirements. An Armour Technology Centre has been set up in the 70 acres of land specially acquired for the purpose of armour development.

There are a few critical factors which I believe have contributed to the above successes in the tryst with indigenous armour development. (i) Full trust, support and freedom provided by the organisation. (ii) Participation by industries in a deeply collaborative way with implicit mutual trust. (iii) From the user's side, the area of protection, being close to the heart of every member from soldier to the chief, which evoked spontaneous, deeply supportive and encouraging responses.

On my part, I kept myself in the excited state throughout for over three decades and spread the excitement to my colleagues. I gave full credit to my colleagues. We used to analyse the less successful results and come up with a solution plan next day itself to work on. I used to keep a notebook and pen on my bed to scribble ideas that often turned up after midnight. Practicing the art of lateral expansion of horizon to generate new ideas and delving deep for robust analysis and experimentation to confirm the utility of new ideas was the method. I used to say Jai Hind before going to bed every day and pray for success of my project every morning.

My heartfelt acknowledgement is to my wife Shantha Kumari, my family and work-related family, whose unfathomable depth of emotions, support and commitment has enabled the attainment of deeply satisfying results in my tryst with destiny in the service of the nation, which reconfirms my childhood feelings that I indeed own the whole country. Jai Hind

Inertial Sensors and Navigation Systems

Suresh Chandra Gupta

Introduction

Highly precise and reliable indigenous inertial sensors and navigation systems, developed ab initio, have contributed immensely to the success of space launch vehicles of India. The venue of development was the erstwhile space science and technology center, since outgrown into, Vikram Sarabhai Space Centre, VSSC, Indian Space Research Organization, ISRO, Trivandrum.

Background

Founding Fathers of the Space Program of India were deeply impressed by the potential of the emerging space-technology-enabled systems to upgrade the infrastructure for knowledge needed for taking India from backwardness to modernity. Their vision encompassed (1) mass communication, telecommunication, and remote sensing, among other vital systems, which existed in the country, but, in a minuscule level. They foresaw the potential much ahead of most of the professionals in advanced countries. They foresaw that space-technology-enabled infrastructure can be set up in much quicker time and at a fraction of cost compared to the then prevalent terrestrial systems. They insisted that India can skip many intervening technology-related steps followed by advanced countries and leap frog safely by going directly to space technology.

At the same time, they stressed compliance with four cardinal conditions, namely

S.C. Gupta (✉)
Vikram Sarabhai Space Centre, Thumba, Trivandrum, Kerala 695 022, India
e-mail: suvidya456@gmail.com

P. Ghosh and B. Raj (eds.), *The Mind of an Engineer*,
DOI 10.1007/978-981-10-0119-2_55

(a) The space-technology-enabled infrastructure, STEI, should encompass the whole territory of the country,
(b) The infrastructure should remain contemporary at world-level despite rapid advancements in systems and technologies,
(c) The program should be affordable, such that return on investment in India ought to be highly competitive, and
(d) The required human resources should be grown largely within the existing Indian educational system. It meant that large-scale foreign training should not be required.

Options

Even with these postulates, two options existed to set up the needed systems, namely (a) buy them out from the developed countries and operate them ourselves. This would be like what India has done in the field of civil aviation. The other option was (b) to develop, produce, and deploy them indigenously, largely through self-reliance. Realizing that the needs of the country were huge and ever expanding and also that the emerging technology will be advancing rapidly, the bought out technology would require frequent upgradation. It precluded the bought out option. So, a courageous decision was taken to take up the challenge to develop, produce, and deploy the whole range of technologies and systems indigenously. The cheering status after nearly fifty years of relentless efforts is that at present the country owns its own indigenously developed satellite-based telecommunication and television system, which are comparable to the best in the world in respect of both the depth of penetration and performance. In addition, the country has satellite-based advanced weather observation and warning system. Moreover, indigenously developed and sustained set up for seamless remote sensing of crops, forestry, floods, and of a host of other natural resources are operational and have acquired great economic significance to the country. A significant piece of data is that just the telecommunication service alone recovers the cost of satellite in about five years. The other services come virtually for free. So, what has been accomplished justifies amply the courageous decision of the founding fathers. Other significant by-products are concurrent emergence and utilization of enormously competent human resources, which were waiting to be employed in meaningful national tasks. Further, the space program has put to use the investments already made and utilized fully by the country in non-space programs. Noteworthy examples are metals and special alloys, metal forming and machining, and polymers and composite materials.

It was in the context of aspiring to achieve total self-reliance in space technology that country ventured into the esoteric field of 'rocket science' and the associated fields of propellants, propulsion systems, and inertial navigation systems. Here, we note that the major components of space systems are satellite launch vehicles, satellites, satellite-borne payloads, and Earth-based tracking and data acquisition systems. The payloads include communication transponders, specialized multi-spectral cameras for observations of diverse natural phenomenon.

Placing a Satellite in Earth Orbit

A satellite launch vehicle, SLV, lifts a satellite mounted on its top from the launch pad to the point of injection into orbit (1). It follows a preplanned trajectory, which involves a vertical take-off, followed by a predetermined set of pitch maneuvers, ending in a near-horizontal path at a required altitude and with the required velocity. Control and Guidance system stabilizes and steers the vehicle. Guidance and navigation system takes it to the destination. Actually, the SLV is a multistage rocket, with the first stage being ignited on the launch pad. When the propellant in the first stage burns out, the stage is jettisoned and the next stage takes over. This sequence continues till the penultimate stage reaches a predetermined altitude and velocity. At this point, the velocity vector is parallel to the local horizon. Generally, after a brief coasting period the final stage gets ignited, imparting enough addition of velocity so as to give the satellite the required orbital velocity. A complex flight dynamics computer program determines and implements the multistage launch sequence. Inertial sensors and systems are integral to implementing these manoeuvres. As inertial sensors and systems are in a highly restricted category under advanced technology control regime practiced by developed countries, their indigenous development was critical to self-reliance in space launch vehicles.

We elaborate that the SLV lifts off vertically and reaches the desired point of injection in orbit following a predetermined steering program along a trajectory. To be sure, the SLV experiences destabilizing forces and moments originating from aerodynamics and propulsion systems. An autopilot, comprising angular rate sensing gyroscopes, stabilizes, and steers it along the trajectory, taking due care about the varying density of the air medium, prevailing winds and unpredictable propulsion-related disturbances. The propulsion system has unpredictable variations requiring in flight corrections in order to reach the destination precisely. An inertial navigation system measuring the instantaneous orientation and sensors for measuring angular rates are integral components of the autopilot. An integrated inertial navigation system conjoined with an on-board digital computer measures the instantaneous altitude and relative velocity of the SLV and estimates its state at the end of the flight. The variations from the intended end conditions are corrected in real time by the closed loop guidance algorithm. Thus, the consolidated inertial system of the SLV works in conjunction with its multistage propulsion system to impart the desired orbit injection conditions and orbital velocity with the required accuracy.

Significance of Orbital Parameters

It is clear that the function of a satellite launch vehicle is to place its passenger satellite in a desired orbit of Earth. The Earth orbits are generally elliptical in shape. The main parameters of an orbit are its apogee, perigee, orbital period, and the inclination angle. The apogee and the perigee are, respectively, the farthest and the nearest points to Earth in the orbit. Orbital period is the time taken to complete one orbit around the Earth. Inclination angle is the angle between the plane of the orbit and the equatorial plane of the Earth. These as well as orbital parameters play crucial role in the utilization of satellite as explained in the following.

A satellite in orbit traverses Earth below it. Inclination angle of the orbit determines the portions of Earth it visits during an orbit. The footprint of such visit is limited by the swath coverage of the camera. Through successive visits, the satellite covers the entire area of Earth, accessible for a given inclination angle. For polar orbits, the inclination angle is 90° enabling access to the entire surface of Earth. Thus, it is the preferred inclination angle for remote sensing application. Further point to be noted is that a satellite in polar sun-synchronous orbit returns at the same time of the day. It covers the whole of Earth in 22 days enabling repeated collection of data over the same place. This characteristic helps in keeping track of changes in target features on Earth.

When the inclination angle is zero, it becomes an equatorial orbit and with orbital period of 24 h it becomes the geosynchronous orbit with the property that it appears stationary with respect to Earth. This property is used as a technical simplification for direct-to-earth, D2H, connection to fixed orientation antennas on Earth, making them simple and inexpensive and accordingly widely employed.

A satellite-borne camera gets an exceptionally large field of view, the size being related to the altitude of the satellite. The camera collects gigabytes of photographic data in selected bands of the spectrum. Additionally, depending upon the inclination angle the satellite visits nearly the whole of Earth during its orbits. With the inclination angle of 90°, a polar orbit provides view of the entire Earth. Thus, gigabytes of data on the state of plant, water, and or minerals on Earth are collected within a few days, aiding scientific analysis of state of agriculture, forestry, and soil conditions. In comparison, an airborne camera will take several months to collect similar data on a small portion of Earth.

Likewise, a geosynchronous satellite-borne communication transponder can communicate with points located on nearly one-third of Earth, replacing thousands of terrestrial repeaters. Three properly located geosynchronous satellites can cover nearly the whole of globe. Only the regions located near the North Pole and South Pole and in the high-latitude zones are inaccessible from the equatorial plane due to the spheroid geometry of the Earth. The satellite-borne communication links can be set up as soon as the satellites are placed in orbit and a suitable Earth station is established. Consequently, the speed of setting up a space-technology-based system is incredibly high, demonstrating the validity of the expectations of the speedy establishment.

Satellite Launch Vehicles of India

The Polar Satellite Launch Vehicle, PSLV, and Geosynchronous Satellite Launch Vehicle, GSLV, are the current operational vehicles of India. A more powerful vehicle under development is GSLV Mark 3.

Owing to its reliability and orbit imparting accuracy the PSLV has become the work horse of ISRO. The launch on April 4, 2014, of the second Indian Regional Navigation Satellite—IRNSS-1B was its 25th consecutive successful mission. The injection was so precise that the subsequent insertion in the final orbit by the satellite-borne liquid apogee engine required 7 kg less fuel, extending the service life of the satellite by more than 6 months. The precision came from the performance of the indigenous on-board inertial navigation system.

Some of the other missions accomplished by PSLV are IRNSS-1A, Chandrayaan-1, and MARS Orbiter. PSLV has placed in its 26th mission on June 30, 2014, the French Earth Observation Satellite SPOT-7 and four more commercial passenger satellites in the intended orbits. Thus, far PSLV has launched 40 commercial satellites, signifying its reliability and cost effectiveness. Microsatellites built in Indian academic institutions have received piggy back rides. Its inertial guidance and navigation system is programmable such that it can guide satellites to polar orbits as well as geosynchronous transfer orbits. Thus, PSLV is a versatile vehicle.

Anatomy of Inertial Navigation Systems

A cluster of gyroscopes to detect the movements in three inertial aspect angles, three accelerometers to measure orthogonal components of rectilinear accelerations and an on-board digital computer to process data make up the basic inertial navigation and guidance system (2). Currently, two numbers of two-axis dynamically tuned gyroscopes, DTG, serve as angular position sensors. DTG have replaced the single-axis rate integrating gyroscope, RIG, as the former does not require as stringent precision in its parts as does an RIG. Heart of both types of gyros is an in-house developed hysteresis synchronous motor, which maintains constant speed under severe launch vehicle environment of high acceleration and vigorous vibration, thus assuring the accuracy of the gyro. Participation of specialists in material science enabled development of the hysteresis synchronous motor. Extremely low internal loss materials for flexures employed in Accelerometers and DTGs contribute to precision in measurements. Ultrahigh precision, of the order of one micron, in machining and assembly in super clean environment assure stability in the characteristics of errors in sensing. This stability in the characteristics of hardware permits further reduction of final errors by sophisticated software compensation. The outcome is the world-class precision in orbital injection of satellites in their desired orbits, thereby enhancing the class of PSLV (3) among the satellite launch vehicles. Chandrayaan and Mars orbital missions depend on the precision of the inertial measurement system. Thus, the ab initio development has fulfilled its mission.

Further Readings

Gupta SC. Growing rocket systems and the team. Prism Books; 2006. p. 22–9 and 102–115.
Bose A, Puri S, Banerjee P. Modern inertial sensors and systems, India: Prentice Hall of India; 2008. p. 6–12
Manoranjan Rao PV, Radhakrishnan P. A brief history of rocketry in ISRO; Universities Press. 2012. p. 235–9 and 249–252.

Perspectives on Technology Trends in Biopharmaceuticals Manufacture

K.V. Subramaniam

Biopharmaceuticals and Society

Over the last 40 years, starting from being nowhere, biopharmaceutical drugs have come to occupy a dominant presence in the list of top ten selling drugs globally. Today, worldwide biopharmaceutical revenues are estimated at USD 288 billion, or 23 % of global pharmaceutical industry revenues, and comprise seven of the top ten list of drugs (Table 1). Many of these drugs, particularly monoclonal antibodies, offer cures for difficult-to-treat diseases, mainly in cancer and rheumatoid arthritis.

However, these drugs cater only to approximately 15 % of global population, primarily because of affordability and accessibility issues. In order to expand the reach of these drugs, it is imperative for their prices to come down significantly. This can happen in two ways: expansion of biosimilars, which are to large proteins what generics are to small chemical molecules, and reduction in costs.

Till recently, the big biopharmaceutical manufacturers had little motivation to reduce costs, given that total costs of manufacture were a small proportion of the net selling price. With the emergence of cost-competitive, high-quality, and biosimilar products with comparable safety and efficacy, big biopharmaceutical manufacturers would have to reduce price and therefore focus on reducing manufacturing costs to reduce impact on margins.

Governments in some developed countries have brought these products under the tender system and have therefore created another pressure point for a focus on manufacturing costs. Governments in many countries have also now placed biologics and biosimilars in particular, as thrust areas for domestic production and self-reliance. Several policies, ranging from thrust area status, local manufacturing support, fast track product registration, import duty benefits, and reimbursements

K.V. Subramaniam (✉)
Reliance Life Sciences, Dhirubhai Ambani Life Sciences Centre,
Thane–Belapur Road, Rabale, Navi Mumbai 400 701, India
e-mail: k.v.subramaniam@ril.com

P. Ghosh and B. Raj (eds.), *The Mind of an Engineer*,
DOI 10.1007/978-981-10-0119-2_56

Table 1 List of top-selling drugs worldwide

Sr. No.	Product	Pharma or Biopharma	Indication/segment	2014 sales (USD billion)
1.	Humira (Adalimumab)	Biopharma	Anti-inflammatory	12.54
2.	Sovaldi (Sofosbuvir)	Pharma	Hepatitis C	10.28
3.	Remicade (Infliximab)	Biopharma	Anti-inflammatory	9.24
4.	Enbrel (Etanercept)	Biopharma	Anti-inflammatory	8.54
5.	Lantus (Insulin glargine)	Biopharma	Diabetes	8.43
6.	Rituxan (Rituximab)	Biopharma	Oncology	7.55
7.	Avastin (Bevacizumab)	Biopharma	Oncology	7.02
8.	Seretide/Advair (fluticasone + salmeterol)	Pharma	COPD	6.97
9.	Herceptin (Trastuzumab)	Biopharma	Oncology	6.87
10.	Januvia (Sitagliptin)	Pharma	Diabetes	6.00

under health coverage to preference in government tenders, have been instituted. While these measures are undoubtedly important, manufacturing costs have to necessarily come down.

It is here that technology can play a big role, in addition to improving product quality and reducing time to market, scalability, and better processing.

Biopharmaceutical Technology

Small molecule drugs are inorganic and typically less than 900 daltons (Da) in size. In contrast, large molecule biologics, such as proteins, are made up of several amino acids and are typically more than 1500 daltons (kDa) in size (for reference 1 Da is equal to 1 atomic mass unit).

While small molecule drugs can be synthesized by a series of chemical reaction processes, proteins are far more complicated to make. The intermediate size molecules, between small molecules and proteins, are peptides and typically have 6–20 amino acids.

Proteins are made by engineering cells using recombinant DNA technology. This technology involves integrating genes that express proteins of interest into the genome of mammalian or microbial organisms. These engineered cell lines are grown in chemically defined media. As these cells grow and fed on oxygen, nutrients, and vitamins in a medium, they produce proteins of interest, either within the cells or outside the cells. These proteins are harvested and purified. In monoclonal antibody technology, antibodies are generated by sensitizing cells to antigens. The genes of the antibodies are sequenced and used to create engineered cell lines by recombinant technology. These cell lines are used to produce monoclonal antibodies.

Cell lines can be microbial (e.g., *Escherichia coli*), mammalian (e.g., Chinese hamster ovary or CHO), yeast (e.g., *Pichia pastoris, Saccharomyces cerevisiae*), and insect cells (e.g., Sf9). One of the desirable characters of recombinant cell lines is to express product stably at higher volumes, such as 2500 L.

Biopharmaceutical Manufacture

Manufacture of biopharmaceuticals follows a process that can broadly be divided into four parts:

(a) upstream cell culture in bioreactors or fermenters, depending on whether the protein is expressed in mammalian cells or microbial cells, respectively.
(b) downstream separation of expressed protein from within or outside the cells, depending on where the protein is expressed, by centrifugation or filtration.
(c) purification, which involves a series of chromatographic steps, such as ion-exchange chromatography and gel filtration chromatography, as also other steps such as tangential flow filtration, ultrafiltration, and nanofiltration
(d) fill–finish for formulating and filling the proteins in single-dose vials, multi-dose vials, pre-filled syringes, and cartridges,

These processes are carried out in clean rooms conforming to the regulations and subject to rigorous audits. Typically, mammalian cell cultures grow slowly, while microbial cell cultures grow relatively faster. As a result, a batch of a mammalian cell culture process will take several weeks to harvest proteins from the cells, while microbial cell lines can grow in a few days' time.

Technology Trends

As with most technology sectors, biotechnology has also seen a rapid pace of development of new technologies.

Product Technologies

Science has come up with several new product domains in biopharmaceuticals.

Leading this, facet of product delivery technologies is small interference ribonucleic acid (siRNA) or antisense technology. Another prominent development in this category is antibody-drug conjugates (ADC) for the targeted delivery of anticancer drugs which are otherwise highly toxic, such as Kadcyla and Adcetris.

However, the promise of this technology has not played out in practice. Issues around product safety for patients and long-term efficacy have resulted in their inability to meet targeted treatment outcomes.

Payload Delivery Technologies

Many biopharmaceutical products require delivery of proteins in a targeted manner to the diseased cells or tissues. This is analogous to monoclonal antibody technology. There have been several developments here such as delivery of proteins into cells using virus-like particles (VLPs) and nanoparticle vehicles for intracellular protein delivery.

Ambient Temperature Protein Storage Technologies

Proteins are temperature-sensitive and have to be stored at temperatures in the range of 2–8 °C. This necessitates systems to continuously monitor storage temperatures, from manufacture to product administration into patients at the bedside. Excursions in temperature beyond the range mean non-compliance and naturally create product integrity questions. The cold chain infrastructure, continuous monitoring and periodic validation required, stability studies under different conditions and for different geographic zones collectively raise costs and management effort and intensity across the value chain.

Some technologies have evolved in making proteins thermostable, i.e., robust to wider temperature. These include formulations using novel excipients and encapsulating in polymers. However, these technologies are faraway from commercialization, given that a long, capital-intensive process would be involved in proving their clinical safety and efficacy.

Product Expression Technologies

Improved expression systems with higher titers lead to reduced manufacturing capacity requirements. When it comes to higher levels of expression, several technologies are underway.

Chinese hamster ovary (CHO) cells are commonly used for the production of mammalian cell culture-based, recombinant therapeutic proteins. With increase in understanding of handling and engineering of Chinese hamster ovary (CHO) cells, expressions levels of up to 5 g/L have been achieved from 1 g/L earlier.

At the same time, alternative expression systems are being developed. Avian lines (e.g., duck embryo quail sarcoma and chick embryo fibroblasts) have been reported to transfect well, have promoters that work with mammalian genes, and grow faster. These cell lines also promise higher levels of cell density and specific expression. Human cell line Per C6 has been proposed to be effective as well as beneficial for the production of biopharma products.

Baculoviral insect cell systems have also been gaining popularity as a substitute for the production of recombinant proteins and have been effective vectors for large-scale production of monoclonal antibodies.

Glycosylation plays a decisive role in ensuring proper protein function. Glyco Express is a novel expression platform that has emerged. This platform allows the generation of proteins with full human glycosylation and optimized sialylation. The technology is based on a glycol engineered human cell line and enables the production of large amounts of improved recombinant proteins. GlycoFi's technology is another novel technology used to control the glycosylation of recombinant therapeutic proteins.

Single-Use Disposable Manufacturing Technologies

The focus on single-use or disposable systems in manufacturing increased with the need for facilities to produce multiple products in parallel. Single use can now be seen in laboratory, pilot, clinical, and commercial manufacturing operations. Single-use systems have been shown to reduce capital cost by 40–50 %, reduce operating costs by 20–30 %, and reduce time-to-build by 30 % when compared with traditional stainless steel technology.

Single-use technologies are becoming an industry standard for the manufacturing of clinical and production batches in biopharmaceuticals. They do away with high levels of capital expenditure upfront, give flexibility on operating scale, and obviate large steam requirements, cleaning-in-process, and sterilization-in-process steps, there by reducing product contamination issues. Above all, manufacturing facilities can be more compact with attendant benefits in capital costs or lease rents.

With their growing adoption, especially in clinical grade material and small volume product or country-specific manufacture, this technology is expected to find widespread use.

Protein Purification Technologies

Protein purification is an expensive proposition. Protein A is currently the most expensive chromatographic step. Some, lower-cost alternatives have emerged. These include precipitation, crystallization, cation exchange, and mixed-mode chromatography. Synthetic protein A resins are also available but have showed lower selectivity and affinity.

Microbial Contamination Detection Technologies

Rapid and reliable detection of microbial contamination is paramount for product safety. Conventionally, these tests are done manually using standard plate count methods. Rapid methods help reduce assay time using automated systems. High-throughput methods for detection of microbes will improve environmental control. Several new technologies in this area are emerging.

Bioluminescence, to measure light output produced by a reaction that is dependent on adenosine triphosphate (ATP) released from microbial cells, is one such technology. Flow cytometry, to detect microorganisms in liquid sample either in flow or captured on a solid surface by membrane filtration, is another.

Industrial Big Data Analytics

Just as with many industrial sectors, big data analytics have been making a beginning in biopharmaceuticals and have tremendous scope. Hybrid platforms are emerging, which are a combination of high-performance operational data management and data storage, in order to leverage technologies of real-time analytics. This promises to lead to designing of improved ligands and matrices, which allow shorter bioreactor residence times, higher flow rates, and longer life cycles.

Robotics and Automation Technologies

Likewise, robotics and automation have started to make a mark in biopharmaceutical technology. High-throughput robotics in both upstream and downstream process developments can be vital to screen and improve cell lines, media compositions, and chromatography media. High degree of automation degree and decreased scale (microtiter plates and tube bioreactors) will allow broader screens.

Continuous Process Manufacturing

Biopharmaceuticals are produced in batch processes or fed-batch processes. With newer technologies, continuous manufacturing is becoming a possibility. This would encompass developments in perfusion bioreactors instead of fed-batch bioreactors and smaller scale modular chromatographic systems.

A series of small columns have been demonstrated to mimic one single large column with a diameter and a bed height equal to the total bed height of the smaller columns. Translated into practice, this could be imagined as an array of different types of columns, controlled by a microprocessor that assigns different types of chromatographic columns to different chromatographic cycle sequences; based on the manufacturing process specific for a product.

An analogy of this can be found in semi-conductor processing where machines specializing in a unit operation are assigned to carry them out for a bunch of materials, delivered to them by a materials handling and transport system; all controlled by microprocessors.

Process Intensification and Integration

With process analytical technology (PAT), it is becoming feasible to merge characterization, release and online product quality assays. In addition, techniques in modeling unit operations by computational fluid dynamics (CFD) and modeling, monitoring, measurement, and control (M3C) of bioprocesses are emerging.

CFD is an engineering tool for virtual experiments and can predict fluid flow, heat, and mass transfers. Results from CFD are relevant for (a) process understanding, (b) detailed product and process design, and (c) conceptual study on cause and effect with a parametric operating conditions.

Conclusion

To conclude, it is fair to say that there have been no new radical technologies that have come about in the realm of expression of proteins, akin to recombinant technology and monoclonal antibody technology. However, there are several relatively less radical technologies coming into play in the near term and medium term that promise to change the face of the biopharmaceutical industry globally.

Collectively, these technologies will impact the industry through better protein quality, higher expression of proteins in cells, higher yields per liter of bioreactor or fermenter volumes, and higher levels of capture of proteins in the downstream separation and purification stages and better use of resources of biopharmaceuticals manufacture.

All this would translate into the desired outcome and imperative of lower costs of manufacture without compromising product quality. In turn, these technologies would have found a much deeper and intensive social engagement.

Acknowledgement The Author would like to thankfully acknowledge inputs from his colleagues at Reliance Life Sciences—Ravishankar Kasturi, Dr. Venkata Ramana, Dr. Sachin Tendulkar, and Rahul Padhye—in the preparation of this article.

The Hydropower Development in India—Challenges and Way Forward

M.M. Madan

It is now a well-recognized fact that availability of electric power is an essential requirement for the development of a nation and the fact is very much applicable to India. As on 31 July 2015, the total installed generating capacity of power in India is 275911.62 MW out of which hydropower accounts for approx. 15.22 % (41997.42 MW). The percentage of hydropower has come down from 45 % in 1970 to less than 16 % in 2014. Today, the ratio is about 15:85 which is too less as to have efficient and balance grid for operation of all types of plants. The hydropower and Renewable sources of power are the best form of energy security. The water and other forms of renewable sources are getting wasted every day by not utilizing them, as these sources cannot be stored for future use. Whereas with the continuous extraction of fossil fuels, the sources are getting depleted fast. These sources could be saved for the future generations to come and could be made use of at the time of need and emergency.

Indian rivers carry more than two-thirds of their annual flow in three monsoon months. We are wasting this huge amount of water flowing continuously which can be converted to energy. Can we afford to waste the bulk of the energy of water by letting it flow down the drain unutilized in the monsoon months?

Dams and reservoirs are designed to last over hundred years. Which other type of energy source has such a long life? They provide us inexpensive energy.

Therefore, there is a need to increase and shift dependence on hydropower for the development and prosperity of the nation. Therefore, to meet country's energy demand at a faster pace and make up for the lost time, development of mega hydropower projects is essentially required.

This also warrants the need of sustainable development of water resources to ensure the continuous availability of water for hydropower generation and other activities and also for the flood moderation for the population living downstream.

Nevertheless, there are number of practical obstacles to the realization of this potential which needs to be tackled through socio-political intervention.

M.M. Madan (✉)
GVK Power & Infrastructure Limited, 2nd Floor, Building No. 10-C,
DLF Cyber City Phase-II, Gurgaon 122 002, Haryana, India
e-mail: mm.madan@gvk.com

P. Ghosh and B. Raj (eds.), *The Mind of an Engineer*,
DOI 10.1007/978-981-10-0119-2_57

This paper addresses the current scenario of hydropower development in India, the obstacles which this sector is facing and the need for future action plan for encouraging development of hydropower sector.

Hydropower Potential in India

India has a vast untapped resource of HEP. The total hydro potential assessed by CEA is approximately 148,701 MW, of which economic potential works out to 84,044 MW at a PLF of 60 % (Table 1).

The above hydropower potential has been assessed by CEA from 845 economically feasible schemes in different river basins with likely annual generation of 600 billion units including seasonal energy variation. As on 30.10.2014, about 49 schemes having an aggregate potential of about 14,287 MW are under execution for 12th Plan and beyond (excluding projects <25 MW) (Fig. 1).

India is a power-starved nation, for the month of July 2015 the energy deficit (MU) in India was 2.1 % and peak power deficit (MW) was 2.6 %. On one hand, Indian government is building ultra-mega thermal power projects (UMPP) and encouraging wind and solar power projects; but on the other hand, a step-child treatment is being meted out to hydropower plants. Following chart gives the status of source-wise installed capacity in India (Fig. 2).

The following graph shows the trend of capacity addition from 1st Five-Year Plan onwards (Fig. 3).

The percentage of hydro has come down from 42 % in 1970 to 16 % in 2014. Decline in hydropower capacity ratio has been distressingly continuous since its peak in 1962–1963 (45.68 %). At the beginning of liberalization of the sector in 1990s, its share in the power position was 28.77 % which reduced to 25.51 % by the end of 8th Five-Year Plan (1992–1997) and 25.40 % by the end of 9th Five-Year Plan, the capacity addition of hydropower was only to the tune of

Table 1 Basin-wise hydropower potential in India

Basin/river	Economic potential at 60 % load factor (MW)	Probable installed capacity (MW)
Indus	19,988.00	33,832
Ganga	10,715.00	20,711
Central Indian rivers	2740.00	4152
West flowing rivers	6149.00	9430
East flowing rivers	9532.00	14,511
Brahmaputra	34,920.00	66,065
Total	84,044.00	148,701

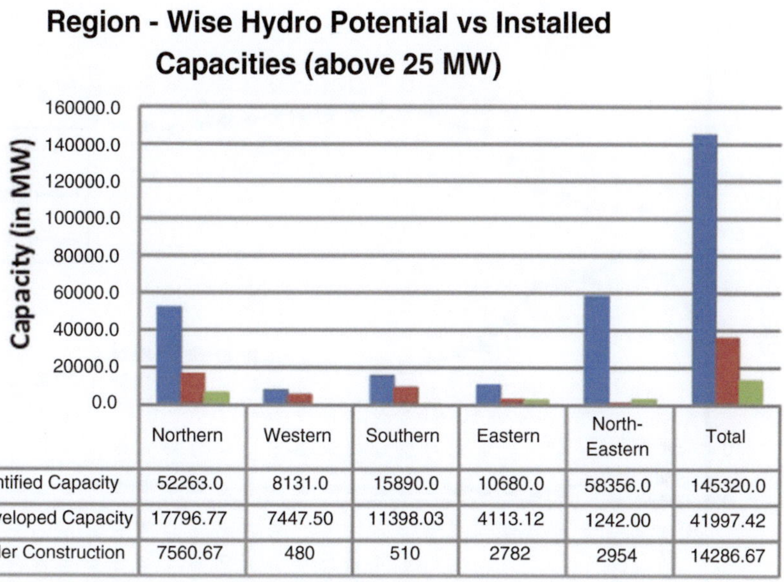

	Northern	Western	Southern	Eastern	North-Eastern	Total
Identified Capacity	52263.0	8131.0	15890.0	10680.0	58356.0	145320.0
Developed Capacity	17796.77	7447.50	11398.03	4113.12	1242.00	41997.42
Under Construction	7560.67	480	510	2782	2954	14286.67

Fig. 1 Region-wise hydro potential versus installed capacities (above 25 MW)

4538 MW against the target of 9818 MW. During 10th Five-Year Plan, a capacity addition of 41,110 MW was envisaged out of which 14,393 MW was planned from hydro-sector. It is generally accepted that while the central sector will remain a key developer, a significant amount of contribution needs to come from the private sector. In the Eleventh Five-Year Plan target was 15,627 MW, out of which achievement was 5544 MW, 35.48 % of the planned. In line with an optimistic GDP growth of 9–10 %, the plans for the subsequent two five-year periods have been revised as (Table 2).

With planned addition of 10,897–12,000 MW by the end of 12th and 13th Five-Year Plans, respectively, the share of hydro is expected to increase to 25.14 %. The long-term goal is to increase the share of hydropower capacity from the present level of 20 to 40 % of total installed capacity.

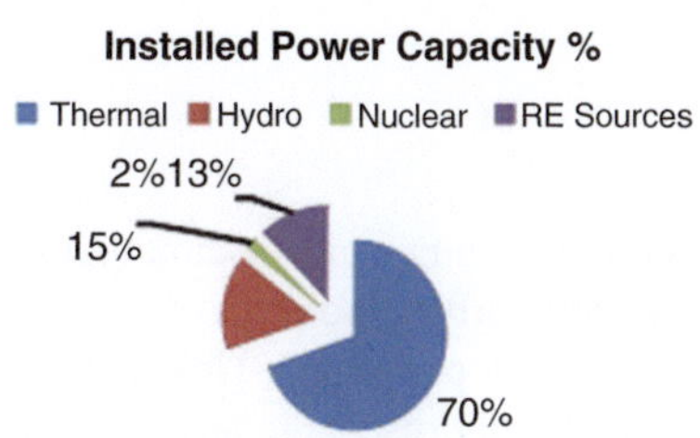

Fig. 2 All India generating installed capacity (MW) (as on 31-07-2015)

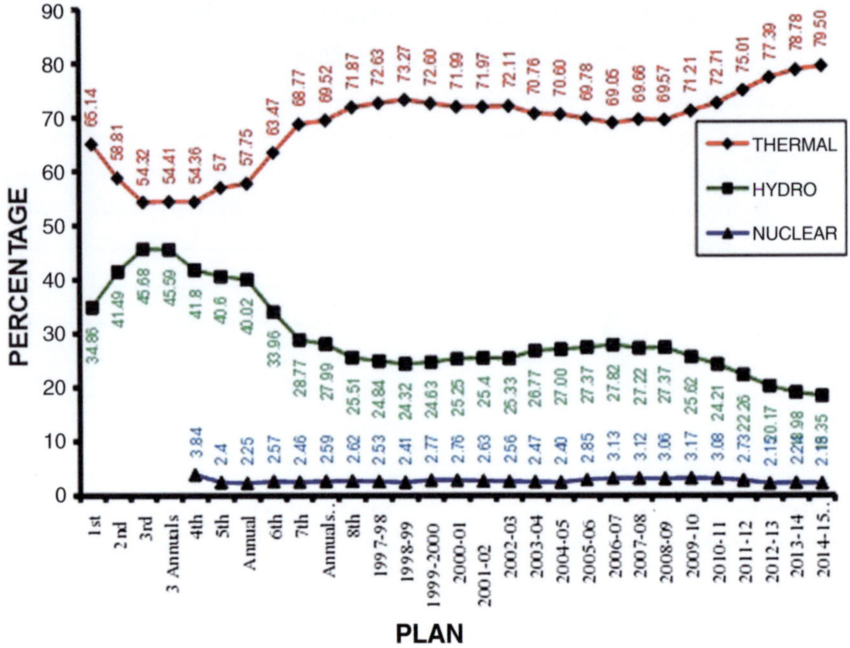

Fig. 3 Trend of capacity addition from 1st Five-Year Plan onwards as % of total (hydro thermal mix)

Table 2 Twelfth and thirteenth Five-Year Plans

Sector	12th Five-Year Plan (up to July 2015)	13th Five-Year Plan (target)
Hydro (target/achievement)	10,897/3025.02	12,000
Nuclear (target/achievement)	5300/1000.00	18,000
Thermal (target/achievement)	72,340/60,769.10	49,200
Total	88,537/64,794.12	79,200
Wind	15,000	11,000
Solar	10,000	16,000
Other RES	5,000	3,500
RES-Total	30,000	30,500

Advantages of Hydropower

- Helps in the fighting climate change and sustainability issues, helps in bringing down carbon emissions and carbon footprint.
- Environment-friendly and non-polluting—Unlike nuclear and thermal power plants, hydropower plants hardly discharge any form of solid, liquid or gaseous wastes into the ecosystem.

- Relatively longer useful plant life.
- Run-of-river (ROR) projects have comparatively low environmental damage.
- Helps in providing inexpensive power, especially once the project achieves financial breakeven/depreciates its assets as operations and maintenance costs are much lower and consistent (independent of cost escalations in fuel) as compared to thermal power plants.
- Offloads the pressure on current account deficit as it helps in lowering fuel Import bills for the country.
- Helps in sustainable usage of coal for power demands of the country.
- Helps in meeting the peak power demand in the country, enhancing power system stability.
- Improved plant load factor of thermal units—grid stability and peak load management.
- Spontaneous starting, stopping and load variation ability.
- Hydropower provides high level of service to power system (reliability, flexibility, efficiency).
- Many international studies consider hydropower as the best available option for reducing GHG emissions and present as a good candidate for CDM benefits.
- Remote area development—improves infrastructure of interior areas and local population gets manifold advantages including better economy, job opportunities, education and healthcare facilities and connectivity and communication facilities.
- Multipurpose hydropower projects—help in flood moderation, irrigation, navigation and drinking water requirement.
- Catchments of river basins get developed as catchment area are treated by the hydro project developers. This helps in curtailing massive soil erosion and checking landslides in mountainous regions.
- With relative independence from international market like oil prices, hydropower involves no extra foreign exchange outgo.
- Hydropower is a no-inflation power as water, the "raw material", for power generation is free of inflation.

Challenges in Hydropower Development

1. Hydropower projects are site specific.
2. Location disadvantage—Projects are located in far flung areas having very little infrastructure and communication facilities. For instance, in Arunachal Pradesh, it takes at least 2 days for a normal passenger to reach project sites and if there are multiple landslides, hundreds of vehicles could easily get stuck for days together without access to basic amenities such as food and water. In such conditions, one could easily imagine how difficult it would be to transport project equipment, machinery, etc., via such routes. And if by chance there are

not much of landslides, locals/student bodies of neighbouring districts and states normally calls bandhs and block the roads at their whim.

3. Lengthy process of preparation of DPR and clearances having uncertainty of timeline and shortage of people with clearing agencies, for example:

   - land acquisition, environment, forest and wildlife clearances and forest rights settlement
   - Project features, layout plan, land requirement and certain environment management plans are required for forest clearance. Details for a hydro electric project are known after investigations and preparation of DPR and EIA/EMP reports.
   - The biggest problem with such clearances is the dynamic nature of changing requirements for obtaining these clearances.
   - Improper circulation of such dynamic changes to the project proponents and general public. Improper circulation/notification of Intermittent changes in law and new notifications.
   - Non-adherence to a fixed guideline/checklist by government personnel at the lower levels due to their overconfidence of knowledge of clearance process and document requirements, which causes rework in the clearance process.
   - One of the most complex problems experienced in Arunachal Pradesh is non-availability of land record with the land and revenue department. Project developer has to pay land compensation amount to the local community as well as state forest department, thereby bearing double financial burden for same piece of land.
   - Another major issue is the shortage of manpower with the State Government of Arunachal Pradesh. It has been observed that number of government officials at lower level are very few. A project can be checked and cleared by the official only after clearing previous projects in the pipeline. This results in time and cost overrun. Many a times, certain time bound activities get lapsed leading to start whole process afresh.
   - Fulfilling statutory requirements of central and state government.
   - Identification of land for compensatory afforestation (CA)—In accordance with Forest Conservation Act, 1980, CA shall be done over double the area for diversion on degraded or non-forest land. It is a well known fact that the State of Arunachal Pradesh has the second largest forest cover in the country. Out of the total geographical area of 83,743 km^2, the forest covers accounts for 68,000 km^2 (approx.) making 81.20 % of the state under forest cover. Thus, finding a large non-forest area or degraded forest area for CA in the State is very difficult and time consuming for the developer.

4. Scarcity of result-oriented reputed contractors/skilled technicians/workers.
5. Non-availability of grid power during construction phase of project.
6. Creation of new sanctuaries and national parks by forest department without consulting the hydropower department of state.

7. Security concerns—Vast hydro potential of the country is available in the areas affected by insurgency and militant problems. The law and order problem in such areas lead to delay in execution of the project as well as cost over runs.
8. Inadequate infrastructure/lack of communication system—As hydro projects are located in interior far flung areas, hilly terrain, landslides, hill slope collapses, road blocks particularly during monsoon season because of heavy rains and unprecedented floods cause severe setbacks in construction leading to time and cost overruns. Non-availability of approach road to project site—the cost of approach road, if included within the project infrastructure, results in increase in overall project cost. Apart from poor road connectivity, the region has extremely poor or no mobile connectivity.
9. Evacuation of power from remote generation sites.
10. Hydropower projects located in far flung areas of Arunachal Pradesh also suffers from the local mindsets. Project developers frequently encounter many types of demands from the locals of these areas. This is obviously over and above the various compensations planned under various regulations. Such illegitimate demands if not met, by the project developers, may lead to a total unrest between the local population and the project head.
11. Interstate disputes

    - Under Indian Constitution, water is a state subject. No objection certificate is required from each downstream state for getting sanction even for run-of-the-river projects which is a time consuming job. If all major rivers are made national resources and its water is distributed by centre keeping requirement of states in mind, the time could be saved.
    - Interstate issues are not limited to state government but also the local populous. For instance, in case of hydropower projects in Arunachal Pradesh which has rich hydropower potential with nearly 50,000 MW of potential, compared to Assam which has significantly low hydropower potential of merely 650 MW, any hydropower development in Arunachal Pradesh is an object of envy by the neighbouring states such as Assam. Any development happening in Arunachal Pradesh has to make its way through Assam. Whenever trucks/trailers carrying equipment or other essentials pass through Assam, they are blocked enroute, trucks and payloads destroyed and in some cases drivers beaten up and sent back to their starting points, for example, Lower Subansiri project.

12. Public awareness—There is inadequate public involvement during the project planning stage conducted by the government agencies and limited or no effort is taken to gain public acceptance through public involvement and transparency.
13. Intermittent stoppage of projects due to local agitations and frequent bandhs in entire north-eastern states resulting in time and cost overrun.
14. Geological surprises: The features of the hydro electric projects, being site specific, depend on the geology, topography and hydrology at the site. The construction time of a hydro project is greatly influenced by the geology of the area and its accessibility. It is, therefore, essential that state-of-the-art

investigation and construction techniques are adopted to minimize geological risks as well as the overall gestation period of hydel projects. Even if, extensive investigation using new techniques of investigations are undertaken, an element of uncertainty remains in the subsurface geology and the geological surprises during actual construction cannot be ruled out.

15. Hydrological challenges—River discharge observations are made available to the developers on pretext of confidentiality to the concerned government department only after the approval of the Ministry of Water Resources, GoI. Considerable time is lost in getting the approvals and the data.
16. Storage versus ROR Projects—There is a lot of controversy in the development of hydro projects as storage versus R-O-R. Most of the distress caused by storage schemes occurs in the hill states, whereas the benefits are largely in the states in the plains which are perceived to be more prosperous. Therefore, hill states prefer R-O-R schemes. For the maximization of benefits in a basin, the judicious blend of both type of schemes (storage/ROR) needs to be considered.
17. Till the late nineties hydropower Development was undertaken by public sector undertaking companies or Government of India/state government owned companies. These companies had relatively easy access to huge debt capital required to develop hydro projects, as these companies had sovereign/state guarantees. But as the hydropower sector has started to open up for the private developers, access to such huge debt capital is not easy as project financing under such risk prone environment for hydro project development is hardly available and these private developers would certainly not like to have such huge debt on their balance sheet as full-recourse funding option. Therefore, debt financing of hydropower projects for private players is a big challenge.

The New Land Acquisition Act (The Right to Fair Compensation and Transparency in Land Acquisition, Rehabilitation and Resettlement Act, 2013). The Act, which replaces the century-old Land Acquisition Act, 1894, proposes a unified legislation for acquisition of land and adequate rehabilitation mechanisms for all affected persons.

- Key Features of the Act
- Land acquisition:
- Consent of 80 % of landowners required in case of land acquired by private companies and 70 % for land acquired under public private partnership (PPP) model for public purpose.
- Compensation up to four times the market value in rural areas and twice in urban areas.
- Mandatory Social Impact Analysis (SIA) to assess nature of public interest and estimation of socio-economic impact prior to acquisition for all projects except irrigation projects.
- Land cannot be vacated until the entire compensation is awarded to the affected parties.
- No irrigated multicropped land shall be acquired under this Act.
- Companies can lease the land instead of purchasing it.

- If the acquired land has been unused for 5 years from the date of the possession, then it shall be returned to the original owner or owners or their legal heirs or to the land bank. If any unused acquired land is transferred to another individual within five years of it being acquired, 40 % of the appreciated land value shall have to be shared amongst the original land owners or their legal heirs.
- The award or agreement made under this Act shall be exempt from stamp duty and income tax and fees.
  - The Act had a retrospective clause saying compensation must be paid in line with the proposed law for ongoing projects (except irrigation projects) where the money has not been disbursed or possession of land has not been taken up.
  - In every project, those losing land and belongings to the SC or ST will be provided land equivalent to land acquired or two and a one-half acres, whichever is lower (this is higher than in the case of non-SC/ST-affected families). The land for land clause has been relaxed for irrigation projects, where land is not available.
  - Government also moved the amendment to specify that either compensation or rehabilitation and resettlement (R&R) will be given to farmers whose land is acquired for irrigation projects
  - Resettlement and rehabilitation (R&R):
- The provisions relating to R&R under this Act shall apply when (a) private companies acquire/purchases land through private negotiations which is equal to or more than such limits in rural and urban areas, as may be prescribed by the appropriate government in accordance with provisions of Sect. 46 and (b) private companies requests appropriate government for acquisition of a part of an area so prescribed for a public purpose.
- Affected families include land owner, farm labour, tenants, sharecroppers and workers on the piece of land for three years prior to the acquisition.
- Compensation includes house, land for land, choice of annuity/employment, subsistence grant for one year, transportation grant, grant against cattle shed/petty shops, one-time grant to artisans/small traders/others, fishing rights and one-time resettlement allowance.

Climate Change and Hydropower Development with Special Emphasis on Recent Floods in Uttarakhand

The discharge in the Himalayan river generally consists of two parts, one which is derived from melting of glaciers and another resulting from the monsoon rainfall in catchment areas. The discharges derived from the melting of snow make the rivers perennial. The excessive melting of glaciers, due to climate change may increase river flow for next few decades and may subsequently boost power generation in these Himalayan rivers. On the contrary if we look at the future prospect, the fast

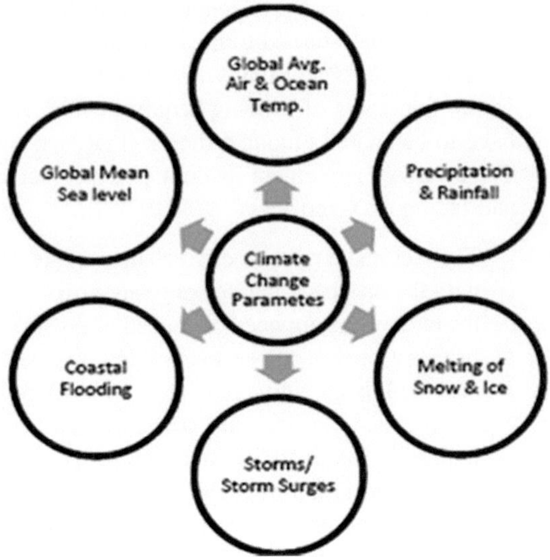

Fig. 4 Climate change parameters

melting of glaciers would lead to decrease in snow cover and subsequently decrease in run-off rate of all rivers. The perennial rivers could be changed to seasonal streams which results in water scarcity.

Further, climate change is evident from the observations (as shown in Fig. 4) of increase in global average air and ocean temperatures, precipitation and extreme rainfall, widespread melting of snow and ice, storms/storm surges/coastal flooding and rising global mean sea level. In future, climate change is expected to increase the frequency and intensity of current extreme weather/hydro-meteorological events, greater monsoon variability and also the emergence of new disaster, i.e., sea level rise and new vulnerabilities with differential spatial and socio-economic impacts on communities. This unprecedented increase is expected to have severe impact on the hydrological cycle, water resource (drought, flood, drinking water, forest and ecosystems, sea level/coastal area/losses of coastal wetlands and mangroves), food security, health and other related areas.

Uttarakhand Floods 2013

During June 2013, there was a deadly flood in the State of Uttarakhand which raised many issues on the development of hydropower. Although the fact remains and accepted by all that hydropower projects were nowhere involved or were responsible for this kind of tragedy, but the entire blame was put on the hydropower development in the state. As a matter of fact, hydropower storage dams and large dams were the saviour of the towns and cities in the downstream.

Now the recent flood in states of Assam, Bihar and UP is an example that the hydropower projects are not the reasons for these kinds of natural disasters since there are no hydropower projects in Nepal which are responsible for floods in Assam, Bihar and UP states.

Some of the identified reasons for this kind of disaster could be due to the following:

- Unseasonal heavy and torrential rain unexpected increase in the water level in the rivers.
- Cloud burst causing a sudden increase in water level.
- Besides rain fall, huge quantity of water was released into the river from melting of ice and glaciers due to high temperatures during the months of May and June.
- The surface air temperatures of Indian Himalayas have increased by one degree celsius in the past decade.
- Consequently, some of the Himalayan glaciers are rapidly melting causing glacial lake outburst floods (GLOFs) without any warning and several new glacial lakes are forming.
- A climate change is occurring due to global warming. Evidence of climate change includes the instrumental temperature record, rising sea levels, and decreased snow cover in the Northern Hemisphere. According to the Intergovernmental Panel on Climate Change (IPCC), most of the observed increase in global average temperatures since the mid-twentieth century is very likely due to the observed increase in human greenhouse gas concentrations.
- Constructions of hotels and guest house in and around the river beds and near river banks without proper planning.
- Rise in uncontrolled tourist traffic, travelling with AC vehicles to hill shrines causing rise in temperature of the area.

Sadly, there is a widespread misunderstanding that hydro projects only contributed to this kind of disaster. Whereas the fact is that the hydropower projects not only provide protection from floods, it also assures a steady supply of water for irrigation and drinking and controls the floods. In India, the rivers carry more than two-thirds of their annual flow in three monsoon months. Can we afford to waste the bulk of the water power by letting the water of the rivers flow unutilized in the monsoon months?

Dams and reservoirs are designed to last over hundred years. Which other type of energy source has such a long life? They provide us inexpensive energy. Even in current times of inflation and high prices, the State of Uttarakhand is getting energy at an average rate of 72.0 paisa/unit, from its old power stations.

The fact that hydro projects help mitigate the fury of floods can be gauged from the fact that had Shrinagar and Tehri Dam not withheld 90 % of the flood inflows during the June, 2013 floods, the Rishikesh and Hardwar would have been washed away. Had there been a dam of the size of Tehri on Alakananda, the people downstream of the dam would not have suffered any harm at all.

Countries, whose leaders and people had vision enough to develop their hydro resources before going for other options, enjoy the highest standard of living in the world (e.g. Canada, Norway, Sweden and Switzerland).

Our neighbours like Bhutan too has realized the importance of converting hydropower potential into catalyst for economic development and the results are here to see. They have higher per capita income than India.

The Way Forward

1. The growing power demand in India warrants the need of increase in power generation which can be fulfilled by development of reliable energy sources such as hydropower. Presently, the existing scenario of hydropower in India demands an urgent need of accelerated hydropower development.
2. The public at large must be made aware that there is no alternative to the increased availability of energy in generating economic growth commensurate with the target of welfare and up-liftment desired by them and the sources of power that are capable of providing them without perpetual cost liabilities in terms of environmental degradation.
3. The involvement of private sector and joint ventures with the neighbouring countries can go a long way towards achieving the goal of "power to all" in the coming years.
4. To counter the problem of location disadvantage, The GoI in collaboration with the state governments and selected project developers should prioritize stretches of approach roads to select priority project sites. These roads could be constructed on PPP or BOOT model, with viability gap funding or grants. Better security forces should be deployed in priority-based interior areas, so as to nullify the impacts of local extremist groups.
5. Amendments in acts, rules, regulations, etc. should be methodically publicized by the various Ministries of Government of India/State Government. Strict adherence to checklists of various clearances should be mandated from the Ministries, thereby keeping a check on the processes followed and documents required by the Clearing agencies for various clearances. Various Ministries, in collaboration with state government departments and project developers could organize regular workshops on new changes and amendments in acts, rules and regulations so that all stakeholders have the same understanding of the clearance procedures and processes. The lowest level government officials actually working on clearance process and proposal files should be given good incentives on timely delivery of intermediate and final clearances. Such incentives could be funded by clearance fees to be submitted by project developers at intermediate and final clearances.
6. Interstate issues could be solved by conducting relevant stakeholder dialogues, understanding their core issues and addressing these issues through various modes of discussions, negotiations, arbitrations or at last legal proceedings.

7. The stringent evaluation process for clearance of a selected project has to be accepted by all authorities and stake holders. Even though the process of TEC and EAC and FAC is stringent, but still if more aspects are required to be added for clearance shall be added to satisfy all the stake holders.
8. Bankable DPR should be examined considering all aspects so that it is accepted by financial institutions. Financial institutions may be consulted before hand for examination so that it becomes a valid document pre-examined for getting debt.
9. After considering all technical and geological aspects at the time of TEC and consultation with stake holders at the time of public hearing, no midway stoppage of the project on any account shall be permitted. Once construction starts, the project shall be treated as project of national importance.
10. In principal, approval on the feasibility report of the project must be given by MOEF so that later on project is not rejected on account of number of trees to be felled. MOEF may even give a formula every state- and district-wise about what number of trees would be allowed to be felled per hectare considering the density of the forest.
11. Wildlife areas and national parks or national reserves should be identified and eco-sensitive zones well defined by respective state governments before handing over of the project to the developer.
12. There should be two public hearings; one before the main public hearing where all doubts are cleared before hand, so that it is clearly known that public is in favour of or against the project. The second public hearing to be conducted on the due date when all the presentations are made and public views are heard.
13. Tariff determination for hydro project must be relooked to provide adequate return to the developer.
14. Hydropower obligation (HPO) of 10–15 % should be provided as incentive so that it becomes mandatory for the large industrial consumers procure power from hydro sources.
15. For regulatory and control aspect, strict monitoring of all conditions imposed during clearances to be monitored strictly.
16. Respective government to provide support and subsidy for development and up gradation of roads and bridges, railway sidings, river jetties, etc. or develop infrastructure on cost sharing basis so that one project in the basin does not feel the pressure.
17. Construction of balancing reservoirs intermittently in the large river basins so as to provide continuous water flow required for the biodiversity mechanism.
18. Construction of storage dams for creating buffer storage in the technically feasible areas.
19. Sustainability analysis shall form a part of the project development and should be carried out by identified and experienced agency before start of the project so as strengthen confidence with respective government and stake holders.
20. PPA has become another difficult area for private developers. Central/state government should extend all necessary help in signing of PPA for private hydro projects.

Further Readings

Is Hydro a Step Child for India? (2014, Feb). Energetica India, p. 72–76.
Ramesh C. Bottlenecks in hydro power development.
Srivastava ABL. Hydro power in India: the bottlenecks., NHPC Ltd.
Madan MM. "Hydro Power"—A Bane or a Boon.
Madan MM. Hydropower- an environmental disaster or climate saver. Indian J Power River Valley Dev Kolkata. 2014;64(9–10):126–31.

Integrated Coastal Zone Management of Chilika Lake

R. Sundaravadivelu

The Chilika Lake in Orissa is one of the largest wetlands in the world (1000 km^2 area) habited by migratory birds and by a special type of dolphin. This lake is the livelihood for about 100,000 fishermen and also acts as drainage for Mahanadhi River Basin. The old mouth on the northern side of lake was not flushing properly, and hence, the tidal exchange was not sufficient to maintain the quality of brackish water.

The project on Integrated coastal zone management of Chilika Lake (opening of a new mouth based on numerical model study and dredging of channels linking the north east and western sides costing Rs. 50 crores) was coordinated by me in association with Chilika Development Authority (CDA), CWPRS, Poona, and NIO, Goa.

The seawater interaction with lagoon has significantly diminished and the maximum salinity has dropped (22.31 ppt in 1957–1958 to around 10 ppt in 1961–64, but appears to have stabilized after the rapid drop). The decrease of salinity level has a great adverse impact on the biodiversity as well as fisheries of the Chilika Lake. This is due to

- Occurrence of sedimentation in the Chilika Lake and its coastal vicinity.
- Narrowing down of the Chilika Lake mouth and decrease of its depth.
- No major effort has been made to understand comprehensively the sedimentation process, lake–sea water interactions, and tidal flushing in the Chilika Lake.
- The salinity levels show remarkable variations, both temporally and spatially. A complex combination of freshwater discharge, evaporation, wind condition, and tidal inflow of seawater governs the seasonal changes in salinity levels. Salinity levels in Chilika Lake vary along the north–south gradient.
- Due to shallow depth of the lagoon, the salinity concentration fluctuates between 1 and 3 ppt between the surface and the bottom. The lagoon is brackish over

R. Sundaravadivelu (✉)
Department of Ocean Engineering, Indian Institute of Technology Madras,
Chennai 600 036, India
e-mail: rsun@iitm.ac.in

P. Ghosh and B. Raj (eds.), *The Mind of an Engineer*,
DOI 10.1007/978-981-10-0119-2_58

most of its area and varies from freshwater (0 ppt) adjacent to the Daya River mouth to hyper—saline level (42 ppt) in the outlet channel during the dry period.

CWPRS has carried out the numerical model study to estimate the tidal prism and determining the width and depth of the mouth. IIT Madras has proposed the dredging methodology and NIO Goa has carried out the EIA study.

The artificial mouth was opened on September 23, 2000, which reduced the length of the outflow channel by 18 km. The dredging of the channel was completed before opening the new mouth.

Author Index

P. Ghosh and B. Raj (eds.), *The Mind of an Engineer*,
DOI 10.1007/978-981-10-0119-2

MIX
Papier aus verantwortungsvollen Quellen
Paper from responsible sources
FSC® C105338

If you have any concerns about our products,
you can contact us on
ProductSafety@springernature.com

In case Publisher is established outside the EU,
the EU authorized representative is:
Springer Nature Customer Service Center GmbH
Europaplatz 3, 69115 Heidelberg, Germany

Printed by Libri Plureos GmbH
in Hamburg, Germany